RURAL GROUNDWATER CONTAMINATION

Frank M. D'Itri
Lois G. Wolfson

LEWIS PUBLISHERS, INC.

Library of Congress Cataloging-in-Publication Data
Rural groundwater contamination.

   Bibliography: p.
   Includes index.
   1. Water, Underground--Pollution--
United States.   2. Water-supply, Rural--
United States.   I. D'Itri, Frank M.
II. Wolfson, Lois G.
TD223.R87   1987   363.7'394   87-3085
ISBN 0-87371-100-9

Second Printing 1988

LEWIS PUBLISHERS, INC.
121 South Main Street, Chelsea, Michigan 48118

PRINTED IN THE UNITED STATES OF AMERICA

PREFACE

Groundwater is one of the most precious natural resources of the United States. However, contaminants from natural and man made sources are causing an ever more increasing decline in the quality of this resource. While the quality of groundwater resources in the United States is still relatively good, now is the time to take action to prevent and resolve groundwater contamination problems.

Everyone has the responsibility to assist with the improvement and protection of water for the benefit of both urban and rural populations. Potable water is needed for domestic consumption while industry and agriculture require sufficiently good quality water to permit sustained production. These consumption demands are not mutually exclusive, and can be maintained if proper precautions are taken.

Public concern has risen as increasing population and industrial growth have placed heavier demands on the existing water supply system, especially groundwater. In excess of 95 percent of this nation's rural inhabitants and over 50 percent of its urban residents depend on groundwater for drinking water.

Pollution of groundwater from industrial, domestic and agricultural chemicals presents a serious threat. For example, there are an estimated 181,000 industrial surface lagoons or impoundments in addition to more than 16,000 industrial landfill sites containing hazardous wastes, 18,500 active municipal landfills, and 20 million septic systems. Each year in the United States, farmers apply approximately 11.5 million tons of commercial nitrogen fertilizers and 1.1 billion tons of animal manure. In addition, approximately 700 million pounds of pesticides are used in the United States each year, and between 3.5 and 21 million pounds (0.5 to 3 percent) are estimated to gravitate to the ground or surface water resources.

Current debate among agricultural, geohydrological and/or environmental scientists does not so much relate to whether or not industrial, domestic, and agricultural practices contribute to groundwater contamination but how to deal with the causes and consequences of the problem. Thirty-four states have identified agricultural nonpoint source pollution as a major cause of their failure to achieve state water quality goals. Therefore, groundwater contamination is perceived as the most critical environmental problem of the 1980s, and the data suggest that the problem could reach crisis levels by the twenty-first century.

iii

One of the reasons that some water quality crises have arisen is because potential problems went undetected under existing monitoring programs. Consequently, aquifer restorative/remedial action programs must be developed even though they can be very expensive. In addition, considerable time is required to develop information on the problem and the technically feasible measures needed to resolve it. Prevention is definitely preferable.

Research has revealed many aspects of groundwater pollution and the chemical interactions between contaminants and the subsurface matrix, but much more needs to be done. The public must also be educated on the causes and prevention of groundwater contamination. Education and research are the basic tools needed to protect the nation's groundwater resources.

Over the long run, water resource protection and control require recognition of a myriad of socio-economic and technological factors. Because water policy involves tradeoffs between water use and quality, leading experts suggest a systems approach which includes consideration of the physical, biological, and economic factors. Government, private institutions, and domestic water users must all take into account the physical characteristics and limits of the atmospheric and hydrologic structures in order to provide a total system approach to determine the economic value and role of water.

While the focus of this book is on groundwater contamination from agricultural sources, it is not limited to this. Rather, the reader is offered a general overview of the sources, impacts, assessments, methods, and health and risk implications of groundwater pollution as well as an explanation of the necessary remedial action program strategies.

The book is organized into six major sections: (1) overview; (2) sources and impacts; (3) assessment and modeling; (4) drinking water standards, health implications and risk considerations; (5) regulation and remedial action; and (6) strategies and assistance. The first two chapters provide an overview with respect to agricultural issues. Chapters 3 through 10 present information on the losses, transport, and abatement of nitrogen from soils, nitrates in groundwater, and the impact of chemigation systems, large scale animal production, landfills, underground storage tanks and septic systems as sources of groundwater contamination. Chapters 11 through 13 describe methods to access, monitor and model the movement of contaminants in soil. Chapters 14 through 19 summarize drinking water standards, pesticide contamination of well waters, well construction, treatment of contaminated groundwater as well as pesticide risk considerations and health implications of groundwater contaminants. Chapters 20 through 23 address regulation and remedial action including groundwater law, liability, economics, land use planning, zoning, public policy and sociological considerations. The major emphasis of these chapters is on a systems approach for total resource management. It considers all or most interrelated issues and concerns such as economics, public health, ecological protection, long-term resource conservation and resource use priorities.

Chapters 24 to 26 represent the authors' collective judgement on research priorities and strategies for the future role of groundwater resources to meet human needs. These chapters identify and describe long-range planning methods and priorities regarding the uses, management conservation, and protection of the nation's groundwater resources.

Frank M. D'Itri
Lois G. Wolfson

East Lansing, Michigan

Frank M. D'Itri is a Professor of Water Chemistry in the Institute of Water Research and Department of Fisheries and Wildlife at Michigan State University in East Lansing, Michigan. He holds a Ph.D. in analytical chemistry with special emphasis on the transformation and translocation of phosphorus, nitrogen, heavy metals and hazardous organic chemicals in the environment.

Professor D'Itri is listed in *American Men of Science, Physical and Biological Science.* He is the author of *The Environmental Mercury Problem,* and co-author of *Mercury Contamination* and *An Assessment of Mercury in the Environment.* He is the editor of *Wastewater Renovation and Reuse, Land Treatment of Municipal Wastewater: Vegetation Selection and Management, Municipal Wastewater in Agriculture, Acid Precipitation: Effects on Ecological Systems, Artificial Reefs: Marine and Freshwater Applications* and *A Systems Approach to Conservation Tillage* and co-editor of *PCBs: Human and Environmental Hazards* and *Coastal Wetlands.* In addition, he is the author of more than forty scientific articles on a variety of environmental topics.

Dr. D'Itri has served as chairperson for the National Research Council's panel reviewing the environmental effects of mercury and has conducted numerous symposia on the analytical problems related to environmental pollution. He has been the recipient of fellowships from Socony-Mobil, the National Institutes of Health, the Rockefeller Foundation and the Japan Society for the Promotion of Science.

Lois G. Wolfson, an aquatic ecologist, is the technology transfer specialist and educational programs coordinator at the Institute of Water Research, Michigan State University, East Lansing, Michigan. She is responsible for the development, interpretation and dissemination of scientific information to the non-scientific community. Ms. Wolfson received her M.S. in Botany from Michigan State University with an emphasis in nutrient cycling in aquatic plants, and is currently engaged in a Ph.D. program in the Department of Fisheries and Wildlife at Michigan State. Her current research efforts are directed toward determining the interacting physical and chemical parameters controlling bluegreen algal growth and buoyancy.

Ms. Wolfson has had over eight years of experience in the field of water quality management and has authored or co-authored several articles dealing with impacts of nonpoint sources on inland water quality. She is also the editor of *Water Impacts,* a monthly publication on water quality and quantity issues.

ACKNOWLEDGMENTS

The success of this endeavor was due to the interest and dedication of the participating individuals and organizations. Primary among these were the Michigan State University College of Agriculture and Natural Resources, the Agricultural Experiment Station, and the United States Geological Survey which sponsored and contributed to the necessary financial support for the conference and publication of this book. Our special thanks are extended to Mr. Homer R. Hilner, State Conservationist, USDA Soil Conservation Service, for his encouragement and support during the preliminary planning as well as his direct participation in the conference itself. We would also like to thank Dr. Harvey J. Liss, Extension Program Leader, along with the many other staff members of the Kellogg Biological Station Conference Center, not only for their hospitality and assistance, but also for providing the conferees with an atmosphere that was free from distractions and conducive to the discussion of the many diverse aspects related to groundwater quality. Last, but certainly not least, a grateful acknowledgment also goes to all of the authors whose research, in addition to contributions of time, efforts, and counsel made this volume possible.

In the Institute of Water Research at Michigan State University, special thanks are extended to its Director, Dr. Jon F. Bartholic, for his support and encouragement. In the preparation of this manuscript, we acknowledge and thank Ms. Dara J. Philipsen for many editorial suggestions that contributed materially to the overall style and Ms. Terry Heineman-Baker for help with the figures.

CONTRIBUTORS

James H. Anderson, College of Agriculture and Natural Resources, 104 Agriculture Hall, Michigan State University, East Lansing, MI 48824 (3).

Charles S. Annett, Keck Consulting Services, 1099 West Grand River, Williamston, MI 48895 (201).

Ann Beaujean, Legislative Coordinator, Michigan Association of Counties, 319 West Lenawee Street, Lansing, MI 48933 (383).

Jon F. Bartholic, Institute of Water Research, 334 Natural Resources, Michigan State University, East Lansing, MI 48824-1222 (Presenter).

Alfred M. Blackmer, Department of Agronomy, Iowa State University, Ames, IA 50010 (85).

Donald J. Brown, Science For Citizens Center, Western Michigan University, Kalamazoo, MI 49008 (339).

Michael Brown, Legislative Assistant, Michigan Association of Counties, 319 West Lenawee Street, Lansing, MI 48933 (Presenter).

George Carpenter, Site Assessment Unit, Groundwater Quality Division, Michigan Department of Natural Resources, Lansing, MI 48909 (147).

Adger Carroll, Natural Resources and Public Policy, 108 Agriculture Hall, Michigan State University, East Lansing, MI 48824 (Presenter).

Robert M. Clark, Drinking Water Research Division, U.S. Environmental Protection Agency, 26 W. St. Clair Street, Cincinnati, OH 45268 (235).

Steven J. Corak, Department of Agronomy, University of Kentucky, Lexington, KY 40506 (161).

Joseph Cotruvo, Criteria and Standards Division, Office of Drinking Water, Environmental Protection Agency, Washington, DC 20460 (213).

Thomas J. Dawson, Wisconsin Public Intervenor, Wisconsin Department of Justice, Post Office Box 7857, Madison, WI 53707-7857 (261 and 323).

Frank M. D'Itri, Institute of Water Research, 334 Natural Resources, Michigan State University, East Lansing, MI 48824-1222 (Presenter).

Joe Ervin, Institute of Water Reserach 334 Natural Resources, Michigan State University, East Lansing, MI 48824-1222 (Presenter).

Edward E. Everett, Keck Consulting Service, 1099 West Grand River, Williamston, MI 48895 (201).

Walter A. Feige, Drinking Water Research Division, U.S. Environmental Protection Agency, 26 West St. Clair Street, Cincinnati, OH 45268 (235).

Carol Ann Fronk, Drinking Water Research Division, U.S. Environmental Protection Agency, 26 West St. Clair Street, Cincinnati, OH 45268 (235).

Wilbur W. Frye, Department of Agronomy, University of Kentucky, Lexington, KY 40506 (161).

Thomas J. Gilding, National Agricultural Chemicals Association, 1155 15th Street, NW, Washington, DC 20005 (379).

Ray J. Gillespie, Cooperative Extension Service, 48 Agriculture Hall, Michigan State University, East Lansing, MI 48824 (Presenter).

Norman Grannemann, U.S. Geological Survey, 6520 Mercantile Way, Lansing, MI 48910 (393).

Denise Gruben, Site Assessment Unit, Groundwater Quality Division, Michigan Department of Natural Resources, Lansing, MI 48909 (147).

Gordon E. Guyer, Michigan Department of Natural Resources, P.O. Box 30028, Lansing, MI 48909 (Presenter).

George R. Hallberg, Geological Survey, Bureau of the Iowa Department of Natural Resources, 123 North Capitol Street, Iowa City, IA 52242 (23).

DeLynn R. Hay, Agricultural Engineering Department, Institute of Agriculture and Natural Resources, University of Nebraska-Lincoln, Lincoln, NE 68583-0726 (105).

Alan G. Herceg, U.S. Soil Conservation Service, 610 West Burr Oak Street, Centreville, MI 49032 (399).

Homer R. Hilner, U.S. Soil Conservation Service, 1405 South Harrison Road, Room 101, East Lansing, MI 48823 (15).

Arthur G. Hornsby, Soil Science Department, 2169 McCarty Hall, University of Florida, Gainesville, FL 32611 (193).

James Jenkins, Department of Crop and Soil Sciences, Michigan State University, East Lansing Michigan, 48824 (Presenter).

Richard S. Johns, Groundwater Division, Michigan Department of Natural Resources, P.O. Box 30028, Lansing, MI 48909 (Presenter).

Maurice Kaercher, Cooperative Extension Service, 201 West Kalamazoo Street, Kalamazoo, MI 49000 (Presenter).

Michael A. Kamrin, Center for Environmental Toxicology, C231 Holden Hall, Michigan State University, East Lansing, MI 48824 (225).

Donald K. Keech, Michigan Department of Public Health, 3500 North Logan, P.O. Box 30035, Lansing, MI 48909 (291 and 396).

Kyle M. Kittleson, Institute for Water Research and Center for Remote Sensing, 302 Berkey, Michigan State University, East Lansing, MI 48824 (69).

Gary Klepper, Groundwater Quality Division, Michigan Department of Natural Resources, P.O. Box 30028, Lansing, MI 48909 (147).

James A. Koski, Great Lakes Planning Commission and Drain Commissioner, Courthouse, 111 South Michigan, Saginaw, MI 48602 (384).

Jessica T. Kovan, Department of Agricultural Economics, Michigan State University, East Lansing, MI 48824 (351).

James N. Krider, National Environmental Engineer, U.S. Soil Conservation Service, P.O. Box 2890, Washington, DC 20013 (115).

Leighton L. Leighty, Department of Resource Development, Michigan State University, East Lansing, MI 48824 (305).

Lawrence W. Libby, Department of Food and Resource Economics, 1157 McCarty Hall, University of Florida, Gainesville, FL 32611 (351).

Christine Lietzau, Environmental Division, Michigan Department of Agriculture, Box 30017, Ottawa Tower North, 4th Floor, Lansing, MI 48909 (395).

Benjamin W. Lykins, Jr., Drinking Water Research Division, U.S. Environmental Protection Agency, 26 W. St. Clair Street, Cincinnati, OH 45268 (235).

Edward Martin, Department of Agricultural Engineering, Michigan State University, East Lansing, Michigan 48824 (Presenter).

Bob Minning, Keck Consulting Services, 1099 West Grand River, Williamston, MI 48895 (Presenter).

David R. Mumford, Michigan Department of Public Health, Bureau of Environmental and Occupational Health, Division of Occupational Health, State Office Building, 411H E. Genesee, Saginaw, MI 48607 (399).

Dan O'Neill, Land Applications Unit, Groundwater Quality Division, Michigan Department of Natural Resources, P.O. Box 30028, Lansing, MI 48909 (395).

Phyllis A. Reed, Pesticide and Toxic Substances Branch, U.S. Environmental Protection Agency, 230 S. Dearborn St, Chicago, IL 60604 (253).

Joe T. Ritchie, Professor and Homer Nowlin Chair, A-570 Plant and Soil Sciences, Michigan State University, East Lansing, MI 48824-1325 (179).

Patricia M. Ryan, Department of Resource Development, Michigan State University, East Lansing, MI 48824 (305).

Jack Sage, Soil Conservation Society of America, 17230 Robbins Road, Grand Haven, MI 49417 (Presenter).

James S. Schepers, Agricultural Research Service, U.S. Department of Agriculture, University of Nebraska, 113 Keim Hall, Lincoln, NE 68583-0915 (105).

Ruth Shaffer, Legislative Science Office, Legislative Service Bureau, Lansing, MI 48909 (129).

Daniel A. Smith, U.S. Soil Conservation Service, 1405 South Harrison Road, Room 101, East Lansing, MI 48823 (394).

M. Scott Smith, Department of Agronomy, University of Kentucky, Lexington, KY 40506 (161).

James G. Truchan, Environmental Enforcement Division, Michigan Department of Natural Resources, P.O. Box 30028, Lansing, MI 48909 (Presenter).

Joe VanderMeulen, Legislative Science Office, Legislative Service Bureau, P.O. Box 30036, Lansing, Michigan 48909 (129).

Jac J. Vargo, Department of Agronomy, Mississippi State University, Mississippi State, MS 39762 (161).

Pete Vergot, Cooperative Extension Service, 226 East Michigan Avenue, Paw Paw, MI 49079 (399).

Craig Vogt, U.S. Environmental Protection Agency, 401 M Street, SW; WH 550, Washington, DC 20460 (213).

Lois G. Wolfson, Institute of Water Research, 334 Natural Resources Building, Michigan State University, East Lansing, MI 48824-1222 (399).

Numbers in parentheses indicate the page on which the author's contribution begins.

CONTENTS

PART I

OVERVIEW

PART II

SOURCES AND IMPACTS

PART III

ASSESSMENT AND MODELING

PART IV

DRINKING WATER STANDARDS, HEALTH IMPLICATIONS
AND RISK CONSIDERATIONS

PART I

OVERVIEW

CHAPTER 1

AGRICULTURE AND NATURAL RESOURCES: THE BROADENING HORIZON

James H. Anderson
College of Agriculture
and Natural Resources
Michigan State University
East Lansing, Michigan 48824

INTRODUCTION

On space ship earth, agriculture sustains most of life because of a thin veneer of soil, nourished by the fresh waters of nature, warmed and cooled by the gentle breezes that sweep across the land and energized by the soft rays of the sun. Through countless centuries nature has developed this delicate balance and life support system. However, in the quest to exploit and utilize the earth's resources to produce food for an ever expanding population, man has threatened to upset that delicate balance. On a global scale, millions of acres of prime agricultural lands are being irreversibly converted to other uses. At the same time, through misuse, countless acres are being severely eroded by wind and water. This not only destroys the land but deposits sediments in streams, lakes and estuaries. In many parts of the world, deforestation and mismanagement of land speed desertification.

As the land resources are continually driven harder in the developing nations, productivity may decline because of reduced organic matter, soil compaction and mismanagement unless appropriate technology is used. In some places underground water resources are literally being mined and in others, rivers and streams are becoming closed systems. These are global problems and demand attention, because survival of the human species depends upon maintaining the soil and water resource base so that future generations may live in abundance and prosperity.

Profitability, productivity and efficiency remain priorities for American agriculture. However, additional emphasis must be placed on conserving and protecting natural resources. The term "steward-ship of the land" has attained popular status, and it is imperative that commitments in this regard be seriously examined. Furthermore, all agricultural practices must be re-evaluated in light of the need to protect and enhance the total natural resource base. This demands evaluation of the benefits as well as the cost of all future tech-nological developments in agriculture. Immediate benefits alone cannot be the sole criteria for determining the suitability of an agricultural practice in the future.

Pollution of surface and groundwater mandates examining both point and nonpoint sources of discharges from agricultural operations. Pollution and contamination of water resources are becoming increas-ingly critical. Presently, 34 states blame nonpoint discharges from agriculture for their failure to achieve water quality goals under Public Law 92-500, the Federal Water Pollution Control Act passed in 1972; and 29 states have identified nonpoint discharges from agricul-ture to groundwater as a major concern. Thus, new clean water legislation is being contemplated to deal with this problem, and it is obvious that ground and surface waters will be monitored more carefully in the future.

There is no question that groundwater contamination is a serious problem; however, the public's perception of it is often shaped by what they see on television or read in the newspapers. These accounts often convince people that government and other officials do not have the ability to protect the public effectively. This is certainly the perception in California where in November, 1986, voters adopted a new environmental statute entitled, "Safe Drinking Water and Toxic Enforcement Acts--Proposition 65." This new law proposes an arbitrary and simplistic approach mandating zero discharge of any chemical that causes cancer, birth defects or sterility. It holds industries accountable for both point and nonpoint releases with respect to both keeping any significant quantity from reaching a drinking water source and making a reasonable effort to warn anyone who could be exposed to a significant level. However, legally establishing the amount of the chemical that constitutes a "significant quantity" will be extremely difficult because toxicologists cannot presently agree on how much of a contaminant in drinking water is harmful. Consequently, Proposition 65 legislation may severely restrict beneficial uses of chemicals in California. Moreover, like Proposition 13, similar legislation could also be implemented in other states.

To meet the environmental challenge, new methods must be devised to cope with old problems as well as new ones that emerge. Soil depletion is an old problem, yet one to which more attention must be given to conserve the top soil in all parts of the country. Some of the most fragile soils are eroding at five to ten times the acceptable rate of three to five tons per acre per year, and it is estimated that, in Michigan alone, 42 million tons of soil are lost annually due to erosion (MSU, 1985). In addition to the loss of fertility, erosion contributes excessive sediment, nutrient and pesticide loadings to surface waters, decreases stream flow, threatens fish and other aquatic species and accelerates eutrophication. Therefore, new conservation tillage techniques must be devised or old ones revived which minimize the soil losses that also often trigger other major problems.

One of these is that when the soil is depleted due to erosion, increased fertilizer application usually follows in order to maintain yields. Since 1945, the use of both nitrogen fertilizers and pesticides has dramatically increased. For example, during the period 1960-85, annual fertilizer applications increased as follows:

> 1960 -- 2.7 million tons
> 1970 -- 7.5 million tons
> 1980 -- 11.4 million tons
> 1985 -- 11.5 million tons.

This is more than a four-fold increase during two decades (USDA, 1985). In 1985, about 44 percent of the total nitrogen applied in the United States was on 74 million acres of corn, and the average rate per acre was 136 pounds (Welch, 1986).

Besides chemical fertilizers, animal manure provides about 8 percent of the nitrogen required for crop productivity and 20 percent of the phosphorus and potassium. An estimated 1.1 billion tons of animal manure are produced each year, about 420 million tons in feedlots and other confined rearing situations (Miller and McCormack, 1978; VanDyne and Gilbertson, 1978). More than 70 percent of this is applied to the land to enhance crop productivity. About 39 percent of the 28 million tons of nitrogen available for food production in the United States comes from fertilizers (Giese, 1985). Major grain crops typically have the highest rate of utilization and may consume up to 50 percent of all applied nitrogen. A three-year study by the USDA showed that, on an average, only 36 percent of the nitrogen applied in fertilizers, crop residues, manure and other organic wastes was recovered by crops (Power, 1985).

It is obvious that major changes need to be made in the way crops are fertilized and that the Extension Service needs to redouble its effort to properly educate farmers relative to fertilizer application.

As fields are overfertilized, water courses become green with plants that then die and support bacteria that consume oxygen and turn the waterways eutrophic. Agricultural nonpoint discharges are the main source of pollution for surface water in the United States, including 64 percent of stream miles and 57 percent of lake surface areas (ASIWPCA, 1986). Nutrients are the prime culprits in the impairment of lake water, and nonpoint pollution has had an impact on or threatens 11 percent of the rivers, 30 percent of the lakes and 17 percent of the estuaries (ASIWPCA, 1986). In fact, water quality in 43 percent of the 404,000 miles of streams examined in one study was threatened by or impaired to some degree by nonpoint source pollution (ASIWPCA, 1986). Moreover, of the 15 million acres of lakes studied, 53 percent were labeled as threatened or impaired as well as 28 percent of the estuaries (ASIWPCA, 1986).

EFFECTS ON GROUNDWATER

Overfertilization not only affects surface waters but also groundwater. Much of the residual nitrogen remains in the soil where it percolates into the groundwater with rain and snow melt. Not much is known yet about the extent or rate of percolation, partly because of previous misconceptions. One long held belief was that excess fertilizer added to the soils simply stayed in place. This was replaced by the assumption that nonporous soils retained nutrients and pesticides more readily than sandy soils. New studies have shown that fissures and even minute tunnels made by plant roots, insects and animals open the way for water to seep through nonporous soils more readily than was previously thought and to carry the fertilizer and pesticides with it. Excessive irrigation in combination with rains and snow may expedite the process. Thus, the degradation of groundwater quality as well as soil erosion and surface water pollution must now be addressed when considering proper productive farming technology.

The most immediate concern is the impact on drinking water in rural areas. Presently, the U.S. Environmental Protection Agency estimates that approximately 603,000 (2.7%) of the drinking water supplies for the 22 million rural households in the United States exceed the maximum nitrate contaminant limit of 10 mg per liter as nitrogen or 45 mg per liter as nitrate (Francis, 1982).

Half of the population and 90 percent of those who live in rural areas use groundwater for drinking. Shallow wells, especially those in alluvial aquifers, are easily contaminated with nitrogen fertilizers and pesticides. These wells may pose a major health risk to farm families, and evidence is mounting that this is occurring. Recently, it was reported that public water supplies in between 40 and 50 rural Iowa towns normally exceeded the maximum nitrate levels set to protect the health of infants. About one in five rural wells tested showed excessive levels of nitrates ranging up to 20 times the recommended limit (Fruhling, 1986a). The possible consequences were demonstrated too graphically in July, 1986, when a two month old infant died of blue baby syndrome caused by drinking formula mixed with water from a rural well containing 152 mg nitrate per liter; 3.4 times the allowable limit (Fruhling, 1986b).

If these problems are to be prevented in the future, the first step is to encourage all rural farm and nonfarm families to have their drinking water checked for nitrate contamination on a regular basis. This goes counter to the general assumption by farm families that well water is natural, pure and healthy. It also is difficult to accept because contaminated water may substantially decrease the value of property and its related potential to obtain FHA and conventional loans. In spite of this, the quality of life and the health of families must be the most immediate concern.

PESTICIDES AND HERBICIDES

In addition to recognizing and counteracting the problems caused by fertilizers, the potential damage from pesticides and herbicides must also be realistically appraised. The annual total output of pesticides in the United States is more than 2.5 billion pounds with 1500 active ingredients in some 45,000 products (Eichers, 1981). Each year, approximately one billion pounds of pesticides, containing more than 600 active ingredients, are applied to farms and gardens (Maddy, 1983). Of this quantity, an estimated 0.5 to 3.0 percent (3.5 to 21 million pounds) reach surface waters before degrading (ECOP, 1985). The World Health Organization has reported that many thousands of people suffer illness from pesticide exposure, and some of them die each year (WHO, 1985a; 1985b). In the United States, atrazine, alachlor and other herbicides that were once thought to dissipate in the soil have been found in groundwater in at least ten states and have contaminated some wells. In 1985, of 3000 drinking water wells tested in California, 185 were contaminated with organic chemicals (Anon., 1986). In a recent national survey, 17 different pesticides were discovered in the groundwaters of 23 states (Cohen et al., 1986). More are likely to be found as additional tests are conducted.

In addition to pesticides, the extent of damage from other types of pollutants, such as leakages from underground fuel storage containers, must be recognized and dealt with. Very small quantities of gasoline have the potential to contaminate large groundwater resources; consequently, farmers must pay more attention to this source of contamination than they have in the past.

THE RESEARCH AND EXTENSION CHALLENGE

Agricultural scientists and engineers must move rapidly to develop, adopt and incorporate new technologies so that the land can be preserved and its productivity enhanced, desertification impeded in the arid areas, salinity minimized in the irrigated areas, compaction reduced in the highly mechanized areas and irreversible soil deterioration minimized everywhere. More support should be forthcoming to do this as the public becomes increasingly concerned about the pollution of land and water resources and objects to letting the environment continue to deteriorate because of man's activities. Agriculturalists have a special responsibility to weigh the tradeoffs and determine the long-range consequences of their actions. These factors must be evaluated by an informed citizenry and not a few do-gooders acting on emotions and devoid of the necessary knowledge to make intelligent decisions. Everyone must continually evaluate both the pros and cons of change brought about by scientific developments and engineering applications in agriculture and related fields, but this is particularly true for degradation and contamination of the natural resource base.

It is imperative that the State Agricultural Experiment Stations and the Cooperative Extension Service increase their efforts to develop and disseminate information to farmers as well as others who potentially impact the natural resource base. Efforts must be intensified to inform farmers that fertilization, pesticide application and supplemental irrigation beyond the optimum rate will not necessarily produce better yields or higher profits without serious side effects. Furthermore, special care must be taken to ensure that agriculture and natural resources unite to conserve and protect the environment so that farmers can remain competitive in a global economy. There is an urgent need to coordinate, at the county level, the nonpoint source pollution information and assistance provided to farmers and their families by state, local and university agents. This is necessary to broaden the scope of assistance and, at the same time, minimize overlapping programs.

It is also imperative to intensify research efforts to develop procedures to utilize water more efficiently for agriculture. In 1980, the total groundwater withdrawals in the United States were esti-mated at 87 billion gallons per day, and two-thirds of this was used for irrigation in agriculture (NAS, 1986). Presently, the efficiency of water application in the United States is approximately 45 to 50 percent, so there is great potential for increases in the efficient utilization of water for irrigation. Although groundwater withdrawals for agricultural irrigation in Michigan are not a serious problem overall at the present time, the lowering of groundwater tables does create a problem in certain locations. Unless constructive measures are taken now, the future groundwater supplies are likely to diminish further because of "mining," increased demand and impairment. The advantages of change besides conserving water are that alternative technologies such as conservation tillage coupled with the optimum utilization of fertilizer, pesticides and irrigation can result in substantial savings to the farmer. For example, instead of irrigating, fertilizing and applying pesticides according to a prearranged schedule, techniques are available to determine how much and when these procedures are needed. Obviously, there is no complete answer for all situations, and efforts must be intensified to make sure that recommendations are sound and produce results if farmers are expected to adapt them.

In addition to developing no-till cultivation and other ways to minimize nonpoint pollution, research must be intensified to identify areas with highly permeable and susceptible soils where fertilization and pesticide application should be most carefully managed. More information is needed on effective application schedules, placement, rate and time of application and speed of release for specific crop requirements. Simple management techniques such as applying only enough fertilizer to meet crop requirements for realistic production goals, planning the time of application, and using slow-release fertilizers can substantially reduce the amount lost. For irrigation systems, effective scheduling of water and nitrogen application according to the need for optimal crop yields can result in less irrigation water being applied and also less supplemental nitrogen. The work of Dr. Joe Ritchie, Nowlin Chair Professor at Michigan State University, on agricultural simulation models to predict the interacting effects of weather, hydrology, nutrient cycling and movement and soil temperature on crop growth and yield has and will continue to contribute substantially to this agricultural advancement. These models enable researchers to plan fertilizer and supplemental irrigation application strategies to minimize groundwater contamination. More accurate research is also needed on crop requirements for the best growth cycle, and better testing devices are needed to monitor the crops both for water and fertilizer requirements as well as runoff and percolation losses. Dr. Ritchie and his colleagues are presently pursuing a whole host of these issues at the Kellogg Biological Station, and the results of their activities certainly are promising.

Biotechnology research also is poised to revolutionize the ancient calling of farming. Molecular biologists and geneticists worldwide are now investigating mechanisms to produce heartier strains of plants. Their being more disease resistant thereby reduces the need for pesticides. Plants, genetically engineered to produce natural molecules that repel insects could also be several orders of magnitude more cost effective than traditional chemical approaches. These biologically engineered plants will be better able to not only tolerate and resist the ill effects of insects, bacteria, fungi and viruses, but also to assimilate nutrients more efficiently.

A variety of kinds of research is also underway to extend the nitrogen fixing capacity of the legumes and transfer this ability to other crops as well. Researchers are also investigating new, natural, biological pesticide controls such as genetically engineered bacteria to control black cutworms on corn. The future possibilities are also promising for fungi that kill specific types of weeds bacteria that attack vegetable worms and viruses that neutralize insects. Genetic engineers are also currently working to alter beneficial insects to make them more persistent and able to kill more than one specific pest.

Plans are also underway for soil bacteria that can produce a naturally occurring insecticide capable of protecting plant roots against soil-dwelling insects. New vegetable species that are disease resistant and superior in growth and uniformity are now being produced as well through cell culture methods rather than seeds grown in the conventional manner. Obviously, these developments will have a significant impact on farming in the future.

These advancements are not without serious concern; and care must be taken that genetically engineered microbes, fungi, viruses and insects that may be released into the environment do not create bigger problems than those they were designed to eliminate. Few data are available on how genetically altered organisms would behave in nature; therefore, concerns have been raised that they could have ecologically disruptive effects such as killing beneficial insects, eliciting resistance in target organisms and persisting in the environment. Obviously, such possibilities need additional study, and research must proceed with great caution.

Education is the most cost effective way to deal with many of these problems. Education not only provides information, technical assistance and motivation to farmers but also information to the rural nonfarmers and urban inhabitants as well as local and state lawmakers. This is necessary to neutralize emotional opposition to agricultural operations as well as to maintain crop goals and protect the environment. The most pressing needs are to educate farmers on the differences between the uses and misuses of agrichemicals so that they can be applied more judiciously. Federal, state and local officials as well as the general public must also be educated regarding the benefits and liabilities of agrichemicals to society. Therefore, coordinated educational programs must be developed and implemented that:

1.   stress the importance of the nation's water resources to agriculture, industry and society and the strategies necessary to maintain a continuing adequate supply of safe water for all sectors.

2.   describe the impacts of agricultural and nonagricultural chemicals relative to the environment and especially groundwater quality; their use, rate, proper handling and disposal; the significance of soil variability to groundwater impact; the health risks associated with small concentrations of hazardous chemicals in drinking water; and the need to test rural groundwater drinking supplies for contamination.

3.   increase the understanding and awareness not only of federal, state and local lawmakers but also of farmers and the general public on the interaction between land use, agrichemical needs and groundwater quality, and the options available to address these issues.

CONCLUSIONS

Agriculture's present system of voluntary effort to deal with nonpoint pollution problems is being seriously questioned by the public, regulatory agencies and legislators. The debate continues on the development of new legislative initiatives, regulatory and enforcement programs to deal with agrichemicals and nonpoint discharges. It must be acknowledged that, as long as chemicals are applied within the various climatic conditions, losses to the environment cannot be completely eliminated; but they can clearly be minimized.

The cooperation of government agencies, state agricultural experiment station researchers, Cooperative Extension Service personnel and farmers is needed to design relevant research and demonstration projects and to implement strategies for change. Agribusiness and industry must also become involved. A coordinated effort is needed to define cost effective management alternatives, new technology, educational programs and agricultural policies that can be more flexibly applied. This will require many changes in priorities at a time when the main concern for many farmers is simply surviving. The burden cannot simply be placed on the back of the farmer alone.

Resolving these problems requires an integrated effort by all segments of the agricultural community to gain the experience and data necessary to effect a satisfactory balance between efficient agricultural production and the protection of the greatest resources-- water and soil. To accomplish this, efforts must be coordinated among community participants, both rural and urban, by various government agencies such as the United States Department of Agriculture, the United States Environmental Protection Agency, the Michigan Department of Natural Resources, the Michigan Department of Agriculture, the Michigan Department of Public Health and the universities. The latter have a particular mandate to conduct appropriate research to be able to educate and advise not only farmers but legislators who increasingly pass more restrictive laws to regulate nonpoint discharges of fertilizers, pesticides and herbicides into the environment.

In conclusion, then, chemicals have played an important role in the growth of agricultural productivity during the last three decades, but they must be applied more efficiently in order to retain profitability in agriculture and protect the environment at the same time. The judicious and economical use of agrichemicals can still play an important role; but more holistic approaches to agricultural management, coupling efficient crop production and soil conservation with surface and groundwater protection, must be developed. Therefore, it is imperative to broaden the scope of research and the dissemination of knowledge with greater sensitivity to community concerns regarding the protection of the environment.

LITERATURE CITED

Anon. 1986. Of 3,000 drinking-water wells in California, 18% were found to be contaminated. Environ. Sci. Technol. 20(9):845.

ASIWPCA. 1986. America's Clean Water: The State's Nonpoint Source Assessment 1985. Association of State and Interstate Water Pollution Control Administrators, 440 North Capitol Street, N.W., Suite 330, Washington, DC 20001.

Cohen, S.Z., D. Eiden and M.N. Larber. 1986. Monitoring groundwater for pesticides. In: W.Y. Garner, R.C. Honeycutt and H.N. Nigg (Eds.), Evaluation of Pesticides in Groundwater. ACS Symposium Series 315, American Chemical Society, Washington, DC, pp. 170-196.

Eichers, T.R. 1981. Agricultural Economic Report Number 464, Farm Evaluation: 1981. U.S. Department of Agriculture, Washington, DC.

ECOP. 1985. Groundwater Education: A Challenge for the Cooperative Extension System. Report of the Groundwater Task Force to the Extension Committee on Organization and Policy (ECOP), p. 10.

Francis, J.D., B.L. Brower, W.F. Graham, O.W. Larson III, J.L. McCaull and H.M. Vigorita. 1982. National Statistical Assessment of Rural Water Conditions. Office of Drinking Water, U.S. Environmental Protection Agency, Washington, DC.

Fruhling, L. 1986a. Water in Iowa tainted by farm chemicals. Des Moines Sunday Register, May 4, 1986, p. 1.

Fruhling, L. 1986b. Nitrates blamed for baby death in South Dakota. Des Moines Register, July 29, 1986, p. 1.

Giese, A. 1985. Testimony of the Fertilizer Institute before the Subcommittee on Toxic Substances and Oversight of the U.S. Senate Committee on Environment and Public Works, December 12, 1985, 13 pp.

Maddy, K.T. 1983. Pesticide usage in California and the United States. Agriculture Ecosystems and Environment 9:159-172.

Miller, R.H. and D.E. McCormack. 1978. Improving soils and organic wastes. USDA Task Force, United States Department of Agriculture, Washington, DC.

MSU. 1985. Water and Land Management, Enriching Michigan's Future: A Planning Document. Agricultural Experiment Station and Cooperative Extension Service, Michigan State University, East Lansing, MI 48824, p. 31.

NAS.   1986.   Groundwater Quality Protection: State and Local Strategies.   National Research Council, National Academy of Sciences, National Academy Press, Washington, DC, 309 pp.

Power, J.F.   1985.   Nitrogen in the cultivated ecosystems.   In: F.E. Clark and T. Rosswall (Eds.), Terrestrial Nitrogen Cycles: Ecosystem Strategies and Management Impacts.   Ecol. Bull. (Stockholm) 33:529-546.

USDA.   1985.   Commercial Fertilizers.   Statistical Reporting Service, U.S. Department of Agriculture, Washington, DC.

VanDyne, D.L. and C.B. Gilbertson.   1978.   Estimating U.S. livestock and poultry manure and nutrient production.   USDA-ESCS Publication No. ESCS-12, March.   United States Department of Agriculture, Washington, D.C.

Welch, L.F.   1986.   Nitrogen, Ag's old standby.   Solutions 30(5):18-21.

WHO.   1985a.   Environmental Pollution Control in Relation to Development.   World Health Organization, Technical Report Series, No. 718, Geneva, Switzerland, 63 pp.

WHO.   1985b.   Safe Use of Pesticides.   World Health Organization, Technical Report Series, No. 720, Geneva, Switzerland, 60 pp.

CHAPTER 2

PERSPECTIVE ON SOIL AND WATER CONSERVATION
AND
AGRICULTURALLY RELATED GROUNDWATER CONTAMINATION

Homer R. Hilner
U.S. Department of Agriculture
Soil Conservation Service
East Lansing, Michigan 48823

The Soil Conservation Service (SCS) participated actively in the planning and implementation of this conference to raise awareness and transfer current technology and information. The primary purpose is to enable Soil Conservation Service and Cooperative Extension Service personnel as well as others, notably farmers and other rural citizens, to incorporate the presently available information into current activities and plan future efforts to control potential causes and prevent rural groundwater contamination.

Before livestock was concentrated in large numbers on small areas, and before the widespread use of pesticides and chemical fertilizers, there was little potential for seriously contaminating rural groundwater from agricultural sources. With limited effort and nature's help, the small amounts of human and animal wastes could be controlled. The relatively simple compounds were largely biodegradable and, in many instances, soil served as a "living filter" after minimal planning with regard to locating and applying animals wastes and other potentially polluting materials. However, enough evidence has now been accumulated so that no one would disagree that groundwater contamination is at least a potentially serious problem. Therefore, conservation programs must be adapted to protect this valuable resource.

This need to protect the groundwater is being recognized by a steadily growing proportion of society, both urban and rural. The recently intensified concern over groundwater contamination is part of a trend wherein citizens are becoming more aware of human impacts on the natural environment. Individual instances of soil and water quality deterioration are being recognized as integral parts of the potential decline of the larger ecosystem.   Therefore, soil and water conservation activities can no longer be conducted on individual farms or in locally isolated communities under the old assumption that they interact with many elements in a larger ecosystem.

Maintaining high quality groundwater will require practical approaches to prevent contamination. Soil and water conservation systems that fit into farm enterprises are the ones that have lasting benefits.   Measures to prevent groundwater contamination must complement farm enterprises. Standards and specifications for many conservation practices and other farm management efforts will have to incorporate protection of both surface and groundwater.   Contaminated surface water should be diverted around recharge areas of vulnerable aquifers or be cleaned up before it is released to the groundwater.   One definite need is to assist farmers to adjust applications of fertilizers to what the crops actually use and avoid excess enrichment of the waterways.   A second need is to limit pesticide applications over vulnerable aquifers to those that degrade rapidly or adsorb to soil particles instead of passing through to the groundwater.

With increased rural populations, concentrated livestock production operations, and extensive use of chemicals, including some relatively complex ones, conservation measures must now be planned, designed and installed that will control potentially negative effects. Conservation systems for erosion control and animal waste management typically include measures to handle surface water. However, groundwater levels and quality are often also affected. As water is diverted for farm and other uses, the levels of groundwater may also be raised or lowered. Appropriate measures must be taken in areas with both high and low infiltration rates to protect the groundwater. Typically, erosion control measures slow or reduce the surface flow and increase infiltration. To determine appropriate measures, types of soil and underlying groundwater flow systems must be taken into account.

More sophisticated information, education and technical assistance are needed by farmers and other land users to prevent contamination.   Even this will not assure total success, especially where groundwater is already contaminated or becomes contaminated. In these cases, reclamation measures would likely differ from those required for prevention. To accomplish a complete job will require contacts and working relationships with individuals, groups and larger organizations with whom conservationists previously have had limited contact; geohydrologists and public health agencies, for example. Maintaining effective communications and working relationships is always difficult, and this is likely to be even more true when new ones must be established with individuals in unfamiliar organizations or disciplines.   Nonetheless, these relationships are needed to augment the already complex existing network.

Of course, some efforts directed at change are already underway by individuals and groups, some involving several counties. These should be encouraged, supported, and combined into a comprehensive approach. This is likely to become easier as increasing numbers of people live in rural areas and become aware of and express concern for groundwater contamination, reinforced by the knowledge that more people could also experience negative health effects in the future if proper measures are not taken now. This awareness, not only in Michigan but throughout the country, is already bringing about an encouraging trend toward the use of less persistent and more readily biodegradable materials. This trend should be supported and accelerated along with development and adoption of additional measures.

Soil Conservation Service personnel and others that advise farmers must be provided with the capability to identify agriculturally related potential groundwater quality problems, sources and ways to evaluate alternative solutions. They should be able to recommend cost effective technical solutions, but like everything else, this will still have a price. Resource management decisions on farms are greatly influenced, if not dictated, by economic considerations. Because groundwater management is, and will continue to be, influenced by economics as well as environmental and health considerations, alternative impacts on water quality and methods of resolving problems must be thoroughly examined in planning farm enterprises.

The Soil Conservation Service's role in groundwater protection will largely be in prevention of contamination due to surface activities. With the present and anticipated federal financial resources being as limited as they are, the best that can be done at this time is to incorporate groundwater concerns into programs that deal with erosion control, flood prevention, irrigation and animal waste management. At a minimum, field offices should have groundwater and related information readily accessible and easily manageable that indicates the local groundwater flow patterns, the aquifers that are most vulnerable to contamination and those that are better protected. Currently, the Soil Conservation Service resources are too thinly stretched to initiate new and accelerated programs related exclusively to groundwater.

In the case of groundwater found to be currently seriously contaminated, outside financial and technical assistance will likely be needed, especially when pollution sources are unknown or financial resources of current surface owners are insufficient to conduct cleanup operations. Such assistance or needed regulation would logically be provided by state or federal agencies that have traditionally assumed these responsibilities, not the Soil Conservation Service.

Fortunately, groundwater contamination from agricultural resources does not presently appear to be a widespread, serious problem in Michigan. However, unless preventive measures are taken, it certainly could become serious in a number of places. However, it won't be known how extensive the problem actually is until detailed additional information is made available. Information regarding vulnerable aquifers needs to be collected and made available to conservationists and farmers as well as other land users for planning activities. Individual farm planning may require site-specific information; however, general information is essential, and much of it still remains to be collected and disseminated.

A good start has been made to collect information on the upper five feet of potential water-bearing strata, the soil, in two-thirds of the state. However, about 11 million acres of Michigan are still without a detailed soil survey. Obviously, to become a more effective data source to use in preventing groundwater problems, the completion of these surveys should be accelerated, preferably with automated methods that will make them more readily available.

Maps of vulnerable aquifer recharge areas should be prepared and made readily accessible to conservationists and other advising the agricultural community. Consistent, current, comprehensive information on water resources from the surface down to all usable aquifers is needed statewide. In addition, having the capability to manipulate and evaluate the information is essential. Mathematical computer models which link soils, geology, surface and groundwater information will be necessary to handle the large volume of data. Presently, efforts are underway in government agencies and academic institutions such as Michigan State University, the Michigan Department of Natural Resources and the Soil Conservation Service to develop those tools and data. These efforts certainly deserve support and should make valuable contributions in the future. In the meantime, the Soil Conservation Service is moving ahead with presently available information and human resources to incorporate groundwater concerns into everyday activities. The Soil Conservation Service's role is to provide technical assistance and information to enable land users to protect soil, surface and groundwaters from contamination insofar as this is possible and practical with given technology.

In conclusion, the current situation raises some questions relevant to agriculturally related groundwater concerns in Michigan:

(1)   What is the groundwater quality situation now and the projected situations in the short run and in the long run? How is going to determine the groundwater quality situation and monitor changes, and who is going to pay for this?

(2)   How is relevant information going to be distributed to those who need it?

(3)   If agricultural activities have already contaminated groundwater to the point where it must be cleaned up, who is going to pay?

(4)   With shrinking federal budgets, who is going to pay for prevention programs when agricultural activities pose a potential threat to groundwater?

(5)   What action should be taken by local, state and federal agencies and organizations?

The Soil Conservation Service and the entire agricultural community need information to answer these questions if groundwater concerns are to be properly addressed.

PART II

SOURCES AND IMPACTS

CHAPTER 3

NITRATES IN IOWA GROUNDWATER

George R. Hallberg
Geological Survey
Iowa Department of Natural Resources
Iowa City, Iowa 52242

INTRODUCTION

During the past six years, studies in Iowa have generated considerable interest and concern with the occurrence and increasing concentrations of nitrate (NO_3) in drinking-water supplies, particularly from groundwater. These studies have shown that nitrates and some pesticides are leaching through the soil into shallow groundwater and are now found in many of Iowa's groundwater aquifers. In many areas, these groundwaters, which supply drinking water to both cities and private individuals, now commonly exhibit nitrate concentrations above the maximum contaminant limit (MCL) for public drinking water of 10 mg/l nitrate as nitrogen (45 mg/l as NO_3). Furthermore, trace amounts of many widely used pesticides have also been found in these aquifers in association with elevated nitrate concentrations.

Well construction and poor waste disposal practices contribute to individual well problems, and spills and misuse of chemicals have caused more severe problems locally. However, the regional contamination of aquifers is the result of conventional usage of agricultural chemicals and constitutes a nonpoint source pollution problem. Much of this research, and hence much of the attention to this problem, has been focused in northeast Iowa. Such contamination, however, is clearly occurring throughout Iowa and the corn belt. Currently, this contamination is primarily found in relatively "shallow" aquifers, but this may only be a function of time.

As with all nonpoint source pollution, the scope of the problem is large. For example, approximately 60 percent of the total land area of Iowa is in row crop cultivation which receives agricultural chemical applications. Because of the vast areas involved, the only effective way to treat nonpoint source pollution is with prevention.

There are justifiable concerns for public health because of the potential for long-term and widespread exposure that may occur with chemicals in drinking water. The long-term implications of the coexistence of nitrates, pesticides, and other chemicals is unclear. Some epidemiologic studies suggest patterns of risk, but unfortunately epidemiologic proof often takes a generation of exposure. Though some protest to raise the drinking water MCL (maximum contaminant level) for nitrate, this standard has a far lower factor of safety than standards for many other potentially hazardous chemicals. It is interesting to note also that while in this country there is discussion to raise the MCL from 10 to 20 mg/l, the European Economic Community has recently *lowered* their standard from 22.6 mg NO_3-N/l to 11.3 mg/l (50 mg/l as NO_3) because of these same health concerns (EEC, 1980; Carey and Lloyd, 1985).

The remainder of this paper will: (1) review studies of nitrates in groundwater in Iowa, (2) discuss pertinent agronomic studies and their implications, (3) describe the interaction of groundwater and nitrates within the hydrologic system, (4) discuss the relative contribution of sources of nitrate, and, thus (5) outline why fertilizer nitrogen is the focus of concern.

THE BIG SPRING BASIN STUDY

The data from the Big Spring Basin study in northern Clayton County have been very important in developing current understanding of the relationship between agriculture and groundwater quality. The detail of knowledge about the basin provides answers to many questions with a level of confidence that isn't possible in most areas. It is important to note, though, that research continues in the Big Spring Basin, not because of its unique or severe water quality problems, but because the area affords a unique opportunity to study groundwater in ways simply not possible elsewhere. These details have been reviewed elsewhere (Hallberg et al., 1983b; 1984b), but a few points are worthy of review.

In the Big Spring Basin there is a very responsive hydrologic system where one can quantitatively gage the discharge of groundwater and, hence, begin to deal with the mass balances of water and chemicals in a real world setting (103 square mile area). The area of the groundwater basin has also been defined, affording an accurate understanding of the land use and treatment. The basin is wholly agricultural--there are no industries, landfills, municipalities, or such to complicate interpretations. Having defined the basin area also allows the use of historic records on water quality and farm practices to reconstruct events and look at the patterns of change over time (reviewed by Hallberg, 1984; Hallberg et al., 1983a; 1983b; 1984a; 1984b).

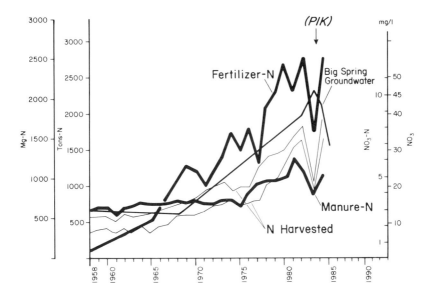

Figure 1.   Mass of fertilizer-N and manure-N applied in the Big Spring Basin and annual average nitrate concentration (dashed line, right axis for scale) in groundwater at Big Spring, and tons of nitrogen harvested in corn grain (from Hallberg et al., 1984a).

**Temporal Changes in Nitrate
Concentrations in Groundwater**

During the 1950s and 1960s, the nitrate concentration in groundwater in the Big Spring Basin averaged about 3 mg/l NO_3-N (12 to 14 mg/l as NO_3).   By the 1980s, the nitrate had increased three times to an annual average of 9 mg/l NO_3-N (39 mg/l as NO_3) in water year 1982 and to 10.1 mg/l NO_3-N (46 mg/l as NO_3) in 1983. Data from a variety of groundwater wells in the area showed the same trend (Hallberg et al., 1984a).   The primary sources of nitrogen in this basin are manure and fertilizer (Hallberg et al., 1983b; 1984b). Manure-N increased only 0.3 times, while fertilizer-N applied increased 2.5 to 3 times as a function of the increasing rate of nitrogen application and the increase in corn acreage.   The increase in nitrate in groundwater directly paralleled the increase in the amount of fertilizer-N applied in the basin (Figure 1).   In addition to the figures for the amount of fertilizer-N applied and manure-N generated in the Big Spring Basin, estimates of the amount of nitrogen removed with the harvested grain can also be made from yield records.   Two different estimates are shown on Figure 1.   The lower line was estimated   using the standard   assumptions of 56 lbs/bu of corn and that

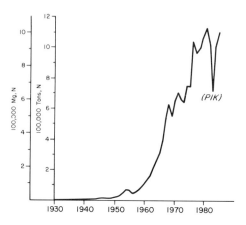

Figure 2. Mass of nitrogen in fertilizer used in Iowa 1930-1984 (from Harmon and Duncan, 1978, and Statistical Reporting Service).

the grain contained 1.5 percent nitrogen. The upper line was estimated using a statistical calculation derived from test plot data (from Dr. A.M. Blackmer, Agronomy Department, Iowa State University, personal communication); the formula is given in Hallberg et al. (1984b, p. 112). This calculation corrects the nitrogen recovery values by subtracting the nitrogen removed in corn grown in check plots without fertilizer. These check data are not available for the Big Spring Basin and thus the unfertilized terms were dropped from the calculations, providing an "uncorrected" or maximum estimate of the amount of nitrogen removed. These estimates simply amplify the prior relationships noted; from the late 1960s to the present, the difference between the amount of nitrogen removed by crops and the amount of fertilizer-N applied increases, and the nitrate concentrations rise in the groundwater at Big Spring.

    This trend in nitrogen use is obviously not unique to the Big Spring Basin. The increase in fertilizer usage in the Big Spring Basin simply parallels the trends for Iowa and the nation as a whole. Across the corn belt, the average nitrogen fertilization rate increased from about 50 kg N/ha (45 lbs N/Ac) in 1965 to 153 kg N/ha (135 lbs N/Ac) in 1982 (Hargett and Berry, 1983). In Iowa, the state average fertilizer-N rate on corn increased from 50 kg N/ha (45 lbs N/Ac) in 1964 to 160 kg N/ha (143 lbs N/Ac) in 1984; on soybeans the rate increased from about 4.5 kg N/ha (4 lbs N/Ac) in 1964 to 26 kg N/ha (23 lbs N/Ac) in 1984. For Iowa as a whole, as shown in Figure 2, the total amount of nitrogen fertilizer used increased over five fold during the time nitrogen use increased 2 to 3 times in the Big Spring Basin. The same timing is evident in other areas. As shown in Figure 3, also since about 1965, more fertilizer-N has been added to corn cropland than was being removed in harvested crops in

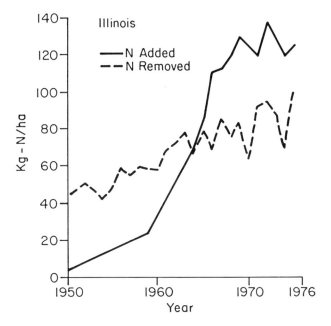

Figure 3. Annual average addition of fertilizer-N and removal of crop-N for corn in central Illinois (NRC, 1978).

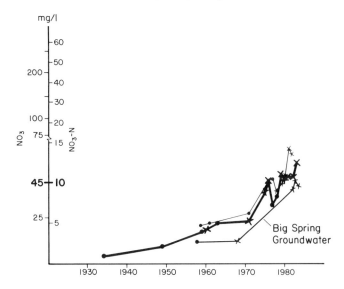

Figure 4. Nitrate concentrations in groundwater with time from two PWS (public water supply) well systems in carbonate aquifers from northeastern Iowa; trend in Big Spring groundwater shown for comparison. X indicates average of three or more samples per year (Hallberg, 1985).

Illinois. During this time, the nitrate concentration in the Kaskaskia River increased concurrently with the increased fertilizer-N use (NRC, 1978).

The Big Spring Basin averages about 60 percent of its area in row crops (essentially all corn). While the basin supports more livestock than many areas of Iowa, it is not as intensively used for row crops. Thus, for perspective, in Figure 4, the nitrate groundwater data from Big Spring are shown with two other sites from northeastern Iowa which have long-term records. All these sites show the same trend of increasing nitrate in a similar time frame; however, the other sites reach greater concentrations of nitrate with time.

Figure 5 shows the change in nitrate concentration with time in the groundwater in two alluvial aquifers in northwestern and west central Iowa. The same timing of major increases is apparent, but the increases in nitrate are considerably higher than in northeastern Iowa. Compare these trends with Figure 2, and the same relationship between fertilizer-N use and nitrate in groundwater, shown directly in Figure 1, is apparent.

For perspective, two other sites are shown in Figure 6, both from alluvial aquifers in western Iowa. Site A is from a well in a rural town, and the water quality is known to be affected by septic tank and cesspool effluents. Site B is from another well system affected by a spill of liquid nitrogen fertilizer in the early 1950s. At site A, high nitrate concentrations are apparent in the 1940s, and they continue to the present with little significant change compared to the other sites described. Likewise, at site B, high nitrate concentrations were noted shortly after the spill, but had seemingly dissipated by the late 1960s. After this time, the nitrate levels rise again, similar in timing and magnitude to the regional changes noted in Figures 4 and 5. Again, there are local, or point sources and factors affecting the nitrate concentration in groundwater, but statewide the regional increases in nitrate in shallow groundwater since the late 1960s occur concurrently with the greatly increased nonpoint source use of nitrate fertilizers, as shown in the Big Spring Basin.

STATEWIDE PERSPECTIVE

Across the country, nitrate problems in groundwater are attributed to various sources such as natural occurrences, mobilization of natural soil nitrogen, septic tanks, other waste disposal, fertilizer-N, and heavy manure usage--often in combination with poor well construction. For individual wells or local areas in Iowa, septic tanks, waste disposal and poor well construction all create problems. In some areas, heavy manure loads create the same problems as fertilizer-N, and manure must be managed properly to avoid such problems. Undoubtedly, some of the nitrate in groundwater is derived

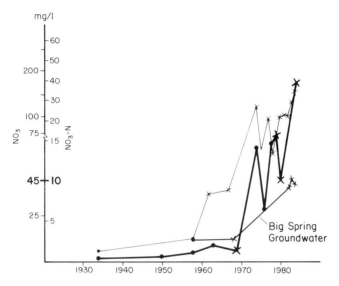

Figure 5. Nitrate concentration in groundwater with time from two PWS well systems in alluvial aquifers; one from northwestern, one from west central Iowa; trend in Big Spring groundwater shown for comparison. X indicates average of three or more samples per year (Hallberg, 1985).

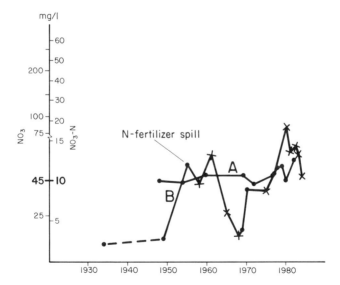

Figure 6. Nitrate concentration in groundwater with time from two PWS well systems in alluvial aquifers from western Iowa. A - affected by septic and cesspool effluent; B - affected by nitrogen fertilizer spill in 1951. X indicates average of three or more samples per year (Hallberg, 1985).

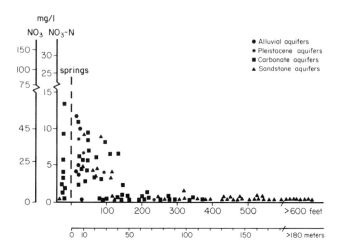

Figure 7.   Nitrate-N concentrations in groundwater vs. depth to groundwater tapped by well, from northeastern Iowa; depth represents casing depth or well depth, where casing depth approximates total well depth (data from Hallberg and Hoyer, 1982; Hallberg et al., 1984b).

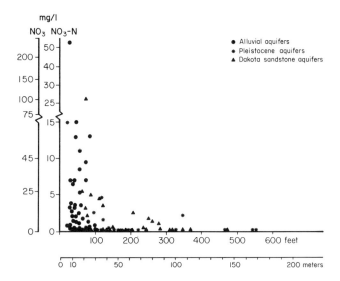

Figure 8.   Nitrate-N concentration in groundwater in relation to depth, as in Figure 3; data from west central Iowa (data from Hunt and Runkle, 1985; Hallberg, 1985).

from all these sources, but nitrate is leaching to groundwater because of the excess amount present in comparison to crop use and removal. In many areas of intensive grain crop production, the largest input of nitrogen is chemical fertilizer. When considering all the sources of nitrogen, from rainfall to soil mineralization, it is also the one variable that can be controlled. This is partly why the regional increases of nitrate in groundwater directly relate to increases in nitrogen fertilizer usage.

As shown in the Big Spring Basin, the nitrate is delivered to groundwater through infiltration recharge, more simply, the percolation of water downward through the soil. Infiltration is the recharge mechanism common to all groundwater and, hence, the leaching losses of nitrate must be expected to occur over large areas in appropriate environments, as indicated in Figures 4 and 5. The Big Spring Basin and other work in northeastern Iowa has concentrated on the regionally important, shallow carbonate aquifers. In these areas, the nitrate concentration generally decreases with increasing depth below the land surface (except in some complex hydrogeologic settings in the karst-carbonate aquifers in northeastern Iowa). This general relationship is shown in Figure 7, from data from the northeastern Iowa studies. Though the work in northeastern Iowa has concentrated on the carbonate aquifers, even these data show that the same effects are apparent in "alluvial aquifers" (sand and/or gravel in stream valleys), "Pleistocene" aquifers (sand and/or gravel buried in glacial deposits), and sandstone aquifers (large sandstone bedrock units, e.g., St. Peter and Jordan Sandstones). As noted in Figures 4 and 5, even though much attention has been focused on northeastern Iowa, nitrate concentrations in groundwater are often greater in other areas. For perspective, Figure 8 shows nitrate groundwater data, similar to Figure 7, but the data are from eight counties in west central Iowa. The data are from wells from alluvial, Pleistocene and Dakota sandstone aquifers, where infiltration is the only mechanism of recharge. The same nitrate to depth relation is apparent. Note that nitrate-N concentrations range higher than in northeastern Iowa data. This is typically the case in western Iowa.

Alluvial aquifer systems are more widely used in western Iowa than in northeastern Iowa because bedrock aquifers are more deeply buried in western Iowa, and the alluvial systems often have had better *natural* water quality. Groundwater in western and north central Iowa is of major concern in relation to agricultural chemicals because of the much greater intensity of row-cropping and attendant chemical use in these areas compared to northeastern Iowa (Figure 9).

Nitrate problems in groundwater/drinking water occur statewide, and affect both public water supplies and private wells. Figure 10 shows locations of public water supplies (from groundwater) where nitrate-N concentrations commonly exceed the MCL. While they are distributed statewide, the majority occur in western Iowa in alluvial aquifers. Figure 11 is a summary of 13,625 groundwater samples from

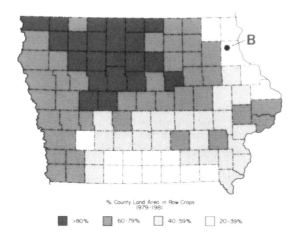

Figure 9. Average percent of county land area in corn and soybeans, 1979-1981 (data from Iowa Department of Agriculture, Crop and Livestock Reporting Service).   B - indicates location of the Big Spring Basin.

private, rural wells less than 30 m deep, which were analyzed for nitrate by University Hygienic Laboratory (UHL) during 1978-1981 (data provided by Dr. R.C. Splinter, UHL).   As in Figure 10, it is apparent that a far greater percentage of groundwater sources exceed the nitrate MCL in western Iowa, particularly northwestern Iowa, than in northeastern Iowa.

The behavior of nitrogen in the environment is complex, as is the distribution of usable groundwater. While problems are apparent in western and northwestern Iowa, Figures 10 and 11 do not indicate any significant problem in north central Iowa, a region of particularly intensive row crop production. In part, this relates to two factors: first, the public water supply wells in this region are deep wells which have not been affected yet; and, second, the groundwater samples from many shallow, private wells and Iowa Geological Survey research wells, in the alluvial aquifers along the upper Des Moines river system do not contain nitrate. Ongoing studies suggest that denitrification is removing the nitrate; pesticide residues do appear in the groundwater, however. Even though denitrification is removing the nitrate, the same leaching losses may still occur.

Nearly all controlled data from susceptible hydrologic environments show the same temporal trend of increases in nitrate concentration noted in Figures 1, 4, and 5 from surface waters, public groundwater supplies less than 30 m (100 feet) deep, or private

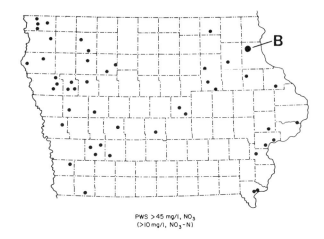

PWS > 45 mg/l, NO₃
(>10 mg/l, NO₃-N)

Figure 10.   Public water supplies which have exceeded the nitrate maximum contaminant level of 10 mg/l nitrate-N (45 mg/l as nitrate) since 1980.   Data from Iowa Department of Water, Air and Waste Management (R. Kelley, personal communication).   B - indicates location of the Big Spring Basin (Hallberg, 1985).

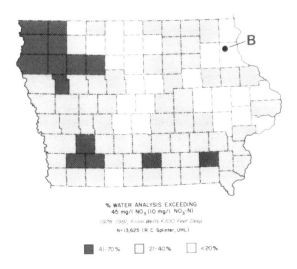

% WATER ANALYSIS EXCEEDING
45 mg/l NO₃ (10 mg/l NO₃-N)
(1978-1981) From Wells <100 Feet Deep
N=13,625 (R. C. Splinter, UHL)

■ 41-70%   □ 21-40%   □ <20%

Figure 11.   Percent of water samples, by county, exceeding the nitrate MCL from private wells less than 30 m (100 ft) deep analyzed by UHL between 1978-1981.   N = 13,625; 28 percent of all samples exceeded the MCL.   (Data from Dr. Roger Splinter, UHL, personal communication).   B - indicates location of the Big Spring Basin (Hallberg, 1985).

groundwater samples less than 30 m (100 feet) deep (see McDonald and Splinter, 1982; Hallberg and Hoyer, 1982; Hallberg et al., 1983a; 1983b). However, it is difficult to evaluate any changes in nitrate problems in relationship to the MCL of 10 mg/l nitrate-N. Approximately 12,000 to 18,000 private well groundwater samples are analyzed by UHL every year. Approximately 18 to 22 percent of those samples have exceeded the nitrate MCL since 1965 (see Morris and Johnson, 1969; Hallberg and Hoyer, 1982). These figures do not represent a clearly significant increase over previous years. While this might make it easy to be complacent over the situation, the story is not that simple. Over time, very shallow, hand dug or bored wells with tile casing were replaced with new deeper, properly constructed wells. As noted by Morris and Johnson (1969), "It is significant to note that the percentage of high nitrate specimens has remained at the 18 to 22 percent level even though the specimens now submitted are from the types of well normally not subject to nitrate contamination." Such is the situation revealed in Figures 4 and 5; municipal wells that had excellent water quality 20 to 30 years ago, now consistently exceed or approach the nitrate MCL.

As noted above, there are many wells which do not exhibit nitrate problems because they are deep wells. This does not mean that one can simply write off shallow groundwater and drill deeper wells and avoid the problem. In part, this depth distribution is a function of time and, in many parts of Iowa, the groundwater aquifers are more deeply buried and there has simply not been enough time for the surficially applied chemicals of the past 10 to 20 years to reach these deeper portions of the groundwater system.

EXPERIMENT FARM STUDIES

Past reports (Baker, 1985a; 1985b; Baker and Laflen, 1983; Hallberg et al., 1983b; 1984b; Libra et al., 1984; Hallberg and Hoyer, 1982), have reviewed a variety of research ranging from water quality studies to agricultural engineering studies, to standard agronomic studies involved with crop yields, that show that losses of nitrogen, particularly as nitrate, below the root zone occur directly as a function of nitrogen fertilization, particularly at high application rates. For example, in a review, Baker and Laflen (1983) note: "NO_3-N losses with subsurface drainage related in nearly linear fashion to nitrogen application for rates exceeding 50 kilograms per hectare." The intent here is not to verbally review the details of these studies again, but to graphically display the results for comparison with the trends in groundwater quality changes and fertilizer rate increases noted in the Big Spring area and other parts of Iowa.

This review will only use data from experiment farm studies from the immediate midwestern states: Iowa, Minnesota and Wisconsin. These studies all measured the relationship between the amount of fertilizer-N applied to the soil during continuous corn cropping and the amount of nitrate-N that was stored in the soil profile to various (specified) depths, and/or the amounts of nitrate-N lost in tile drainage over some period of treatment. The horizontal (x) axis is the same on each diagram and shows the annual rate of fertilizer-N applied in kg N/ha. The vertical (y) axis shows: (1) the amount of nitrate-N (in kg N/ha) that was measured in the soil profile, to the specified depth at the end of the specified number of years of continuous treatment; or (2) the amount of nitrate-N lost in tile drainage water (and, on some, the concentration of nitrate-N in the tile drainage water). The scale of the vertical axis varies from graph to graph because the soil depths and/or years of treatment vary among the studies. Note that most of these studies used an unfertilized, or very low rate, fertilizer-N check plot to correct/adjust the data from the fertilizer-N plots. Also, some of these studies used high annual fertilizer-N rates of 336 to 448 kg N/ha (300 to 400 lbs N/Ac) to define the curves. While these are clearly in excess of recommended and optimum rates, such rates are in practice at least locally.

Fertilizer-N Lost to Tile Drainage Water or Left in the Soil

Figure 12 is a good example of such studies from Minnesota. These studies were conducted on poorly drained Webster soils (Randall, 1985a; 1986). Figure 12A shows the amount of nitrate-N lost in the tile drainage water and the average nitrate-N concentration after five years of treatment (1973-1979). Tile drainage water is shallow groundwater. Note that the same linear response between fertilizer-N amount (rate on Figure 12) and nitrate-N leaching losses to groundwater, reflected as nitrate-N losses in tile water, that appears at Big Spring (Figure 1) and elsewhere (Figures 2, 4 and 5) is apparent. In addition to the nitrogen lost in the tile water, a considerable amount of nitrate-N was left stored in the soil, which is shown on Figure 12B. The authors note that most of the nitrate-N was accumulated in the upper 2 m (6.5 ft) of the soil profile; or to a depth just slightly below the depth of tile drainage. Figure 12B also shows data from a companion study (Buzicky et al., 1983) which, using isotopically labeled fertilizer-N, directly calculated the percentage of fertilizer-N lost through leaching into tile effluents.

Similar data on nitrate-N losses in tile effluent in relation to fertilizer-N have been compiled in Iowa (Baker and Johnson, 1981; Kanwar et al., 1983; Baker and Austin, 1982; Hallberg et al., 1983b). Kanwar et al., (1983, p. 1457) summarize:

Measured as well as predicted data indicate that an equivalent of nearly half of the applied fertilizer nitrogen is being discharged with tile drainage water. As farmers decide to apply more fertilizers to obtain higher yields, large leaching losses of nitrates can be expected to occur, an economic as well as an environmental concern.

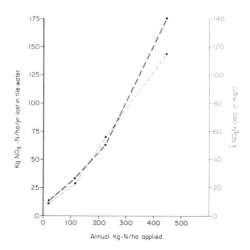

Figure 12A.   Losses of NO$_3$-N/ha (kg) in tile water and average nitrate-N concentration.

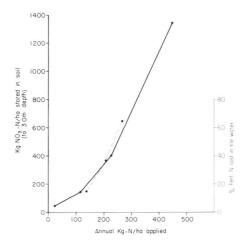

Figure 12B.   NO$_3$-N/ha (kg) stored in soil to 3.0 m (10 ft) depth and percent fertilizer-N lost in tile water from companion N-15 study.

Figure 12.   Minnesota experiment farm studies showing relationship between fertilizer rate (kg N/ha) and the amount (kg N/ha) of nitrate-N lost in tile drainage and the flow weighted average nitrate concentration in the tile drainage water (light dashed lines).   Webster CL soils; after 5 years treatment, 1973-1979; data from Gast et al., 1978; Nelson and Randall, 1983; Buzicky et al., 1983; Randall, 1985a; 1986.   (To convert to lbs NO$_3$-N/Ac, multiply by 0.9.)

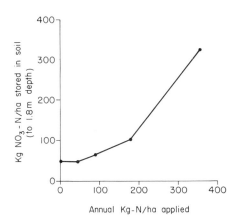

Figure 13A. Webster SiCl; 6 years of treatment (Jolley, 1976).

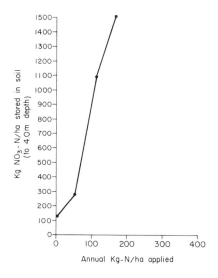

Figure 13B. Moody SiCL; 17 years (Jolley, 1974).

Figure 13. Iowa experiment farm studies showing relationship between nitrogen fertilization rate (kg N/ha) and the amount of nitrate-N (kg N/ha) stored in the soil to a given depth after specified
time of treatment. The following are noted under each figure: the
soil type, the years of treatment, and the reference.

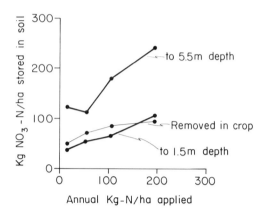

Figure 14.   Relationship among fertilizer-N rate and amount of nitrate-N stored in soil to 1.5 m (5 ft) and 5.5 m (18 ft) and nitrogen removed in crop: Minnesota; Webster L; 10 years (Nelson and MacGregor, 1973).

Other experiment farm work in Iowa shows the same trends in the amount of nitrate-N left stored in the soil profile also.   Figure 13A illustrates this for a Webster soil in Iowa.   In comparison to Figure 12, Figure 13A is for six years of treatment, but only to about 1.8 m (5 ft) depth, whereas Figure 12B sums up the nitrate-N in the soil to about 3.0 m (10 ft).   Figure 13B shows data for well to moderately well drained loess derived soils in western Iowa, over fertilizer-N application rates only reaching 168 kg N/ha (150 lbs N/Ac).   The important point to note in all these results is the nearly linear or curvilinear increase in the amount of nitrate-N stored in the soil after corn harvest in relationship to the increased rates of fertilization.

Crop Uptake of Fertilizer-N

Figure 14 shows similar trends from another study in Minnesota.   In this study (Nelson and MacGregor, 1973), the amount of nitrogen removed in the corn grain was also measured.   Note that the amount of nitrogen removed with the grain increases very little between application rates of about 100 to 200 kg N/ha, while the amount of nitrate-N accumulating in the soil continues to increase. Nelson and MacGregor also note that: (1) additional nitrate-N was lost in tile drainage water (an amount equivalent to about 35 percent of the fertilizer-N at the 200 kg N/ha rate may be estimated from their data); (2) little evidence indicated significant volatilization or gaseous losses of nitrogen; (3) greater nitrogen losses and slightly

Table 1. Estimated percentage recovery of fertilizer-N in corn grain from continuous corn production in the rotation fertility experiments in Iowa (Blackmer, 1984).

Location	Maximum Yield	Economic Optimum
	- - - % Recovery Fertilizer-N - - -	
Northwest Iowa	36	39
North Central Iowa	37	40
Northeast Iowa	25	28

lower corn yield were obtained with fall applied ammonium nitrate; and (4) corn nitrogen uptake may be somewhat greater with urea-N, than ammonium-nitrate-N.

Figure 15 shows the percentage of fertilizer-N remaining in the soil and removed in the grain for the 17 year Moody farm study by Jolley (1974; also see Figure 13B). In the range of 112 to 168 kg N/ha (100 to 150 lbs N/Ac) only 11 to 17 percent of the fertilizer-N was recovered in the grain; about 50 percent was left stored in the soil, and another 30 percent was not recovered. Jolley (1974, p. 126) notes that   30 percent "was possibly lost from the soil by NH_3 volatilization, denitrification or surface runoff or was immobilized by NH_4 fixation." In 1974, leaching losses were not considered, but even in the region Jolley was working, it is likely that some of this nitrogen was lost as nitrate in shallow groundwater moving laterally to discharge to streams. Perhaps the main point here is how little nitrogen was actually accounted for in the crop at these very common ranges of fertilization.

In this regard, Keeney (1982, p. 632), in a review of nitrogen in agriculture, notes: "N recovery by agronomic crops is seldom more than 70%, and the average value is probably nearer to 50%. . . ." However, as noted above for Jolley's study (1974), many published results for corn suggest that nitrogen recovery may be much lower. While the total fertilizer-N uptake of corn may be around 50 percent, the stover typically remains in the field. The amount of fertilizer-N removed in the harvested grain would more typically be in the range of 35 percent or less, particularly for continuous corn (e.g., Owens, 1960; Olsen et al., 1970; Jolley and Pierre, 1977; Chichester and Smith, 1978; Gerwing et al., 1979; Cooper et al., 1984; Meisinger et al., 1985).

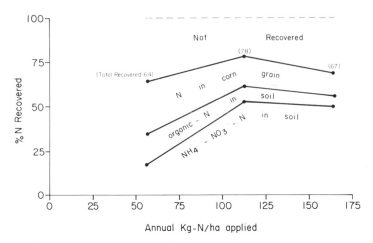

Figure 15. Percentage of fertilizer-N recovered in soil and grain (corrected for no-fertilizer-nitrogen check plot); Iowa; Moody SiCL; 17 years treatment (Jolley, 1974).

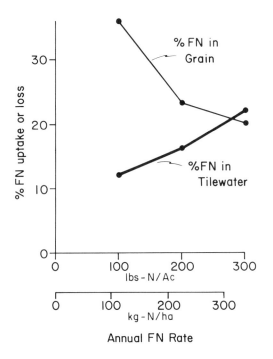

Figure 16. Relationship among fertilizer-N rate and the percentage fertilizer-N accounted for in grain and the percentage lost in tile drainage water.

Table 1 shows the estimated fertilizer-N recovery by corn grain from long-term continuous corn, fertility experiments in Iowa (Blackmer, 1984). The fertilizer-N recovery is only 25 to 37 percent at maximum yield, and 28 to 40 percent at optimum. These data are also derived by comparison with plots that are "cropped down" in respect to nitrogen; i.e., compared to check plots which have received no fertilizer-N during the long-term experiments. Blackmer (1985) notes even lower fertilizer-N recovery from labeled fertilizer-N studies in continuously fertilized plots. He notes a three year mean recovery of fertilizer-N in grain of only 19 percent during the year of fertilization. The two year mean for recovery of fertilizer-N in grain for the year following fertilization was only 0.8 percent. Further analysis of soil samples from the plots indicated only 15 to 20 percent fertilizer-N in the soil one year after application. This indicates that 60 to 65 percent of the fertilizer-N was lost by processes other than grain harvest.

Figure 16 graphically summarizes some of these short-term experiment farm results. As described in the preceding figures and text, as the rate of fertilizer-N goes up, the percentage recovery of fertilizer-N in grain generally goes down (above 100-150 kg fertilizer-N, at least), and the percentage of fertilizer-N lost in tile water and to other processes goes up.

Management Effects

Chemical leaching losses can also be affected by management differences. Burwell and others (1976) compared the effects of contour tillage and terracing in western Iowa at Treynor. While terracing reduced surface runoff, sheet-rill erosion, and associated chemical losses in runoff, it substantially increased infiltration and subsurface water flow. Most nitrate-N losses occur in the infiltration component, and hence the increased infiltration in these terraced areas greatly increased the losses of nitrate-N. In an adjacent watershed, Alberts and Spomer (1985) also note that contour, till-plant conservation tillage also increased the leaching losses of nitrate-N because of increased infiltration (less runoff). As Burwell and others (1976) note, appropriate chemical management must be combined with conservation practices to minimize soil erosion and runoff as well as chemical losses in "deep percolation" or groundwater.

Figure 17 shows the relationships among nitrogen fertilization rate, nitrate-N remaining in the soil and rotation effects for four end members in the rotation study of Olsen and others (1970). These authors found that the total amount of nitrate-N in the soil profile, the distribution of nitrate-N in the profile, and the amount of nitrate-N below typical corn rooting depth was directly related to the rate of fertilizer-N application on corn, the number of years of corn in the rotation, and to some extent the length of time since harvest

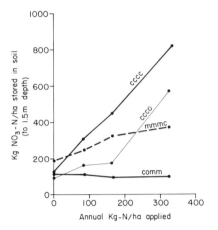

Figure 17.   Relationships among fertilizer-N rate, the amount of nitrate-N stored in the soil profile, and crop rotations; Wisconsin; Rosetta SiL; after 4 years rotation treatment (Olsen et al., 1970).

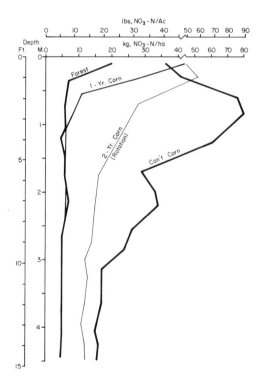

Figure 18.   Nitrate-N stored in soils under different land use/treatment in the Big Spring Basin; after harvest, fall of 1982.

of the last corn crop. Olsen and others (1970, p. 448) note that effective methods for limiting the amount of nitrate-N leached to groundwater are: "limiting rates of N fertilizer to approximately that required by the crop, reducing the acreage and frequency of corn or other crops that receive fertilizer N in the rotation, and maintaining a crop cover on the land. . . ."

Data from soil cores from the Big Spring Basin also illustrate this. Figure 18 shows the nitrate-N content from replicate soil cores under different land use from the fall (post-harvest) of 1982. The land use represents: forest such as woodlots that have been in timber for 40 years or more (some never cleared); first year corn, after forest (168 kg fertilizer-N/ha); second year corn after alfalfa, under high fertilizer-N management (average 1st yr--145 kg fertilizer-N/ha, 2nd yr--170 kg fertilizer-N/ha); and continuous corn (at least 4 to 6 yrs corn, average of 175 kg fertilizer-N/ha). The forested areas show relatively low nitrate-N concentrations, as expected. In many natural ecosystems such as prairie or forest the nitrate-N lost in leaching often has to be measured in parts per billion (McArthur et al., 1985). As noted by Hallberg et al. (1983a; 1983b; 1984b), the natural background nitrate-N concentration in Iowa groundwater was generally less than 1 mg/l.

As noted by Olsen et al. (1970) and shown in Figure 18, the nitrate-N content in the soil, to depth, is directly related to the number of years of fertilization. Proceeding to the continuous corn data, multiple "bulges" of the nitrate-N are apparent, and the nitrate-N content is still increased at depths of 4.5 m (15 ft)!

Beyond these notes on management, the important point of all these figures in relationship to the groundwater data is that the linear response between the increased application of fertilizer-N and the increase in nitrate concentrations in "shallow" groundwater (Figure 1) is exactly what should be predicted from these standard agronomic studies. All of the studies reviewed were conducted in areas where aquifers are at substantial depth. If an aquifer is inserted at depths of 1 to 5 m into each of these studies, as is the case in extensive areas in northeastern Iowa, the nitrate-N being 'stored' in the soil or lost in tile drainage is now translocated into the aquifer. This is particularly true for areas where carbonate or sand and gravel aquifers occur, not only where they are at very shallow depth, but because they promote deep and rapid percolation of soil groundwater, even in areas with a thicker mantle of soil and quaternary deposits. The thicker soil mantle may simply increase the time involved to leach the nitrate into the groundwater. Olsen et al. (1970) note that for silt loam soils, the bulk of the residual nitrate-N moved downward through the soil at a rate of about 0.3 to 0.5 m per year. They concluded that with excess amounts of fertilizer-N on continuous corn, there was a good probability that the groundwater would be contaminated, but because of the slow rate of movement, it may not be apparent for many years.

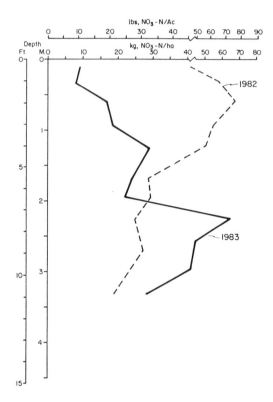

Figure 19. Comparison of nitrate-N in soil profiles, under continuous corn; from fall 1982 and fall 1983, Big Spring Basin.

Nitrate-N can move more rapidly even in these medium textured soils, however. Figure 19 shows the nitrate-N content from replicate soil cores from a continuous corn field on a loess derived Downs silt loam from the Big Spring Basin. The fertilizer-N rate has averaged 190 kg fertilizer-N/ha for at least six years. Data from 1982 and 1983 are shown. The fall of 1983 was quite wet and considerable nitrate-N was leached through the soil as noted by Hallberg et al. (1984b) from soil cores, groundwater and tile line data. In the fall of 1982, the major concentration of nitrate-N occurred between 0.5 and 1.0 m, but after harvest in the wet fall of 1983, the major concentration was between 2 and 3 m, well beyond the root zone.

Timing of Fertilizer-N Losses

There are many long- and short-term aspects of the leaching of nitrate which are important. As Olsen and others noted (above), groundwater contamination may not be evident for many years after excess fertilizer-N use begins in areas with a thick, medium to fine textured soil mantle, or in areas with a substantial thickness of unsaturated zone. Even in sandy soils where the water table is deep some studies suggest it will take 10 to 50 years for the leaching nitrate-N to reach the groundwater (Pratt et al., 1972; Adriano et al., 1972; Pratt, 1984; Adelman et al., 1985; Carey and Lloyd, 1985).

The temporal increases in nitrate in groundwater, nearly concurrent with increases in fertilizer-N use, shown in Figures 1, 2, 4, and 5 are from very responsive hydrogeologic settings. In less responsive settings only slight increases have occurred, and in many areas no effects are yet obvious (in aquifers or groundwater/drinking water supplies). Carey and Lloyd (1985) modeled areas with moderate thicknesses of glacial deposits, analogous to much of Iowa and noted that the full impact of today's excess fertilizer-N use will not be noted in these groundwater aquifers for possibly 30 to 40 years! Thus, the depth distribution of nitrate noted in Figures 7 and 8 are likely only a temporary status quo. For many groundwater supplies the problems are just beginning.

In this regard, over most of Iowa the water table is not very deep; although groundwater that is used for drinking water may be. However, the leaching losses of fertilizer-N go on and are of concern, even where there are no apparent effects on the groundwater used for drinking water. The experiment farm studies cited in the previous sections were conducted in southern Minnesota, north central and northwestern Iowa. These studies clearly indicate the economic concern with fertilizer-N losses, but they were conducted in areas where the shallow groundwater is not a drinking water source. Even though in these areas the leaching of agricultural chemicals probably has not directly affected groundwater/drinking water supplies, the same chemical inefficiencies are apparent.

While these studies clearly show that leaching losses of nitrate-N to groundwater occur in response to large applications of fertilizer-N, this does not mean that all the nitrate-N delivered to groundwater comes from fertilizer-N. From some view points only this cause and effect relation is important. However, to begin to know what to do about the problem, more details on the processes are important.

Labeled-nitrogen studies are providing many insights into these processes. Many of these studies leave little doubt that rapid losses of fertilizer-N occur (Cerrato et al., 1985; Priebe et al., 1983; Randall, 1985a; 1985b; White, 1985; Sanchez and Blackmer, 1985; Owens, 1960). Rice and Smith (1983) noted 30 to 38 percent fertilizer-N lost from the upper part of the soil within two weeks of application. Blackmer (1985) has noted that in some studies more than 50 percent of the fertilizer-N was lost from the rooting zone within a few weeks of application.

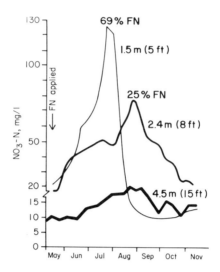

Figure 20. Nitrate-N concentration in water at various depths, with time after application of labeled fertilizer-N (FN); percent fertilizer-N at peak concentration from N-15 (after Gerwing et al., 1979).

Figure 20 shows the concentration of nitrate-N in soil water and groundwater for one growing season beneath test plots fertilized with N-15 labeled fertilizer-N. The lesser increases in nitrate-N with greater depth show the effects of dispersion, dilution, and storage within the soil water system. The main point, however, is to note that at the peak nitrate-N concentration, 69 percent of the nitrate-N at 1.5 m was derived from that season's N-15 fertilizer-N; 25 percent of the nitrate-N at 2.4 m was fertilizer-N. Substantial quantities of fertilizer-N can be readily leached, in part because chemical fertilizer-N forms may often be readily converted to nitrate.

The amount of fertilizer-N stored in the soil also plays a role that must be considered when evaluating the importance of various studies. This can be illustrated by returning to data from the Lamberton, Minnesota, experiments (Figure 12; Randall, 2985a; 1986). Figure 21 shows the amount of nitrate-N lost in tile waters from the plots versus the annual rate of fertilizer-N. Treatments were continuous from 1973-1979. In the early phases of the study (1973-1975) nitrate-N losses were relatively small, but a substantial amount of nitrate-N had been stored in the soil (Figure 22). From 1975 to 1979, soil storage only changed appreciably for the highest fertilizer-N rate (Figure 22), but nitrate-N lost in the tile water increased 3 to 4 fold. In the fall of 1979, the equivalent of 14, 17 and 21 percent of the fertilizer-N applied was lost in the tile water from the 112, 224 and 448 kg fertilizer-N/ha treatments, respectively. When combined with the

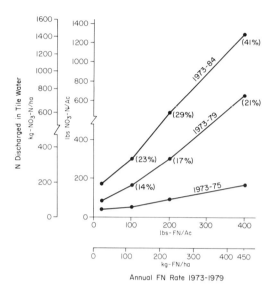

Figure 21. Nitrate-N losses in tile effluent, in relation to fertilizer-N (FN) rate at various times. From Lamberton, Minnesota, experiment farm studies (after Gast et al., 1978; Nelson and Randall, 1983; Randall, 1985a; 1986).

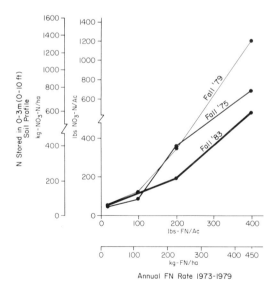

Figure 22. Nitrate-N remaining stored in soil to 6 m depth at various times in relation to fertilizer-N (FN) rate. From Lamberton, Minnesota, experiment farm studies (after Gast et al., 1978; Nelson and Randall, 1983; Randall, 1985a; 1986).

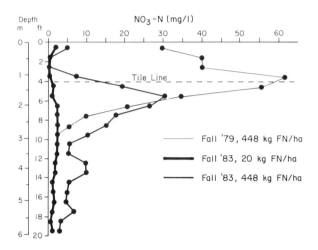

Figure 23. Nitrate-N in soil after seven (1979) and eleven (1983) years of experiment; annual fertilizer-N (FN) rate from 1972 to 1979 (Lamberton, Minnesota; from Randall, 1986).

amount of nitrate-N stored in the soil (1979, Figures 21 and 22), this amounts to a 26, 40 and 62 percent equivalent of fertilizer-N applied.

Of interest is that fertilizer-N application was stopped in 1979, but corn was still grown on the plots through 1984 to evaluate residual effects. From 1979 to 1984, a considerable portion of this previously applied nitrogen continued to be leached to the tile water (and below tile lines; Randall, 1985a; 1985b; 1986). Average annual nitrate-N loss rates were actually higher the five years after fertilizer-N application (1979-1984) than during the seven years of fertilizer-N treatment. This is likely related, in part, to climatic differences. It also illustrates how storage in the soil system, and resultant time lags in nitrate movement and losses, can affect interpretations. Fertilizer-N losses could be highly underestimated using only the tile data over only the first three years of the study, for example.

At Lamberton, five years after fertilizer-N treatment stopped, nitrate-N concentrations in the tile water still remain above back-ground for at least the two highest fertilizer-N treatments. Other studies have also noted such multiyear influences on groundwater quality (e.g., Baker and Johnson, 1981; Owens et al., 1985; Hallberg et al., 1986). Figure 23 graphically shows these differences in nitrate-N in the soil for the fall of 1979 and fall of 1983 (from Randall, 1985a; 1986). Note the downward movement of the peak of nitrate-N and the increase in nitrate-N concentration from 3 to 6 m in depth. The shape of the nitrate-N profile is suggestive of the various processes that go on in the soil. The appearance of the increase in nitrate-N to depth is what would likely occur from the preferential flow of water and nitrate-N down macropores in the soil. Whereas the gradual downward movement of the peak of nitrate-N occurs in response to the slow flux of water and nitrate-N through the matrix of the soil.

NITRATE AND THE HYDROLOGIC CYCLE

As noted above, studies of tile drainage water help to define relationships among chemical application rates and leaching responses to groundwater. It is clear from many studies that what is applied at the soil surface affects the quality of the tile drainage water.

In a responsive hydrologic system such as the Big Spring Basin, an even more direct link between tile drainage water and groundwater in an aquifer can be illustrated. Figure 24 shows nitrate concentrations from tile lines under fertilized, continuous corn, from a perennial stream within the Basin, and from the Big Spring groundwater itself. The same general seasonal trends and even pronounced short-term variations in nitrate are clearly coincident between these sites and Big Spring. This clearly shows that the tile line data are indicative of the quality of the infiltrating groundwater (from fertilized nitrate contributing areas) which recharge the aquifer. What is done at the land surface affects the quality of the tile waters, and these same effects are propagated through the system as stream baseflow and on to the larger groundwater system as a whole.

Surface Water Problems

In Iowa, groundwater discharges into master streams provide perennial flow. This is why streams continue to flow, even after many weeks with no runoff producing events. This interconnection also affects surface water or stream water quality.

In Iowa, many municipal water supplies and other public water supplies rely solely or partially on surface water. Over the past years, a great deal of money and energy has gone into resolving some of the surface water quality problems. Great strides have been made in Iowa, and the nation as a whole, in developing sewage treatment facilities and working with industry to control point source discharges into streams and lakes. Yet even where waste treatment effects have been essentially eliminated, high nitrate loads have continued or increased in many streams.

This is where the interconnection between groundwater and surface water again plays a role. The high nitrate concentrations that appear in Iowa streams are related to periods of high recharge from shallow groundwater into the streams. The nitrate is mobilized by water infiltrating through soil, recharging shallow groundwater which then discharges into surface waters.

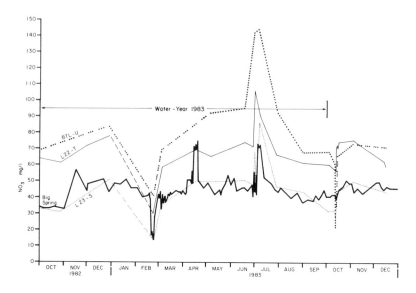

Figure 24.   Nitrate concentrations over time for tile line drainage water (L22-T, BTL-U), surface water (L23-S) and groundwater at Big Spring (Hallberg et al., 1984b).

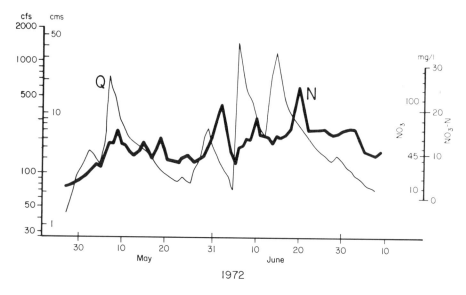

Figure 25.   Discharge (Q) and nitrate concentration for the Skunk River near Ames (Johnson and Baker, 1973; Baker and Johnson, 1977).

This is illustrated in Figure 25, which shows the discharge of the Skunk River and the nitrate-N concentration of the water over time. Note that as the discharge rises the nitrate-N concentration is either stable or actually decreases. That is because the overland runoff water that forms these discharge peaks is actually quite low in nitrate-N. Nitrate forms in the soil and hence is picked up in the water moving through the soil (i.e., groundwater), not moving over it (i.e., runoff or surface water). In Figure 25, note that the peak of nitrate-N occurs as the river discharge is receding. The recession of discharge events are periods of enhanced groundwater discharge into the streams (from tile lines and other shallow groundwater movement). Thus, after the runoff water peak has rapidly moved by, the slower moving groundwater, added by the rain which generated the event (and bank storage, etc.) moves through the soil, mobilizing nitrate and discharges into the stream producing the increases in nitrate concentration.

Thus, the timing of nitrate fluctuations is related to seasonal recharge, not to the timing of seasonal agricultural practices. This is why nitrate concentrations in streams (and wells) increase during spring recharge, often several weeks before nitrogen fertilizers are applied.

MAGNITUDE OF NITROGEN LOSSES

The magnitude of the nitrogen losses measured in the Big Spring Basin are also enlightening. For three years of monitoring, the amount of nitrate-N lost from the basin in surface water and groundwater varied from 820,000 to 1,300,000 kilograms. These losses are equivalent to about 33 to 55 percent of the average amount of fertilizer-N applied during the preceding years (Table 2). Furthermore, these are minimum figures for the amount of nitrogen lost because only the nitrate-N losses were computed. Other forms of nitrogen are also discharged with the water, and losses by denitrification and in-stream consumption cannot be estimated. As with the studies noted above, this is not meant to imply that all of this nitrate-N is derived directly from the fertilizer-N. However, the large losses occur in response to the large amounts applied.

As discussed, these figures are readily compatible with experiment farm data which show that only 35 percent of fertilizer-N applied is likely removed over the long term and that substantial losses occur, directly with water, and to storage in the soil system. Such losses were "built-in" to nitrogen fertilizer yield response models before the inefficiencies and losses involved were understood. This is also evident from studies noting the wide variance in nitrogen fertilizer recommendations which have not resulted in significant yield differences (e.g., Olson, 1985; Olson et al., 1982; 1985).

Table 2. Nitrate-N losses in groundwater and surface water from the Big Spring Basin for three water years (Hallberg et al., 1985).

Nitrate-N Losses	Water Year		
	1982	1983	1984
NO₃-N discharged in water (thousands kg)	821	1300	817
Equivalent kg-N/ha total basin	31	49	31
Equivalent kg-N/ha, long-term corn acreage*	52	83	52
Equivalent of average fertilizer-N applied (percent)*	33%	55%	33%

*five year average, excluding PIK

THOSE OTHER SOURCES OF NITROGEN

When there are so many sources of nitrogen in the environment, why has all the focus turned to fertilizer-N? Some perspectives can be helpful. Even for the United States as a whole, if all the livestock and poultry wastes were reclaimed and land applied, they would still (ideally) supply only 40 percent of the nitrogen currently applied as fertilizer-N. Even under best management, only about half of this nitrogen can be recovered, thus reducing the figure to 20 percent (Pratt et al., 1975; Olson, 1985). Also, sewage sludge from municipal systems would only constitute 10 percent as much nitrogen as fertilizer-N.

In the Big Spring Basin, for example, all estimates of household sewage-N are less than 0.1 percent of the fertilizer-N currently applied! The relative inputs of manure-N and fertilizer-N are clearly shown in Figure 1. Even for the entire state of Iowa, the total amount of nitrogen in household sewage would only equal about one percent of the fertilizer-N applied.

When considering all the typical sources of nitrogen to cropland in Iowa, an average balance shows fertilizer-N contributed 55 to 60 percent; manure-N 10 to 15 percent; legume-N 8 to 10 percent; rainfall-N 5 to 8 percent; and natural soil nitrogen possibly 15 to 20 percent. Using estimates from total crop harvest statistics, harvested corn grain can only account for about 35 to 60 percent of the nitrogen fertilizer applied to corn in Iowa over the past five years. In the Big Spring Basin, conservative estimates from inventories of agricultural practices show that fertilizer-N, alfalfa-N and manure-N account for at least 50 percent more nitrogen than needed for crop production (Kapp and Padgitt, 1985; Padgitt, 1985). Similarly, for Nebraska as a whole, Olson (1985) estimates that, since 1968, nitrogen fertilization rates have exceeded the annual crop requirements by 20 to 60 percent per year.

Well Construction, Septic System Problems

Nitrate problems in rural wells are often attributed to improper septic systems, poor well construction or well placement, and other local problems such as fertilizer spills or manure handling. As discussed with Figure 6, such situations do occur, but they cannot explain the systematic region-wide nitrate contamination of surficial aquifers that has occurred (see Hallberg and Hoyer, 1982; Hallberg et al., 1983a; 1983b; Libra et al., 1984). Hallberg (1985) discussed cases around local agricultural chemical dealerships, mixing and rinse facilities where nitrate-N levels are seriously increased, likely from many small-scale spills and rinse water discharge. However, several studies in Iowa have found no relationship between nitrate in individual well water and feedlots, barnyards, manure storage or density of septic tanks (Mancl and Beer, 1982; Hallberg et al., 1983; Libra et al., 1984).

About half of the waterborne disease outbreaks in the United States are related to contaminated groundwater; septic systems are the most frequently reported cause of the contamination (Yates, 1985). Where groundwater contamination occurs from septic systems, it is generally related to the density of the systems as well. In a study of shallow groundwater quality in relation to a high density of septic systems (130 units over 58 acres), only 1 percent of the wells in the area exhibited any bacterial contamination (Mancl and Beer, 1982). Loading of nitrogen to the soil from septic systems was less than 30 percent of that from adjacent areas of nitrogen fertilized corn. The highest groundwater nitrate values appeared to relate to groundwater flow from adjacent areas of nitrogen fertilized corn into the subdivision area, and nitrate concentrations actually decreased in the area of dense septic system use.

The effects of poor well construction or placement and septic systems on groundwater quality from individual wells should be randomly distributed across various geologic settings. Indeed, the most obvious cases occur in areas where a bedrock aquifer is deeply buried by an aquitard (fine textured glacial materials or shales), and the well is also cased to substantial depth but is poorly located or constructed, i.e., the well is located or constructed such that it allows seepage or runoff from a septic system or a feedlot to enter the well. In such "Deep Bedrock" settings, nitrate or bacterial contamination of groundwater would not be expected (yet) from regional recharge. In a review of over 16,000 water quality analyses from northeastern Iowa, Hallberg and Hoyer (1982) noted that bacterial problems were randomly distributed in relation to geologic setting and well depth, but nitrate contamination was significantly and systematically related to geologic settings prone to contamination from infiltration recharge. One would expect problems from well construction, septic tank placement, etc., to be randomly distributed. Even in these uncontrolled data sets, less than 15 percent of "Deep Bedrock" wells (greater than 50 ft deep) exhibited any nitrate problems.

In a more detailed field assessment of well characteristics and water quality, Hallberg et al. (1983b) inventoried 271 wells in the Big Spring Basin. A total of 32 percent of the wells showed potential construction or placement problems; in 25 percent of all the wells, bacterial problems could be linked to the use of cisterns, however. Again, these problem wells exhibited significant bacterial contamination, but as a whole, there was no significant difference in nitrate concentrations. The median nitrate concentration for nonproblem wells was 6.7 mg/l as NO_3-N (30 mg/l as NO_3) (with a range of less than 1 to 40 mg/l as NO_3-N) and the median for wells with obvious problems was 8.1 mg/l NO_3-N (37 mg/l as NO_3) (with a range of less than 1 to 62 mg/l as NO_3-N). While the median nitrate concentration and the extreme high values are greater in the problem wells, the differences are not statistically significant. In less than 2 percent of the wells sampled could nitrate contamination be attributed to these local placement/construction problems.

In another study, Libra et al. (1984) found that 6 percent of deep wells had potential problems, but while some of these wells showed bacterial contamination, none exhibited nitrate problems and, in fact, all had less than detectable nitrate concentrations. During the past year, sampling of over 400 wells in Mitchell County showed similar results. In "Deep Bedrock" areas, where monitoring has shown that nitrate contamination from recharge has not reached the aquifer yet, less than 5 percent of the wells exhibit any nitrate contamination (greater than 1 mg/l NO_3-N) potentially from placement/construction problems. In most of these studies, wells with frost pits that could allow seepage tended to have somewhat higher nitrate concentrations, but there were always nearly equal numbers with no problems.

In sum, well placement/construction problems and septic systems indeed cause bacterial contamination problems and locally add to nitrate problems. Such problems are of concern for the safety of drinking water from individual wells, especially because these are problems that can be avoided or cured. But such problems contribute only a very small proportion, less than 5 percent, to the overall regional nitrate contamination that is evident in Iowa. Even if one extrapolates from uncontrolled data sets and suggests that 15 to 18 percent of all wells may exhibit local problems and hence contribute nitrate into the groundwater system, various studies indicate that such problems no longer contribute any significant nitrate contamination. The effect of local well problems must be viewed as a site-specific contribution, superposed on a regional nonpoint source nitrate contamination problem and, in fact, in some settings local "problems" appear to dilute the regional nitrate concentrations.

NITROGEN BALANCE IN THE BIG SPRING BASIN

Many of the points discussed in the preceding sections may be summarized by reviewing the nitrogen balance in the Big Spring Basin. Figure 26 shows the cumulative nitrogen inputs from various sources in the Big Spring Basin. The fertilizer-N and manure-N data are as shown on Figure 1 (derived from Hallberg et al., 1983b; 1984b; Padgitt, 1985). Alfalfa-N contributions were reconstructed from statistical data on acreage and rotations (Hallberg et al., 1983b) and by assuming that the alfalfa being rotated to corn contributed 112 kg N/ha.

The nitrogen deposited by precipitation was also estimated. Nitrate-N and ammonium-N in precipitation has been monitored continuously at Big Spring since early 1984, and intermittently since early 1983. The average nitrogen concentration per inch of precipitation from these records was multiplied times the annual precipitation to derive the values shown on Figure 26. Two lines for precipitation-N are shown: the lower line indicates the nitrogen deposited on the land in row crops; the upper line shows the total for the entire basin.

As noted earlier, there is essentially no excess nitrogen contributed from areas of native vegetation or from sewage--the total nitrogen contributed by septic tanks would equal less than 0.1 percent of the fertilizer-N applied. This leaves one other major source of nitrogen--that released by mineralization of organic-N in the soil--both natural organic-N and that derived from corn stover.

The contribution of mineralized soil-N can be estimated by various methods, and for the soils in the basin may range from 28 to 90 kg N/ha (25 to 80 lbs N/Ac) for the areas of cultivated soils (Hallberg et al., 1983b; Harmon and Duncan, 1978). But the basin is dominated by soils formed under forest, and the average contribution for the basin would be on the low end of this range.

No separate values for soil-N are shown on Figure 26 for two reasons. First, there are many uncertainties in the estimate. Second, and more importantly, is that the value for manure-N actually available in the soil crop system is over-estimated. The amount of manure-N shown assumes a much higher level of manure management than actually in practice in the basin, as noted from surveys of management practices (Padgitt, 1985). With the level of management in use in the basin, likely less than half of the manure-N shown is actually available for crop use or nitrate production in

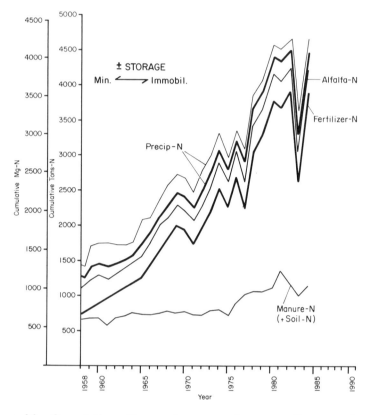

Figure 26. Cumulative nitrogen inputs to the crop-soil system in the Big Spring Basin.

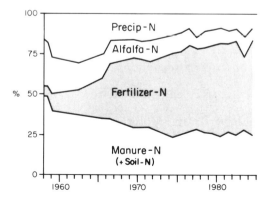

Figure 27. Percentage of nitrogen inputs in Big Spring Basin.

Table 3.   Estimated average percentage of nitrogen inputs from various sources in the Big Spring Basin (see Hallberg et al., 1983b; 1984b; Padgitt, 1985; Kapp and Padgitt, 1985).

Nitrogen Source	Percent
Fertilizer-N	56
Manure-N and Soil-N	25
Legume/Rotation-N	9
Rainfall-N	10

the soil system. The range of soil-N available is essentially equal to (or less than) the excess manure-N shown. Thus, the value shown as manure-N is a good approximation of manure-N plus soil-N actually available. Figure 26 then provides a reasonable estimate of the nitrogen inputs to the crop soil system in the Big Spring Basin over time.

These estimates are summarized on a percentage basis in Figure 27. As apparent on Figures 26 and 27, the total and relative input of fertilizer-N increases dramatically over time. Before 1960, fertilizer-N was often less than 10 percent of total nitrogen and approximately equalled the amount of precipitation-N deposited on corn acreage. Over time, the amount of alfalfa-N has decreased slightly as more land went into continuous corn. The amount of manure-N (soil-N) actually increased, but its percentage contribution decreased because of the very large increase in fertilizer-N.

From 1970 on, fertilizer-N comprises 50 percent or more of total nitrogen; after 1978, fertilizer-N approaches 60 percent. Table 3 summarizes average percentage of nitrogen inputs under current management (1978-1985, exclusive of PIK). Fertilizer-N is the largest input, the most controllable input, as well as the nitrogen input farmers pay for. Hence, it is the input that must be focused upon in any management scheme to mitigate water quality problems.

Figure 28 shows the cumulative nitrogen removed from the basin in relation to the nitrogen input (top line). The nitrogen removed in corn grain shown comes from the two estimates noted on Figure 1. The nitrogen removed in water comes from the values measured in the basin since 1981. Prior to 1981, the values are based on estimates from the relationship between Turkey River discharge and basin discharge (see Hallberg et al., 1983b; 1984b), and the integration of nitrate-N concentrations over time at Big Spring (as shown on Figure 1). The top, light line for water adds an additional 10 percent for ammonium-N lost in water, estimated from the basin monitoring.

The nitrogen "removed" in the stover is estimated assuming that 60 percent of the nitrogen taken up by the corn is in the grain and hence 40 percent is in the stover. However, it is clearly not accurate to call the nitrogen in stover "removed" from the system. In a particular growing year, the nitrogen in the stover has been removed from the soil, but most of the stover is returned to the system (though some is obviously recycled through cattle!). Hence, in the long term much of the stover is also an input, or may be looked at as an equilibrium product, and part of the nitrogen input from soil/organic-N.

It is obvious that even considering the stover there is a substantial amount of nitrogen that cannot be accounted for. This is summarized graphically in Figure 29, which shows the percentage of total input nitrogen removed by grain, water, and stover. The grain percentage was estimated using the average of the two amounts of nitrogen removed in grain (Figure 28), and the stover "plus" values are based on the amount shown for stover on Figure 28, "plus" the remainder of the grain-N estimate.

The large amount unaccounted for during the 1960s results from relatively dry conditions and, hence, relatively low discharges and losses in water. As precipitation and discharge returned to their long-term normal in the 1970s and 1980s, the percent removed in water increased. Prior to 1970, the nitrogen removed in water averaged about 10 percent; after 1970 this averaged about 20 percent of total nitrogen, ranging from 12 to 43 percent. For the entire period, grain-N removal averages 30 percent, but as with water, this increases after 1970, averaging 33 percent because of increased yields (likely related to better moisture and fertilizer-N inputs).

Averaging the past 10 years' data, the nitrogen removed in grain and water only equals about 55 percent of nitrogen inputs; adding the stover "plus" the total only averages about 75 percent! Where has all this nitrogen gone?

First, there are other losses. Some losses from the basin occur as organic-N discharged as particulates with suspended matter in surface water (and some with groundwater). However, much of this nitrogen discharged in major streams relates to eroded sediment from stream beds and stream banks and is not directly related to the short-term nitrogen inputs.

A second process to remove nitrogen is by gaseous losses from the soil. This occurs by both denitrification as well as oxidation of ammonium. Particularly in a well drained, rain fed agricultural environment such as the Big Spring Basin, the bulk of losses are

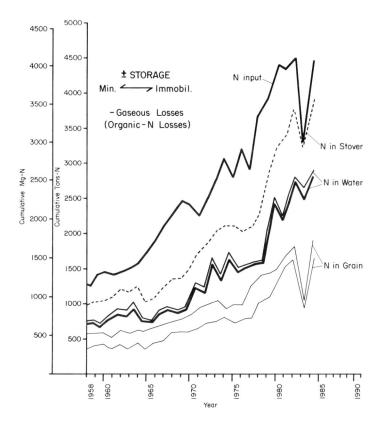

Figure 28. Cumulative nitrogen removed from crop-soil system in the Big Spring Basin. See text for explanation.

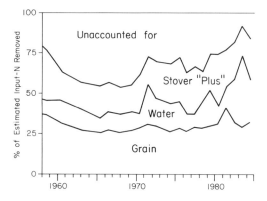

Figure 29. Percentage of total input nitrogen removed in Big Spring Basin. Stover "plus" indicates estimate for stover, plus remainder from range of grain estimate from Figure 28.

likely related to the activity of ammonium oxidizers, or nitrifiers (Breitenbeck et al., 1980; Bremner and Blackmer, 1978; 1981). *Total* rates estimated also depend on the time of year measurements are made but emissions range from less than 1 to 16 kg N/ha for rainfed agricultural systems (Bremner et al., 1980; 1981; Duxbury et al., 1982; Goodroad et al., 1984). The proportion of fertilizer-N evolved as N_2O in a cropping season is generally noted as about 2 percent or less (e.g., Breitenbeck et al., 1980; Cochran et al., 1981; Goodroad et al., 1984). As noted by Blackmer (1985), volatilization and denitrification seem far less important than often assumed.

In apparent contrast to these statements is the fact that the losses in water from the Big Spring Basin are minimum figures because Hallberg et al. (1984b) have documented denitrification losses of nitrogen. However, this denitrification occurs, not at the soil surface but at the top of the water table in particular alluvial soils. Besides denitrification, substantial nitrogen losses occur through in-stream consumption of nitrogen by algae, etc., in surface waters. In total, this may account for an additional 2 to 10 percent nitrogen loss (and perhaps more), depending upon conditions in a particular year.

A major parameter that also comes into play is the soil storage factor. As noted in the experiment farm studies, storage of nitrogen is a major factor, often equal to the amount of nitrogen discharged in water. The storage takes two primary forms: (1) nitrogen immobilized in the soil, either converted to organic-N or as ammonium attached to clay particles; and (2) nitrogen stored as nitrate-N at depth in the soil (and aquifer), some of which has migrated into the slowly mobile, matrix pore water. The magnitude of this effect is apparent during the 1983 PIK year. Note this is the only year the estimated nitrogen removed approximates the amount input (Figures 26 and 28). The reasons for this are: (1) the significant reduction of fertilizer-N input; and (2) the large amount removed in water because it was a wet year. Even though this year had the smallest inputs of recent times, it had the largest outputs in water. The amount removed in water in 1983 equalled 94 percent of the fertilizer-N applied that year, and this would appear to be impossible. The large outputs reflect the removal of nitrogen stored in the system during prior, drier years (see Figure 29). The greater than normal water flux mobilized this nitrogen as nitrate and moved it through the system. Just as noted in the experiment farm studies, excess fertilizer-N may be stored and continue to be leached into the groundwater for many years, even after cessation of fertilizer-N application. This storage/time lag may also be reflected in the sharp drop in nitrate in groundwater in 1985 (Figure 1). This decrease is, in large part, related to dry conditions, but a portion of it may be related to 1983 PIK reductions.

Before 1965, the nitrogen removed in grain and water equalled more than the fertilizer-N input. After 1969, with increasing fertilizer-N rates, grain and water removed an equivalent of between 75 and 100 percent of the fertilizer-N. Hence, with stover uptake considered, an equivalent of over 98 percent of the fertilizer-N applied can always be accounted for. Since 1978, the amount of grain-N was equivalent to an average of 52 percent of fertilizer-N, and measured nitrogen removed in water has equaled from between 25 and 53 percent of fertilizer-N (ignoring PIK). As reviewed in this paper, most of the leaching losses are related to high rates of fertilizer-N inputs, even though the resultant nitrogen leached comes from all sources. The nitrate, from whatever source, is leached to groundwater because the nitrogen inputs are far in excess of nitrogen uptake. As noted, the chemical form of fertilizer-N often allows it to be readily converted to the leachable nitrate form. Fertilizer-N must be the focus of attention in this problem because it is (1) the largest input to the system; (2) the most (only?) controllable input; and (3) the input farmers pay for.

DISCUSSION

The groundwater quality problems related to agricultural fertilizer-N use, that are described here, can only be resolved through a more "holistic" approach to agricultural management. Standard concerns for soil conservation and surface water quality must be coupled with the need to protect groundwater as well. Ways must be found to balance the need for efficient and profitable agricultural production with the need for safe drinking water.

There are legitimate concerns for public health, over the long term, if these water quality problems continue or increase. Beyond the health concerns and environmental impacts, the magnitude of the nitrogen losses are of economic concern as well. When 50 to 70 percent of the nitrogen applied is not going into grain production, there is obvious room for improved efficiency and economic gain. In this regard, the concerns reach much further than northeastern Iowa; the processes and losses documented in this report, for example, go on everywhere even where drinking water quality has not yet been affected.

As additional examples, in a project in Hall County, Nebraska, with irrigated corn, where groundwater nitrate was a growing problem, nitrogen fertilizer application rates were reduced 90 kg N/ha (80 lbs N/Ac) with no reduction in yield over four years, simply with good nitrogen management (Frank et al., 1984; Olson, 1985). Recent work in Pennsylvania shows that some current methods recommend fertilizer nitrogen applications more than 100 kg N/ha (90 lbs N/Ac) greater than rates that would produce economic optimum production (Fox and Piekielek, 1984). Results of this study produced information that allowed more accurate recommendations that, if followed, would reduce nitrogen fertilizer use by over 50 kg N/ha (45 lbs N/Ac) in Pennsylvania, and still achieve economic optimum yields.

While septic tanks, chemical spills, and poor well construction cause local problems, they are no longer a significant factor. Nitrate problems have become regional in scope, resulting from the wide-spread application of fertilizer. As noted, grain production, at best only accounts for about 50 percent of the nitrogen that can be managed.

It must also be made clear that the intent of research programs developed in Iowa is not to simply reduce fertilizer-N inputs. Granted, some reductions will likely play a role. The intent of the on-going efforts is to find the most effective ways to channel the fertilizer-N into grain production and minimize losses into water. This could also result in ways to use current amounts of fertilizer-N and improve yields. Undoubtedly, these efforts will result in improved efficiency of nitrogen use.

These efforts are not only needed, but imperative. Collectively, all of us involved with agriculture must find workable and rational solutions to this problem. Regulatory or mandatory controls on fertilizer-N would be harmful to agriculture at this time but are not beyond the realms of possibility, if groundwater problems increase, and no clear mitigation plan is in progress.

LITERATURE CITED

Adelman, D.D., W.J. Schroeder, R.J. Smaus and G.R. Wallin. 1985. Overview of nitrate in Nebraska's groundwater. Trans. Nebr. Acad. Sci. 13:75-81.

Adriano D.C., P.F. Pratt and F.H. Takatori. 1972. Nitrate in unsaturated zone of an alluvial soil in relation to fertilizer nitrogen rate and irrigation level. J. Environ. Qual. 1:418-422.

Alberts, E.E. and R.G. Spomer. 1985. Dissolved nitrogen and phosphorus in runoff from watersheds in conservation and conventional tillage. J. Soil Water Conserv. 40(1): 153-157.

Baker, J.L. 1985a. Conservation tillage: Water quality considerations. In: F.M. D'Itri (Ed.), A Systems Approach to Conservation Tillage. Chelsea, MI: Lewis Publishing, Inc., pp. 217-238.

Baker, J.L. 1985b. Sources and fates of material influencing water quality in the agricultural midwest. Perspectives on Nonpoint Source Pollution, USEPA, EPA 440/5 85-001, Washington, DC, pp. 467-470.

Baker, J.L. and T.A. Austin. 1985. Impact of agricultural drainage wells on groundwater quality. U.S. Environmental Protection Agency, Contract Report No. G007228010, 126 pp.

Baker, J.L. and H.P. Johnson. 1977. Impact of subsurface drainage on water quality. Proc. Third. National Drainage Symp., Am. Soc. Ag. Eng., St. Joseph, MO.

Baker, J.L. and H.P. Johnson. 1981. Nitrate-nitrogen in tile drainage as affected by fertilization. J. Environ. Qual. 10:519-522.

Baker, J.L. and J.M. Laflen. 1983. Water quality consequences of conservation tillage. J. Soil and Water Conserv. 38:186-193.

Blackmer, A.M. 1984. Losses of fertilizer N from soils. In: Proc. Iowa 37th Ann. Fert. and Ag-Chem. Dealers Conf., Iowa State University, Cooperative Extension Service, CE-2081, 4 p. (20811).

Blackmer, A.M. 1985. Integrated studies of N transformations in soils and corn responses to fertilizer-N. In: Proc. Iowa 38th Ann. Fert. and Ag-Chem. Dealers Conf., Iowa State University, Cooperative Extension Service, CE-2158, 4 p. (2158e).

Breitenbeck, G.A., A.M. Blackmer and J.M. Bremner. 1980. Effects of different nitrogen fertilizers on emission of nitrous oxide from soil. Geophys. Res. Lett. 7:85-88.

Bremner, J.M. and A.M. Blackmer. 1978. Nitrous oxide: emission from soils during nitrification of fertilizer nitrogen. Science 199:295-296.

Bremner, J.M. and A.M. Blackmer. 1981. Terrestrial nitrification as a source of atmospheric nitrous oxide. In: C.C. Delwhiche (Ed.), Denitrification, Nitrification, and Nitrous Oxide. John Wiley, New York, pp. 151-170.

Bremner, J.M., G.A. Breitenbeck and A.M. Blackmer. 1981. Effect of anhydrous ammonia fertilization on emissions of nitrous oxide from soils. J. Environ. Qual. 10:77-80.

Bremner J.M., S.G. Robbins and A.M. Blackmer. 1980. Seasonal variability in emission of nitrous oxide from soil. Geophys. Res. Lett. 7:641-644.

Burwell, R.E., G.E. Schuman, K.E. Saxton and H.E. Heinemann. 1976. Nitrogen in subsurface drainage from agricultural watersheds. J. Environ. Qual. 5:325-329.

Buzicky, G.C., G.W. Randall, R.D. Hauck and A.C. Caldwell. 1983. Fertilizer N losses from a tile drained mollisol as influenced by rate and time of 15-N depleted fertilizer application. Agron. Abstracts, Am. Soc. Agron., Washington, DC, p. 213.

Carey, M.A. and J.W. Lloyd. 1985. Modelling nonpoint sources of nitrate pollution of groundwater in the Great Ouse Chalk, U.K. J. Hydrol. 78:83-106.

Cerrato, M.E., A.M. Blackmer and D.L. Priebe. 1985. Movement of 15-N-labelled nitrate in the rooting zone of Iowa soils. Agron. Abs., 1985 Ann. Meetings, ASA, CSSA, SSSA, Chicago, IL, p. 23.

Chichester, F.W. and S.J. Smith. 1978. Disposition of ^{15}N-labelled fertilizer nitrate applied during corn culture in field lysimeters. J. Environ. Qual. 7:227-233.

Cochran, V.L., L.F. Elliot and R.I. Papendick. 1981. Nitrous oxide emissions from a fallow field fertilized with anhydrous ammonia. Soil Sci. Soc. Am. J. 45:307-310.

Cooper, J.R., R.B. Reneau, Jr., W. Kroontje and G.D. Jones. 1984. Distribution of nitrogenous compounds in a rhodic paleudult following heavy manure application. J. Environ. Qual. 13:189-193.

Duxbury, J.M., D. Bouldin, R.E. Terry and R.L. Tate, III. 1982. Emissions of nitrous oxide from soils. Nature 298:462-464.

EEC (European Economic Community). 1980. Council directive relating to the quality of water intended for human consumption. Off. Jour. Eur. Communities, No. 80/778/EEC, 23:L229.

Fox, R.H. and W.P. Piekielek. 1984. Relationships among anaerobically mineralized nitrogen, chemical indexes, and nitrogen availability to corn. Soil Sci. Soc. Am. J. 48:1087-1090.

Frank, K.D., T. Bockstadter, C. Bourg, G. Buttermore, D. Eisenhauer and D. Krull. 1984. Nitrogen and irrigation management. Hall County Water Quality Project, University of Nebraska Cooperative Extension Service Special Report, Lincoln, NE, 22 p.

Gast, R.G., W.W. Nelson and G.W. Randall. 1978. Nitrate accumulation in soils and loss in tile drainage following nitrogen application to continuous corn. J. Environ. Qual. 7:258-262.

Gerwing, J.R., A.C. Caldwell and L.L. Goodroad. 1979. Fertilizer nitrogen distribution under irrigation between soil, plant and aquifer. J. Environ. Qual. 8:281-284.

Goodroad, L.L., D.R. Keeney and L.A. Peterson. 1984. Nitrous oxide emissions from agricultural soils in Wisconsin. J. Environ. Qual. 13:557-561.

Hallberg, G.R. 1984. Agricultural chemicals and groundwater quality in Iowa. In: Proc. Iowa 37th Ann. Fert. and Ag. Chem. Dealers Conf., Iowa State University, Cooperative Extension Service, CE-2081, 6 p. (2081j).

Hallberg, G.R. 1985. Agricultural chemicals and groundwater quality in Iowa: Status report 1985. In: Proc. Iowa 38th Ann. Fert. and Ag. Chem. Dealers Conf., Iowa State University, Cooperative Extension Service, CE-2158, 11 p. (2158q).

Hallberg, G.R., J.L. Baker and G.W. Randall. 1986. Utility of tileline effluent studies to evaluate the impact of agricultural practices on groundwater. Agricultural Impacts on Groundwater Conference, National Water Well Association, pp. 298-326.

Hallberg, G.R. and B.E. Hoyer. 1982. Sinkholes, hydrogeology and ground-water quality in northeast Iowa. Iowa Geol. Surv., Open-File Rept., 82-3, 120 pp.

Hallberg, G.R., B.E. Hoyer, R.D. Libra, E.A. Bettis, III and G.G. Ressmeyer. 1983a. Additional regional groundwater quality data from the karst-carbonate aquifers of northeast Iowa. Iowa Geol. Surv., Open-File Rept., 83-1, 16 pp.

Hallberg, G.R., B.E. Hoyer, E.A. Bettis, III and R.D. Libra. 1983b. Hydrogeology, water quality, and land management in the Big Spring Basin, Clayton County, Iowa. Iowa Geol. Surv., Open-File Rept., 83-3, 191 pp.

Hallberg, G.R., R.D. Libra, G.G. Ressmeyer, E.A. Bettis, III and B.E. Hoyer. 1984a. Temporal changes in nitrates in groundwater in northeastern Iowa. Iowa Geol. Surv., Open-File Rept., 84-1, 10 pp.

Hallberg, G.R., R.D. Libra, E.A. Bettis, III and B.E. Hoyer. 1984b. Hydrogeologic and water-quality investigations in the Big Spring Basin, Clayton County, Iowa: 1983 Water Year. Iowa Geol. Surv., Open-File Rept., 84-4, 231 pp.

Hallberg, G.R., R.D. Libra and B.E. Hoyer. 1985. Nonpoint source contamination of groundwater in Karst-carbonate aquifers in Iowa. Perspectives on Nonpoint Source Pollution, USEPA, EPA 440/5 85-001, Washington, DC, pp. 109-114.

Hargett, N.L. and J.T. Berry. 1983. 1982 fertilizer summary data. Natl. Fert. Develop. Ctr., TVA, Muscle Shoals, AL, 136 pp.

Harmon, L. and E.R. Duncan. 1978. A technical assessment of non-point pollution in Iowa. Contract Report 77-001 to the Iowa Department of Soil Conservation, College of Agriculture, Iowa State University, 427 pp.

Hunt, P.K.B. and D.L. Runkle. 1985. Groundwater data for the alluvial, buried channel, basal Pleistocene, and Dakota aquifer in west-central Iowa. U.S. Geol Surv., Open File Rept., 84-819, 168 pp.

Johnson, H.P. and J.L. Baker. 1973. Ames reservoir environmental study; Appendix 4, chap. 2 and 3. ISWRRI-60-A4, Iowa State Water Resources Res. Inst., Ames, IA.

Jokela, W.E. and G.W. Randall. 1985. Uptake of N fertilizer by corn as affected by time of application. Agron. Abs., 1985 Ann. Meetings, ASA, CSSA, SSSA, Chicago, IL, p. 175.

Jolley, V.D. 1974. Theoretical and measured soil acidity form N-fertilizer as related to the N recovered in crops and soils. Unpub. M.S. Thesis, Department of Agronomy, Iowa State University, Ames, IA, 166 pp.

Jolley, V.D. 1976. Yields of corn and soybeans and the depth of nitrate removal from the soil as influenced by applied and residual nitrogen. Unpublished Ph.D. dissertation, Department of Agronomy, Iowa State University, Ames, IA, 174 pp. (Diss. Abstr., v. 37, 11, B, p. 5476; University Microfilms, Ann Arbor, MI).

Jolley, V.D. and W.H. Pierre. 1977. Profile accumulation of fertilizer-derived nitrate and total nitrogen recovery in two long term nitrogen-rate experiments with corn. Soil Sci. Soc. Am. J. 41:373-378.

Kanwar, R.S., H.P. Johnson and J.L. Baker. 1983. Comparison of simulated and measured nitrate losses in tile effluent. Trans. Am. Soc. Agric. Eng. 26:1451-1457.

Kapp, J.D. and S.C. Padgitt. 1985. Effects of agronomic practices on groundwater quality: results from a 1984 Iowa survey. Agron. Abs., 1985 Ann. Meetings, ASA, CSSA, SSSA, Chicago, IL, p. 27.

Keeney, D.R. 1982. Nitrogen management for maximum efficiency and minimum pollution. In: F.J. Stevenson (Ed.), Nitrogen in Agricultural Soils, Agronomy Monograph 22, pp. 605-649.

Keeney, D.R. 1986. Sources of nitrate to groundwater. In: Critical Reviews in Environmental Control. CRC 16(3):257-304.

Libra, R.D., G.R. Hallberg, G.R. Ressmeyer and B.E. Hoyer. 1984. Groundwater quality and hydrogeology of Devonian-Carbonate aquifers in Floyd and Mitchell Counties, Iowa. Iowa Geol. Surv., Open File Rept. 84-2, 106 pp.

Mancl, K. and C. Beer. 1982. High-density use of septic systems, Avon Lake, Iowa. Proc. Iowa Acad. Sci. 89(1):1-6.

McArthur, J.V., M.E. Gurtz, C.M. Tate and F.S. Gilliam. 1985. The interaction of biological and hydrological phenomena that mediate the qualities of water draining native tallgrass prairie on the Konza Prairie Research Natural Area. Perspectives on Nonpoint Source Pollution, EPA 440/5-85-001, pp. 478-482.

McDonald, D.B. and R.C. Splinter. 1982. Long-term trends in nitrate concentration in Iowa water supplies. J. Am. Water Works Assoc. 74:437-440.

Meisinger, J.J., V.A. Bandel, G. Standford and J.O. Legg. 1985. Labeled N fertilizer use by corn under minimal tillage and plow tillage culture. Agron. Abs., 1985 Ann. Meetings, ASA, CSSA, SSSA, Chicago, IL, p. 178.

Morris, R.L. and L.G. Johnson. 1969. Pollution problems in Iowa. In: P.J. Horick (Ed.), Water Resources of Iowa. Iowa Acad. Sci., University Printing, Iowa City, IA, pp. 89-110.

National Research Council. 1978. Nitrates: An environmental assessment. Environmental Studies Board, Commission on Natural Resources, Coordinating Committee for Scientific and Technical Assessment of Environmental Pollutants, National Academy of Sciences, Washington, DC.

Nelson, W.W. and J.M. MacGregor. 1973. Twelve years of continuous corn fertilization with ammonium nitrate or urea nitrogen. Soil Sci. Soc. Amer. Proc., 37:583-586.

Nelson, W.W. and G.W. Randall. 1983. Fate of residual nitrate-N in a tiledrained mollisol. Agron. Abstracts, Am. Soc. Agron., Washington, DC, p. 215.

Olsen, R.J., R.F. Hensler, O.J. Attoe, S.A. Witzel and L.A. Peterson. 1970. Fertilizer nitrogen and crop rotation in relation to movement of nitrate nitrogen through soil profiles. Soil Sci. Soc. Amer. Proc. 34:448-452.

Olson, R.A. 1985. Nitrogen problems. In: Plant Nutrient Use and the Environment, The Fertilizer Institute, Washington, DC, pp. 115-138.

Olson, R.A., F.N. Anderson, P.H. Grabouski and C.A. Shapiro. 1985. Soil test interpretation: Sufficiency vs. build-up and maintenance. Agron. Abs. 1985 Ann. Meetings, ASA, CSSA, SSSA, Chicago, IL, p. 180.

Olson, R.A., K.D. Frank, P.H. Grabouski, and G.W Rehm, 1982. Economic and agronomic impacts of varied philosophies of soil testing. Agron. J. 74:492-499.

Owens, L.B., R.W. VanKeuren and W.M. Edwards. 1985. Groundwater quality changes resulting from a surface bromide application to pasture. J. Environ. Qual. 14:543-548.

Owens, L.D. 1960. Nitrogen movement and transformation in soils as evaluated by a lysimeter study utilizing isotopic nitrogen. Soil Sci. Soc. Amer. Proc. 24:372-376.

Padgitt, S. 1985. Farming operations and practices in Big Spring Basin. CRD 229, Cooperative Extension Service, Iowa State University, Ames, IA, 48 pp.

Pratt, P.F.   1984.   Nitrogen use and nitrate leaching in irrigated agriculture.   In: R.D. Hauck (Ed.), Nitrogen in Crop Production.   Am. Soc. Agron., Madison, WI, pp. 319-333.

Pratt, P.F., et al.   1975.   Utilization of animal manures and sewage sludges in food and fiber production.   Council Agric. Sci. Tech., Rept. 11, 96 pp.

Pratt, P.F., W.W. Jones and V.E. Hunsakes.   1972.   Nitrate in deep soil profiles in relation to fertilizer rates and leaching volumes: J. Environ. Qual. 1:97-102.

Preibe, D.L., A.M. Blackmer and J.M. Bremner.   1983. ^{15}N-tracer studies of the fate of surface-applied urea.   Agron. Abstracts, Am. Soc. Agron., Washington, DC, p. 159.

Randall, G.W.   1985a.   Nitrogen Movement.   In: Proceedings from Plant Nutrient Use and the Environment Symposium, Kansas City, MO, pp. 141-152.

Randall, G.W.   1985b.   Nitrogen losses into drainage water and corn production as influenced by fertilizer N rates.   Agron. Abs. 1985 Ann. Meetings, ASA, CSSA, SSSA, Chicago, IL, p. 30.

Randall, G.W.   1986.   Nitrogen research in Minnesota.   Proc. Nitrogen and Groundwater.   Iowa Fertilizer and Chemicals Association, 3/6/86, Ames, IA, 12 pp.

Rice, C.W. and M.S. Smith.   1983.   Nitrification of fertilizer and mineralized ammonium in no-till and plowed soil.   Soil Sci. Soc. Am. J. 47:1125-1129.

Sanchez, C.A. and A.M. Blackmer.   1985.   Recovery of anhydrous ammonia-derived N by corn in Iowa.   Agron. Abs. 1985 Ann. Meetings, ASA, CSSA, SSSA, Chicago, IL, p. 183.

White, R.E.   1985.   A model for nitrate leaching in undisturbed structured clay soil during unsteady flow.   J. Hydrol. 79:37-51.

Yates, M.V.   1985.   Septic tank density and ground-water contamination.   Ground Water 23:586-591.

CHAPTER 4

THE GROUNDWATER PROBLEM IN MICHIGAN: AN OVERVIEW

Kyle M. Kittleson
Institute of Water Research and Center for Remote Sensing
Michigan State University
East Lansing, Michigan 48824

INTRODUCTION

The most serious and fundamental groundwater problem Michigan faces is communication. Specifically, the problem is how to deal with the complex maze of data and information that is available on groundwater quality issues. Obviously, policy makers and the general public must understand those who have the technical expertise in these areas and they, in turn, must make sure their work is directed toward effectively addressing the many serious groundwater problems. In general, these problems and their solutions are extremely complex. They range from mechanisms of chemical and biological transport (which are often simulated with computer programs) to the difficult regulation and policy decisions about groundwater.

The following paper addresses three major topics: (1) a brief summary of the Michigan groundwater problem, (2) a suggestion of a general strategy for responding to this problem, and (3) an example of the proposed strategy.

MICHIGAN GROUNDWATER

Obviously, groundwater contamination results from human activity on the surface of the earth. Mapping land use patterns is one way to aggregate human activity into large enough categories to be significant on a statewide basis. The southern third of the state, roughly speaking, is divided between urban activity and agriculture. The northern half of the Lower Peninsula as well as the Upper

Peninsula are predominantly natural vegetation land cover with relatively small amounts of agriculture and urban activity.

Michigan soils are heavily glaciated as is typical of most states in the Great Lakes area. Repeated glaciation causes a complex mixture of relatively coarse material with many deposits of nearly impervious layers of dense, high clay composition soils. The mixture of textures is particularly important with respect to groundwater contamination because it is difficult to determine whether or not a contaminant that is placed on the surface will ultimately find its way into groundwater resources. The soils are so heterogeneous and thoroughly mixed by repeating glaciation that it is often difficult to know whether or not there is a significant potential for contamination of groundwater resources in Michigan as a result of surface activity.

Approximately one half of the western part of Michigan's Lower Peninsula is characterized by a relatively well developed aquifer system. The eastern half of the Lower Peninsula has less well developed aquifers as well as generally higher clay centers and soil types which protect the aquifer somewhat more from surface contamination. It is important to realize that the unclassified areas may or may not be protected from surface water contamination by impervious layers of soil. It is equally important to understand that because the western half of Michigan's Lower Peninsula contains relatively well developed groundwater aquifers, they are the predominant source of urban and rural drinking water supplies in that area. Unfortunately, the best estimate of known or suspected contamination sites shows a fairly even distribution over the entire state.

To summarize, the land use shows a relatively clear pattern of being predominantly agricultural and urban in the southern half of Michigan's lower peninsula. In general, the majority of the well structured aquifers occur in the western half of the Lower Peninsula. The public water supplies in that area are drawn primarily from the aquifer itself. In terms of the vulnerability of these groundwater resources to surface contamination, the western one half of the lower peninsula is primarily unprotected or unclassified (which for planning purposes must be assumed to be unprotected).

One of the challenges in groundwater quality problems on a statewide basis is successfully dealing with large amounts of data distributed over space and time. The Institute of Water Research and Center for Remote Sensing at Michigan State University have collaborated to develop a method by which these problems can be overcome.

TRENDS IN NITRATE LEVELS OF
PUBLIC DRINKING WATER SUPPLIES

The purpose of the pilot project within which the development of this system occurred was to determine a spatial distribution of the nitrate (NO_3^-) contamination of Michigan's public drinking water supplies. At the same time, the objective was to develop a prototype system for conducting these types of analyses in the future. Nitrate was selected as the target compound because of its high solubility and the fact that it is widely believed to be a good indicator of the presence of other, possibly more serious contaminants.

The data on nitrate contamination of public water supplies was provided by the Michigan Department of Public Health. Their public water supply testing data began in the mid 1930's. The basic strategy of this study was to combine the Public Health Department's data with the geographic information system that had been developed by the Institute of Water Research and the Center for Remote Sensing at Michigan State University (MSU). The MSU Geographic Information System contains computerized maps of land use and cover, soil type, aquifer characteristics, aquifer vulnerability, the location of public water supplies and several other data sets (Krogulecki, 1987). Until the mid 1970s, the predominant nitrate levels observed in public drinking water supplies in Michigan were in the zero to one part per million (ppm) range (Figure 1). Between the mid 1970s and the present, the nitrate contamination levels appear to have increased significantly with many of them approaching the 10 to 11 part per million range (Figure 2). The combination of land use patterns and nitrate contamination observed during the most recent period of the study indicates a variety of land uses in these high nitrate concentration areas (Figure 3). In particular, the southwest corner of the Lower Peninsula of Michigan has a significant amount of agricultural land associated with the high nitrate concentration areas. In fact, much of this land is a combination of agricultural land and rural residential areas.

Two probable causes for elevated nitrate concentrations in these areas have been suggested: a significant amount of nitrogen fertilizer from agricultural activity may be moving from the root zone down to the level where it is being drawn into water supply wells, or a significant amount of nitrate may be leaving septic tank fields and ending in water well supplies. The relative importance of these two sources of nitrates is not well understood. Perhaps the most significant conclusion that can be drawn from the correlation of land use patterns with nitrate contamination is that further study within these high nitrate concentration areas is definitely needed.

In response to this need, the Institute of Water Research in cooperation with the Michigan State University Centers for Environmental Toxicology and Remote Sensing and the Michigan Department of Public Health are conducting a project to test groundwater quality

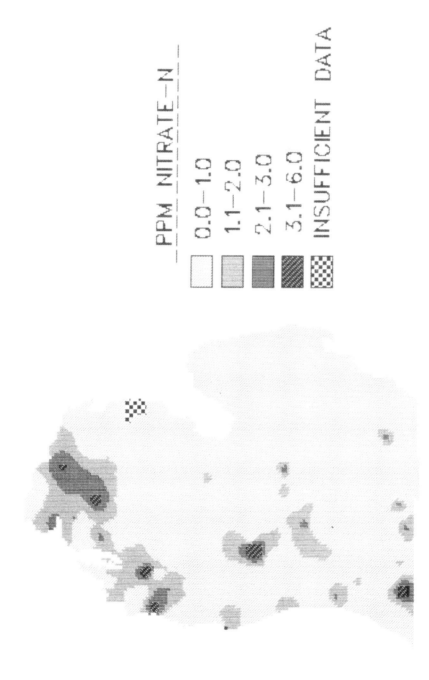

Figure 1. Groundwater nitrate levels based on community public water supply data, 1933–1970.

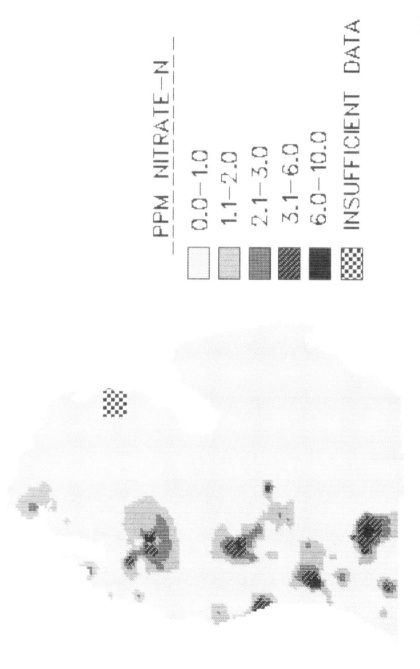

Figure 2.   Groundwater nitrate levels based on community public water supply data, 1975–1984.

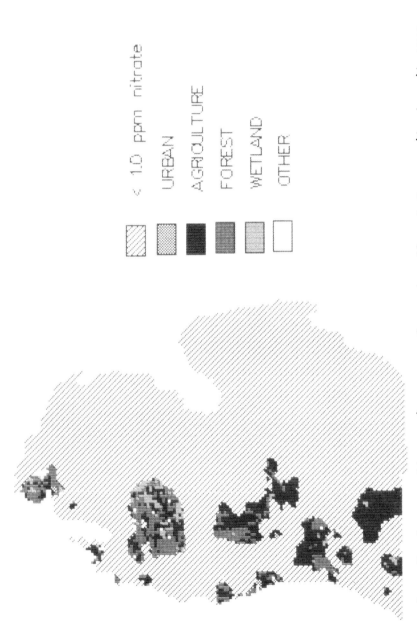

Figure 3.   Land use patterns in areas of greater than 1.0 ppm nitrate nitrogen.

Table 1.   Pesticides and their maximum concentrations detected in groundwater in northeastern Iowa, 1981-1983.

Common Name -- Active Ingredient	Trade Name	Maximum Concentration ug/1 (ppb)
Herbicides		
Alachlor	Lasso	16.6
Atrazine	Atrazine, AAtrex	10.0
Cyhanazine	Bladex	1.2
Metolachlor	Dual	0.6
Metribuzon	Sencor, Lexone	4.4
Insecticides		
Fonofox	Dyfonate	0.1

Hallberg et al., 1983; Hallberg et al., 1984; Libra et al., 1984.

in high nitrate areas for other contaminants. Studies indicate that several common agricultural chemicals are often found in association with nitrates (Table 1). Obviously, these types of data raise serious concerns about public health and safety. One consequence of these concerns is an increased level of public anxiety about this problem. This anxiety has, in turn, produced a shifting of priorities from the general consensus that these problems are essentially isolated, to an awareness that they are serious and relatively pervasive. The important question is how do resource managers, extension agents, scientists, public policy specialists, and elected officials respond to this concern?

One logical place to start is to begin to view the groundwater quality problem as a system. This will help provide a structure for further analysis. Groundwater quality is essentially a function of: (1) land use, (2) soil type, (3) hydrology, and (4) weather. Efficient handling of technical information over large areas of space and time requires an efficient and effective system.

AN INFORMATION SYSTEM FOR
STATEWIDE GROUNDWATER ANALYSIS

The type of system which has been developed to address this problem at Michigan State University is the Spatial Information Management System (SIMS). SIMS consists of the following components: (1) a geographic information system, (2) a tabular data base containing water quality data, and (3) a series of computer models which make generalizations and forecasts based on the geographic information system and tabular data base.

The geographic information system is essentially a set digital data bases which can be presented in a graphic form as it is in the illustrations which accompany this paper. The tabular data base which is used within SIMS is contained in a relational data base. The computer models which run on the tabular and geographic data sets are used to address groundwater flow, chemical transport, crop yield, erosion, sedimentation, and other facets of nonpoint source pollution.

Recent breakthroughs in computer technology make it possible to conduct these types of analyses on microcomputer systems ranging in price from $10,000 to $17,000. These prices are decreasing continuously and are presently within reach of most organizations which are required to manage information of this type. This is extremely important because it effectively removes a major technological barrier from effectively dealing with groundwater contamination data on a significantly large scale of space and over significant periods of time.

This leads to the issue of how to choose to organize space. Major federal agencies such as the United States Geological Survey (USGS) and the Soil Conservation Service (SCS), to name just two, have a well established track record of dealing effectively with spatial data. In part, this results from their charter as national organizations. Many other organizations, notably the Environmental Protection Agency (EPA), have established a strong orientation toward what are often regional level studies. More recently, the state resource management organizations have begun to develop spatial data handling capabilities of their own. In Michigan, the Michigan Resource Information Systems (MIRIS) represents a significant commitment by the state of Michigan to develop a program in this area.

Within the last two to three years, in a large part because of the availability of microcomputer technology, county level organizations have begun to develop their own capability in this area. Michigan has several examples of counties which have developed various types of spatial information handling systems and are already effectively applying them. The important point is that all of the major organizations from federal, to state, to county which are involved in groundwater quality studies have the capability to deal effectively with spatial data. However, the last link in this chain is perhaps the most important.

It is absolutely essential that county and local governments develop the capability to apply the spatial analysis systems themselves. The major reason this is so important is that groundwater contamination problems appear first on a localized basis. Consequently the first public awareness of these problems is normally at the local level. Townships are frequently the point at which the problems begin to move into the public sector. While many of these problems are currently severe enough that they will ultimately require state or possibly federal assistance, many are also small enough that, if they are dealt with effectively in the early stages, the degradation of the resource can be minimized. And, for the first time, the technology and the information is available in most cases to help this local response along.

EXAMPLE OF COUNTY LEVEL GROUNDWATER MANAGEMENT

One example of how this computer technology can be effectively applied on a county and local level is the work that the Institute of Water Research and Center for Remote Sensing at Michigan State University have conducted in Kalamazoo County, Michigan. The purpose of this project was to develop and test a prototype groundwater management system for a county. At this time, the project is approximately one half complete. As with the examples discussed above from the statewide study, the focus of this study was on nitrate contamination because it is a good indicator of other types of possible contamination.

The project was conducted in Kalamazoo County, which is located in the south-central part of Michigan's Lower Peninsula. The data base which was developed for this project contains the results of over 5,000 well water quality tests. Preliminary analysis of the results of this study indicate that approximately 19 percent of the wells which were tested contain nitrate levels over the 10 ppm as nitrogen standard for drinking water.

Perhaps the most important point about this study is that it demonstrates that the technology can shed an enormous amount of new light on the groundwater quality in a convenient to use format. The formulation of an effective groundwater management strategy for any area depends on having the best information available.

A comprehensive data set containing land use and cover information is a very important first step (Figure 4). The next logical step is to develop a geographic information system data base for groundwater concentrations (Figure 5). The combination of these two data sets produces a data set showing land use and cover patterns overlaying groundwater nitrate contaminations of two ppm or more (Figure 6).

Analysis of land use patterns overlaying two ppm of nitrate or more indicates that the majority of this area lies in the eastern half of the township. The question then becomes what characteristics of that portion of the township produces this pattern? One possibility is that the water table might be very near the surface in this area. By incorporating depth to static water level in the data set (Figure

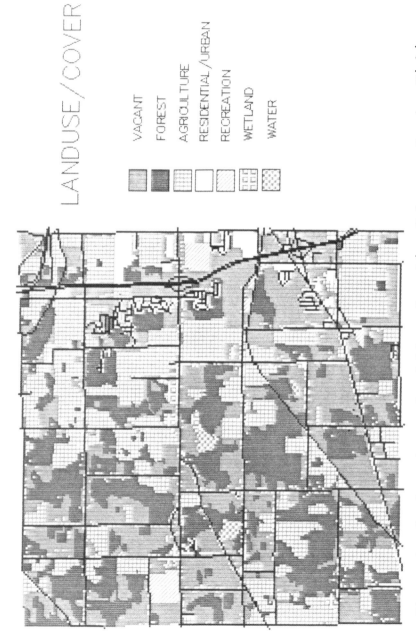

Figure 4.   Land use/cover for Oshtemo Township, Kalamazoo County, Michigan.

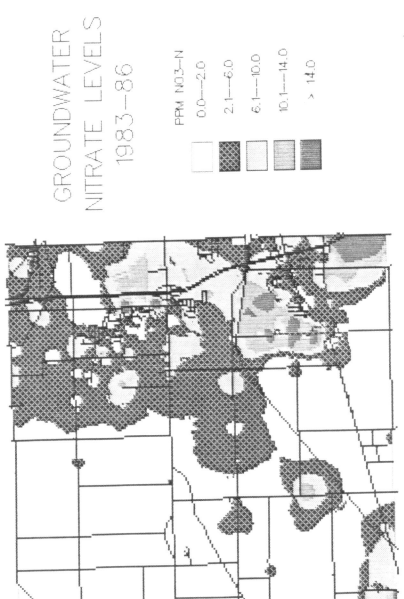

Figure 5. Groundwater nitrate concentrations, 1983–1986, Oshtemo Township, Kalamazoo County, Michigan.

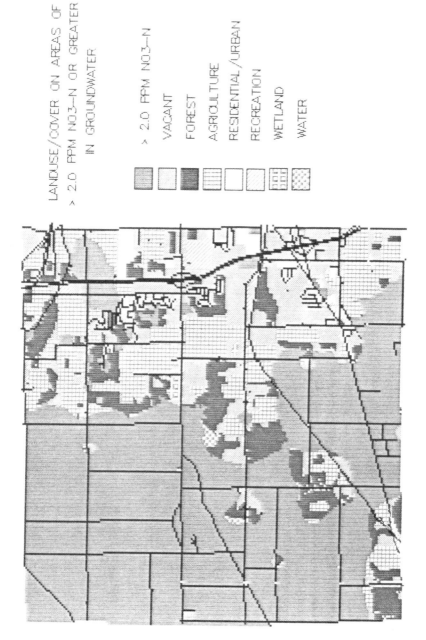

Figure 6. Land use/cover on areas of 2.0 ppm nitrate nitrogen or greater in groundwater, Oshtemo Township, Kalamazoo County, Michigan.

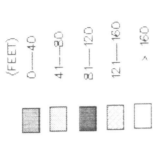

Figure 7.   Depth to static water level in Oshtemo Township, Kalamazoo County, Michigan.

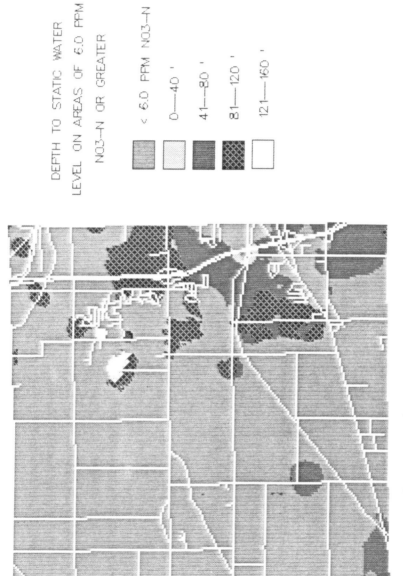

Figure 8.   Depth to static water level on areas of 6.0 ppm nitrate nitrogen or greater in groundwater in Oshtemo Township, Kalamazoo County, Michigan.

7), one can get an idea of whether or not this possibility actually occurs. Examination of Figure 7 indicates that, in fact, many of the areas where the concentrations are above two ppm have a depth to static water level in excess of 80 feet. Therefore, one can conclude that the depth to static water level is probably not the primary factor in determining elevated nitrate concentrations in groundwater in Oshtemo Township.

Some evidence indicates that some of the highest concentrations occur in areas where the depth to static is over 80 feet (Figure 8). These analyses by no means exhaust the possibilities in these data sets. They are merely illustrated here to demonstrate that the geographic information system approach to groundwater study has the potential to supply local decision makers with enough information to begin to do their own analyses and become personally involved in groundwater management in their areas.

These county level data sets provide the basis from which local decision makers can formulate an effective policy to manage this important resource on the county level.

In conclusion, the most important point to be made about the use of the spatial information management systems to solve groundwater contamination problems is that if this type of information is put in the hands of local decision makers, the first real opportunity to formulate effective solutions to these serious problems on a solid informational foundation will be made possible.

By bringing the decision makers together with this information in a format that lends itself to the development of simulation model alternatives to these complex land use problems, they can, for the first time, hope to arrive at long term, economically feasible, management strategies for groundwater contamination. It is equally important to note that the trends in groundwater quality in Michigan are not encouraging. It is absolutely essential that the decision makers at all levels of government have access to the information management systems they need to address these serious problems as soon as possible.

LITERATURE CITED

Hallberg, G.R., B.E. Hoyer, E.A. Bettis, III and R.D. Libra.   1983b. Hydrogeology, water quality, and land management in the Big Spring Basin, Clayton County, Iowa. Iowa Geol. Surv., Open-File Rept., 83-3, 191 pp.

Hallberg, G.R., R.D. Libra, E.A. Bettis, III and B.E. Hoyer.   1984b. Hydrogeologic and water-quality investigations in the Big Spring Basin, Clayton County, Iowa: 1983 Water Year. Iowa Geol. Surv., Open-File Rept., 84-4, 231 pp.

Krogulecki, M. and W. Hudson. 1987. Documentation of the Michigan Geographic Information System. Center for Remote Sensing, Michigan State University, East Lansing, MI, unpublished manuscript.

Libra, R.D., G.R. Hallberg, G.R. Ressmeyer and B.E. Hoyer. 1984. Groundwater quality and hydrogeology of Devonian-Carbonate aquifers in Floyd and Mitchell Counties, Iowa. Iowa Geol. Surv., Open File Rept. 84-2, 106 pp.

CHAPTER 5

LOSSES AND TRANSPORT OF NITROGEN FROM SOILS

Alfred M. Blackmer
Department of Agronomy
Iowa State University
Ames, Iowa 50011

INTRODUCTION

Historically, the ability of soils to supply plant-available forms of nitrogen (N) has been a major factor limiting the agricultural production of food and fiber. This limitation has been alleviated by commercially fixed nitrogen fertilizers, which became widely used only after World War II. Use of fertilizer nitrogen has markedly increased over the past three decades (Table 1) and must be recognized as a

Table 1. Trends in nitrogen fertilizer use and corn production in Iowa.

Period Year	Annual use of fertilizer N[1]		Annual production of corn in Iowa	
	U. S.	Iowa	Area planted	Grain yield
	— — — Gg — — —		Million ha	Mg/ha
1955 - 1959	2,136	64	4.4	3.8
1960 - 1964	5,329	203	4.4	4.7
1965 - 1969	5,897	507	4.4	5.6
1970 - 1974	7,010	605	4.6	6.1
1975 - 1979	8,378	829	5.5	6.4
1980 - 1984	9,916	919	5.2	7.0

[1]919 Gg = 1 million U. S. tons.

major factor contributing to the abundance of food and fiber currently enjoyed by developed nations.

The possibility that some portion of the fertilizer nitrogen applied to soils may leach into groundwater supplies has been recognized for as long as fertilizers have been used. However, concern about the importance of this problem is increasing. This increase in concern is justified in view of (i) increases in amounts of fertilizer nitrogen being used, (ii) increases in awareness of the need to protect the quality of groundwater supplies, and (iii) reports that nitrate concentrations in groundwater supplies are increasing in many areas.

The objective of this paper is to provide an overview of what is, and is not, known about processes that influence the amounts of nitrogen that escape from agricultural soils to groundwater. This objective will be approached by (i) briefly reviewing what happens to fertilizer nitrogen in soils, (ii) describing some of the difficulties associated with determination of the quantities of nitrogen lost from agricultural soils to groundwater, and (iii) summarizing results from some relevant studies in progress at Iowa State University.

WHAT HAPPENS TO FERTILIZER NITROGEN IN SOILS?

As soon as they are applied to soils, nitrogen fertilizers are subjected to numerous chemical, physical, and biological processes. A basic knowledge of these processes is a prerequisite to understanding the difficulties associated with assessing the amounts of nitrogen lost to groundwater and the difficulties associated with preventing these losses. When discussing these processes, it is essential to distinguish between those directly responsible for losses of nitrogen from fields and those not directly responsible for such losses.

Processes Not Directly Responsible for Losses of Nitrogen from Fields

Major processes not directly responsible for losses of fertilizer nitrogen from fields are shown inside the dotted lines in Figure 1. One of the more important is *nitrification*, which refers to the transformation of ammonium (NH_4^+) to nitrate (NO_3^-). This transformation is performed by a very specific group of soil microorganisms and, therefore, can be temporarily controlled by a class of compounds called "nitrification inhibitors." Nitrification is important because most fertilizer nitrogen applied for crop production is applied as ammonium or as forms that are rapidly converted to ammonium. Under conditions usually found when nitrogen fertilizers are applied to soils, nitrification is sufficiently rapid to convert most ammonium or ammonium-yielding fertilizers to nitrate within two to three weeks.

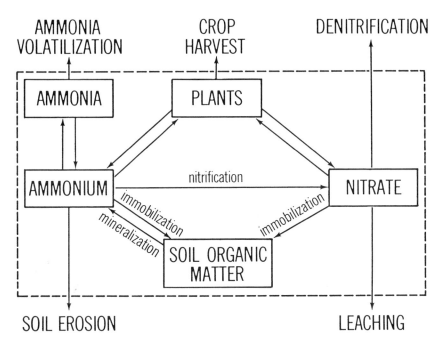

Figure 1.   Schematic diagram of major nitrogen transformations in soils and processes leading to losses of nitrogen from soils.

A notable exception is when fertilizers are applied in the fall after soils are too cold for rapid microbial activity.   Such fertilizers are nitrified after soil temperatures increase during the following spring.

Other important processes are *immobilization*, which refers to the transformation of plant-available forms of nitrogen (ammonium and nitrate) to organic forms that are not available to plants, and *mineralization*, which refers to the reverse transformation.   Immobilization and mineralization are performed by a wide range of soil microorganisms going about their normal life functions within soils. These life functions include growing and building tissues from nutrients obtained by decomposing old tissues of plants, animals, and microorganisms.   Immobilization is favored during decay of organic materials having high carbon to nitrogen ratios (i.e., relatively low percentage nitrogen), whereas mineralization is favored during decay of organic materials having low carbon to nitrogen ratios (i.e., relatively high percentage nitrogen).   The net direction of the immobilization-mineralization transformations largely determines whether the organic matter content of a soil increases or decreases.

Ammonium fixation and *ammonium release* by clay minerals are other transformations that, at least from a functional point of view, are associated closely with immobilization and mineralization. However, these processes do not involve soil microorganisms. During ammonium fixation, ammonium ions just move inside clay structures and become unavailable to plants and soil microorganisms. Ammonium fixation is favored by the presence of high concentrations of ammonium, whereas ammonium release is favored by the presence of low concentrations of ammonium. Ammonium fixation and release by clay minerals usually are considered to be less important than immobilization and mineralization in most agricultural soils and, therefore, are not included in Figure 1.

Another pair of transformations that is similar to immobilization and mineralization in soils is the uptake and release of nitrogen by plants. Even in young plants, roots release nitrogen in the form of exudates, secretions, or sloughed cells. Upon maturation and death, of course, above-ground tissues are returned to the soil and decomposed by soil microorganisms.

Nitrification, immobilization and mineralization, ammonium fixation and release by clay minerals, and plant uptake and release of nitrogen occur simultaneously in soils and result in a mixing of fertilizer-derived nitrogen and soil-derived nitrogen. Because of this mixing, it is usually impossible to distinguish between soil-derived nitrogen and fertilizer-derived nitrogen in soils soon after the fertilizers are applied. It also is impossible to distinguish between nitrogen that was initially applied as commercial fertilizers, plant residues, or animal manures.

**Processes Directly Responsible for
Losses of Nitrogen from Soils**

Major processes that result in losses of nitrogen from fields are shown outside the dotted lines in Figure 1. Soil erosion is an important mechanism by which nitrogen is lost from soils. However, most of the nitrogen lost by this process is as organic matter or ammonium attached to soil particles and usually does not pose a direct threat to groundwater supplies. For this reason, and because the management practices associated with preventing soil erosion involve much broader considerations than merely nitrogen management, soil erosion is not discussed further here.

When ammonium or ammonium-yielding fertilizers are placed on the surface of soils and not incorporated into the soil, nitrogen may be lost by *ammonia volatilization.* Ammonia volatilization occurs because there is always a chemical equilibrium between ammonium (an ion) in the soil solution and ammonia (a gas) in the soil atmosphere. Losses of nitrogen can occur whenever ammonia in the soil atmosphere can escape to the air above the soil. Ammonia volatilization can result in substantial losses of fertilizer nitrogen from soils having a high pH, which promotes the formation of ammonia. Also, some fertilizers tend to intensify this problem by increasing soil pH near the point of application. Although some of the ammonia lost from soils may be sorbed by surface waters, losses of fertilizer nitrogen by ammonia volatilization are not directly related to groundwater problems. For this reason and because losses of nitrogen by ammonia volatilization can be controlled by selecting appropriate methods of fertilizer application, this mechanism is not discussed further here.

Denitrification refers to the biological transformation of nitrogen in the nitrate or nitrite forms to gaseous forms that escape from the soil. The major product of this transformation is nitrogen gas (N_2), but nitrous oxide (N_2O) is formed as an intermediate during denitrification and small amounts of this gas may escape to the atmosphere. Denitrification is strongly inhibited by the presence of oxygen in soils and, therefore, only occurs in soils having sufficient water to fill the pores that normally keep soils well-aerated. Denitrification within the rooting zone is undesirable because it represents a loss of plant available nitrogen. Although there have been some suggestions that the formation of nitrous oxide during denitrification is linked to reactions that result in a depletion of stratospheric ozone (Crutzen and Ehhalt, 1977), there can be no environmental threats from the formation of N_2 by denitrification because this gas is the primary component of air.

When excess water is present in soils, nitrogen also can be lost by *leaching*. Leaching is the mechanism of nitrogen loss that poses the greatest threat to groundwater supplies. Nitrate is the form of nitrogen most susceptible to leaching because it is an anion (negatively charged ion) and, therefore, is not attracted to soil particles. Unless removed from the soil solution by some process such as immobilization, plant uptake, or denitrification, nitrate is free to percolate downward with the water that moves through soils. Ammonium is not leached from soils because it is a cation (positively charged ion) and, therefore, is attracted to soil particles. Organic forms of nitrogen are not leached from soils because they have low solubility in water.

Like excess water that drains from fields, nitrate may be lost from fields through tile drainage systems. It may move below the rooting zone and then laterally to nearby surface waters. It may move nearly straight downward until it reaches groundwater supplies. The major factors affecting the amount of nitrogen lost from a field by leaching are the quantities of water that move through the surface soils and the amount of nitrate available for leaching when this movement occurs.

The most desirable way for nitrogen to be lost from fields is by harvest of plant tissues. Even under ideal conditions, however, the recovery of fertilizer nitrogen during plant harvest must be less than 100 percent because some nitrogen always remains in the field in the form of plant tissues. For this reason, it is important to distinguish between amounts of fertilizer nitrogen taken up by plants and amounts removed from the field during plant harvest. This distinction is made in Figure 1.

PROBLEMS IN ASSESSING LOSSES OF NITRATE FROM FIELDS TO GROUNDWATER

To those who have not made serious attempts to do so, it often seems like a simple problem to assess the amounts of nitrate leached from agricultural soils. There are, however, major problems that have hampered attempts to make such assessments. These problems include difficulties in catching and quantifying the amounts of nitrate that escape from a field, difficulties in distinguishing between leaching and denitrification as mechanisms of nitrogen loss, and difficulties in distinguishing between nitrogen that is lost and nitrogen that is merely hidden within soil organic matter.

Catching Nitrate that Escapes from Fields

Direct measurements of the amounts of nitrate that are leached from fields are impossible to attain under most conditions. It is possible to measure the amounts of nitrate lost in the effluent from tile drainage systems and the results of such measurements have been published (e.g., Hanway and Laflen, 1974; Baker et al., 1975; Gast et al., 1978). Although the amounts of nitrogen lost through tile drainage systems on cropped soils are extremely variable, the results of one three-year study (Gast et al., 1978) can be considered typical of results often observed during continuous corn production. This study showed losses through tile lines to be 41, 53, 93, and 180 kg N/ha when 20, 112, 224, and 448 kg/ha of fertilizer nitrogen were applied. It must be recognized that the results from such studies provide no information concerning the amounts of nitrogen that leach from soils that are not tile drained or the amounts of nitrate that leach between the tiles in soils that have tile drains.

Numerous researchers have assessed the importance of nitrate leaching by using lysimeters, or volumes of soil that are encased by artificial barriers that catch water and nitrate moving through the soil. Although the lysimeters often are placed under field conditions, the installation of the barriers usually involves removing and back-filling the soil within the lysimeter. This disturbance represents a drastic perturbation of both soil structure and soil nitrogen transformations. Also, the placement of a barrier at the bottom of the lysimeter breaks the natural flow of water and nitrate through the soil. Because these perturbations may have major impact on the amounts of nitrate leached, it is difficult to establish the reliability of lysimeter studies for assessing nitrate leaching from soils.

A few locations have been identified that, because of their topography and geology, can be viewed as huge natural lysimeters. One example is the Big Spring Basin located in Clayton County, Iowa

(Hallberg et al., 1984). All of the water that infiltrates though a land area of about 101 square miles is collected by underlying formations and is channeled through Big Spring into the Turkey River. Data from such studies show that substantial amounts of nitrate are leached from soils. In 1982, for example, the amount of nitrate leaving the Big Spring basin was equivalent to 33 percent of the fertilizer nitrogen applied. The major problem with natural catchments like Big Spring basin is that they are difficult to locate and they cannot be used in field plot research designed to identify management practices that minimize losses of nitrogen.

Distinguishing Between Leaching and Denitrification

One of the major problems in assessing the amounts of nitrate leached from soils is that denitrification, as well as leaching, is favored by excessive amounts of water in soils. It is impossible to directly measure the amounts of denitrification that occur under field conditions because the major product of denitrification, N_2, is also the major component of air. However, laboratory studies involving incubation of soil samples under artificial atmospheres in sealed vessels clearly show that microorganisms in soils have the capacity for rapid denitrification when conditions are favorable.

If subjected to conditions ideal for denitrification, soil microorganisms can denitrify within a few days amounts of nitrogen equivalent to annual applications of fertilizer. However, there is little reason to expect that such high rates of denitrification are encountered in most agricultural soils because conditions favorable for denitrification must include the presence of microorganisms, nitrate, and readily available carbonaceous substrate in anoxic (without oxygen) environments having temperatures favorable for microbial growth. Unfortunately, there is a lack of direct evidence showing the frequency at which such favorable conditions occur in fields normally used for crop production. An abundance of laboratory studies showing high denitrification capacity by microorganisms in soils has encouraged many investigators to report large losses of nitrogen by denitrification under field conditions even when the investigators did not have direct evidence to distinguish between leaching and denitrification as mechanisms of nitrogen loss.

It needs to be recognized that laboratory studies of denitrification capacity involve manipulation of most (usually all but one) of the important factors affecting denitrification to be optimal for denitrification. Therefore, these studies have a tendency to exaggerate the importance of denitrification as a mechanism of nitrogen loss. Extreme caution should be taken when using the results of laboratory studies to predict the importance of denitrification under field conditions.

The difficulties associated with measuring denitrification seem even more important when it is realized that nitrate can be denitrified after it leaves the rooting zone. Whether it occurs before or after nitrate reaches groundwater supplies, this denitrification decreases the importance of nitrate leaching from agricultural soils. Although little is known about the amounts of denitrification that occur in subsurface layers of soil or in groundwater, the possibility that such denitrification may be important cannot be ignored.

Because of these and other problems, current knowledge does not enable meaningful and unbiased assessments of the relative importance of denitrification and leaching as mechanisms of nitrogen loss from agricultural soils.

Distinguishing Between Nitrogen Losses and Changes in Organic Nitrogen Levels

There have been many attempts to assess losses of nitrate from fields by measuring changes in the concentrations of nitrate within soils. A major problem with this practice is that transformations such as immobilization and mineralization also influence nitrate concentrations in soils. Especially if the change in concentration is measured over periods lasting a week or more, these transformations may introduce significant errors into assessments of nitrogen losses.

This problem cannot be alleviated merely by measuring changes in the total amount of nitrogen in the soil, because nitrate represents a very small fraction of the total amount of nitrogen in soils at any given time and because the concentration of nitrogen within soils shows marked spatial variability in both the horizontal and vertical directions. Because of these factors, it usually is not possible to determine whether nitrogen as nitrate is lost from a soil or merely hidden within the organic fraction. This problem is especially severe in Corn Belt soils, which tend to have relatively high levels of soil organic matter.

There have been many attempts to study losses of fertilizer nitrogen from soils by measuring the amounts of nitrate plus exchangeable ammonium in soils. This practice is especially useful when studying the fate of ammonium fertilizers, which are rapidly nitrified in soils. However, because the addition of the exchangeable ammonium fraction to the nitrate fraction also adds additional transformations within soils and additional mechanisms of nitrogen loss, this practice usually does not alleviate the problems mentioned in the preceding two paragraphs.

ESTIMATING NITROGEN LOSSES BY MEASURING RECOVERY OF FERTILIZER NITROGEN

Because of the extreme difficulties associated with measuring nitrogen after it has escaped from fields, the most widely used method of studying losses of nitrogen from fields involves measuring recovery of fertilizer nitrogen in soils and (or) crops at various times after fertilization. A major limitation of this method is that it does not indicate the mechanism of loss, although the most probable mechanism often can be inferred from knowledge of how the fertilizer was applied and soil conditions that prevailed. Another major limitation of this method is that it focuses only on losses of fertilizer-derived nitrogen, which may be less than losses induced by addition of the fertilizer or the total amount of nitrogen lost from the field. As noted earlier, the high degree of spatial variability in the concentration of soil nitrogen prohibits measurements of changes in total amounts of soil nitrogen in most studies.

^{15}N-Labeled Fertilizers

Recoveries of fertilizer-derived nitrogen in soil-plant systems can be determined unequivocally only by using ^{15}N-labeled fertilizers. ^{15}N is a naturally occurring stable isotope (most nitrogen in the environment is ^{14}N) and, therefore, does not pose the health and environmental problems associated with radioactive isotopes. ^{15}N-labeled fertilizer materials look exactly like the corresponding ordinary fertilizer materials. They behave exactly like ordinary fertilizer materials in soil-plant systems, and they pose no greater threat to man or the environment than do ordinary fertilizer materials. If samples of soil and plant tissues are analyzed for ^{15}N content at several times after application of labeled fertilizers, it is possible to monitor the transformations and movements of the fertilizer nitrogen within the system. For this reason, these labeled materials are sometimes called "tracers."

A major drawback to the use of ^{15}N-tracers in field situations is the high cost of purchasing labeled fertilizers and of analyzing samples of soil and plant tissues for ^{15}N content. Field plots having areas of 3 to 5 m^2 often require fertilizers costing several hundred dollars if they are sufficiently enriched with ^{15}N to enable accurate determinations of recovery amid the large amounts of nitrogen found within the rooting zone of soils. This high cost prohibits the use of ^{15}N on study areas larger than a few square meters, and it has prevented widespread use of tracers to monitor the transformations and movements of fertilizer nitrogen under field conditions. Analysis of samples for ^{15}N requires an isotope ratio mass spectrometer, an instrument that is very expensive to purchase and operate.

It should be recognized that determinations of ^{15}N recovery in soil-plant systems may underestimate fertilizer-induced losses of nitrogen because of processes called isotope exchange and biological interchange. In each of these processes, an atom of fertilizer nitrogen merely replaces an atom of soil nitrogen and, therefore, there is no net change in the amount of nitrogen in the soil. The ^{15}N incorporated into the soil is considered when recoveries are calculated, but the nitrogen released is not considered in these calculations. For this reason, determinations of fertilizer-induced losses of nitrogen require knowledge of whether recovery of ^{15}N in soils represents a net increase in soil nitrogen content or whether it merely replaced nitrogen that was already present. Such knowledge usually is impossible to obtain under field conditions.

Apparent Recoveries of Fertilizer Nitrogen in Plants

Agronomists have made many measurements of the amounts of fertilizer nitrogen recovered by plants, and the results of these studies often are used when making assessments of the amounts of nitrogen lost from soils. Because of high costs associated with the use of labeled fertilizers, most of these studies have been conducted without the use of ^{15}N. Also, from an agronomic point of view determination of the net effect of the fertilizer on plant growth is more important than is specific knowledge of the fate of the nitrogen applied. Although the distinction is not made in most discussions, it should be recognized that any study that does not use labeled fertilizers involves measurement of "apparent recovery" rather than "recovery" of fertilizer nitrogen.

Apparent recoveries of fertilizer nitrogen often are measured by determining the amounts of nitrogen in plants grown with and without fertilizer. The difference in amounts of nitrogen found is taken to represent the amount of fertilizer nitrogen recovered. This method involves the assumption that similar amounts of soil-derived nitrogen are taken up with and without the addition of fertilizer. This assumption is reasonable if the objective of the study is to identify fertilization practices having the greatest economic benefit for farmers. However, because the addition of nitrogen fertilizer may alter the amounts of soil-derived nitrogen taken up by the plant or lost by leaching, the validity of this assumption is questionable if the objective of the study is to determine the effect of the nitrogen fertilization on total amounts (soil-derived nitrogen plus fertilizer-derived nitrogen) of nitrogen lost from the soil. Current knowledge concerning the effects of nitrogen fertilization on changes in organic nitrogen contents of soils is not adequate to determine the magnitude of error introduced by this assumption.

Apparent recoveries of fertilizer nitrogen often are measured by determining the amounts of nitrogen in the whole plant, rather than only in plant parts that are removed from the field during harvest. From a point of view of crop production, the total amount of nitrogen within the plant is very important because it is the amount of nitrogen needed to grow the crop. However, as indicated in the discussion that accompanies Figure 1, it is important to recognize that the total amount of nitrogen in plant tissues is irrelevant when measuring nitrogen losses from a field. Only the amount of nitrogen that is removed from the field is important when estimating losses of nitrogen from a field.

Overall, studies of apparent recoveries of fertilizer nitrogen by crops can be used to obtain assessments of the amounts of nitrogen lost from soils. However, it must be recognized that, depending on changes in levels of soil-derived nitrogen, losses of nitrogen from soil may be lesser or greater than the amounts of nitrogen not recovered. It also must be recognized that most measurements of fertilizer recovery have not been conducted with the specific objective of assessing the amounts of nitrogen lost from soils and, therefore, these studies may not have been conducted by using methods that provide the best estimates of the amounts of nitrogen lost from soils.

RECENT IOWA STATE UNIVERSITY STUDIES OF NITROGEN LOSSES FROM SOILS

Since 1982, this laboratory has been using ^{15}N-labeled fertilizers in integrated studies of the transformations and movements of fertilizer nitrogen in soils under field conditions. The major difference between these and earlier ^{15}N-tracer studies is in the scope and the intensity of soil and plant tissue sampling and analysis. Most of the data from these studies has not been published in detail at the time of this writing, and detailed discussions of these data are beyond the scope of this paper. However, following is an overview of some important findings relevant to assessments of the amounts of fertilizer-derived nitrogen that are lost from soils to groundwater.

Recovery of ^{15}N-Labeled Fertilizers During Corn Production

The first studies by this laboratory to assess recovery of labeled fertilizers under field conditions were conducted during 1982, 1983, and 1984 on plots located near Ames (central Iowa) and Nashua (northeast Iowa) on soils classified as being somewhat poorly drained. The plots were cropped to corn and were treated with 0, 112, or 224 kg N/ha as spring-applied anhydrous ammonia. These rates represent the lower and upper ends of the range in rates used by farmers in the area. On small (4.6 m^2) "^{15}N plots" located within the larger (4 rows by 10 m) "yield plots," ^{15}N-labeled anhydrous ammonia was substituted for ordinary anhydrous ammonia. The locations of the ^{15}N-plots within the yield plots were moved each year to permit studies of the residual effects of the fertilizer nitrogen. Samples of soil and plant tissues were collected from all ^{15}N-plots at selected times during each year to determine recovery of fertilizer nitrogen.

Analysis of the corn grain showed that, using the mean values over all plots, only about 19 percent of the fertilizer nitrogen was recovered in the corn grain during the first harvest after fertilization (Sanchez and Blackmer, 1985). This recovery is less than was expected, but is within the range of recoveries reported by other researchers conducting similar studies (Chichester and Smith, 1978; Olson, 1980; Kitur et al., 1984; Meisinger et al., 1985). During each of the three years these studies were conducted, the amounts of rainfall that occurred during the two-month period after fertilization were higher than long-term averages and, therefore, the potentials for losses by leaching and denitrification were higher than average. Also, periods of moisture stress during mid-season caused yields of grain to be less than anticipated, but generally consistent with county averages (yields of grain ranged from about 5 to 7.5 mg/ha on fertilized plots).

Analysis of soil samples collected one year after fertilization showed that only about 20 percent of the labeled nitrogen remained in the soil (Sanchez and Blackmer, 1985). This includes nitrogen present in soil microorganisms, plant residues, soil organic matter and other forms of nitrogen found in soils. Much of this nitrogen was in a stable form that still remained in the soil two and three years after fertilization. Overall, these observations indicate that about 61 percent of the labeled nitrogen was lost from the plots during the first year by processes other than grain harvest.

The losses of fertilizer nitrogen observed in these studies are higher than has been reported in most similar studies. Undoubtedly, some of this difference can be attributed to factors related to soil properties and weather conditions. However, a careful review of the literature suggested that much of this difference could be explained by differences in methods used to determine recovery. One of the most important differences related to the time at which fertilizer recovery was determined. Recoveries were determined over a one-year period and, therefore, include all losses that occurred during the year. Previous studies evaluated recovery of fertilizer nitrogen no later than the end of the cropping season and, therefore, did not include estimates of the amounts of fertilizer nitrogen lost between cropping seasons. In the Corn Belt, as in many other regions of the world, large amounts of rainfall produce conditions favorable for losses of nitrate by denitrification and leaching during the fall-to-spring period. During much of this period, soil temperatures are too cold to favor rapid denitrification.

The losses of fertilizer nitrogen observed in these studies may be unusually high due to the weather conditions that prevailed during 1982 through 1984. One way to evaluate this possibility is to calculate apparent recoveries of fertilizer nitrogen from long-term studies where fertilizers have been applied over periods of many years and grain yields have been recorded.

**Apparent Recoveries of Fertilizer
Nitrogen in Long-Term Studies**

Iowa State University is currently maintaining long-term rotation-fertility studies located in north central Iowa (started in 1954 at the Clarion-Webster Research Center near Kanawha), north-west Iowa (started in 1957 on the Galva-Primghar farm near Sutherland), and northeast Iowa (started in 1979 near Nashua). Each of these studies includes several rotations involving corn and is designed so that all crops in each rotation are harvested each year. Fertilizer nitrogen has been applied for the corn crops at several rates and yields of corn at each rate have been measured each year. ^{15}N-tracers were not used in these studies, so it is not possible to directly determine the amounts of fertilizer nitrogen lost from soils.

Apparent recoveries of fertilizer nitrogen by continuous corn and corn in rotation with soybeans are shown in Table 2. These estimates were calculated from the yield data reported by Webb (1982a,b) and Webb and Ross (1982) and from the relationship between relative yields and nitrogen content of corn grain observed by Pierre et al. (1977). Data presented show that significant portions of the fertilizer nitrogen applied to the long-term studies were not harvested with the corn grain. If it is considered that most of the fertilizer nitrogen found in the relatively stable fractions of soil organic matter during the studies using ^{15}N-labeled anhydrous ammonia would be available for plants over long periods of time, it would seem that the apparent recoveries of fertilizer nitrogen observed in the long-term studies are quite similar to the recoveries observed in the 3-year study with labeled anhydrous ammonia. It should be noted that the relative uncertainty associated with a range of 20 to 40 percent recovery does not seem as important when expressed in terms of 60 to 80 percent of the nitrogen being not accounted for.

Table 2. Apparent recovery of nitrogen in corn grain at economic optimum rates of fertilization in long-term studies.[1]

Location	Continuous corn	Corn after soybeans
	— — — — % recovery — — — —	
Galva-Primghar	39	32
Northern Iowa	40	27
Northeast Iowa	28	28

[1]Percentage recovery was calculated for 12-year periods at the Galva-Primghar and Northern Iowa locations and a 4-year period at the Northeast Iowa location.

Apparent Recoveries Calculated
from Fertilizer Recommendations

In practice, recommendations concerning fertilizer needs usually are based on the internal nitrogen requirement of crops and credits for nitrogen that is derived from sources other than fertilizer. The internal nitrogen requirement reflects the amount of nitrogen that is contained in tissues of the mature crop when optimal amounts of nitrogen are present. It generally is expressed in terms of units of nitrogen per unit of yield (e.g., 21.4 kg N/mg grain or 1.2 lb N/bu grain). The rationale for such recommendations has been explained by Stanford (1973). Fertilizer recommendations are obtained by multiplying the internal nitrogen requirement times the yield goal and then subtracting credits for nitrogen derived from other sources (soil, manure, legumes, etc.). This method has great appeal because all farmers have knowledge of yields that are reasonable for their particular farm and management capabilities and, therefore, can easily make their own fertilizer recommendations.

If it is accepted that corn grain has 1.5 percent nitrogen on a dry weight basis at economic optimum rates of nitrogen fertilization (Pierre et al., 1977) and recognized that yields are adjusted to 15.5 percent moisture content, simple calculations show that corn grain contains 12.7 kg N/Mg grain (0.71 lb N/bu). Therefore, the maximum percentage recovery of fertilizer nitrogen in harvested crops is 59 percent when fertilizers are applied according to the concept of internal nitrogen requirement and when it is assumed that the internal nitrogen requirement for corn is 21.4 kg N/Mg grain.

The percentage recovery falls below the maximum of 59 percent when farmers apply fertilizers for a yield goal that is not attained. The maximum percentage recovery of fertilizer nitrogen drops to 54 percent if farmers apply fertilizers for yields that are only 10 percent above yields attained. And farmers usually are advised to fertilize to increase yields by at least this amount because of the potential for high rates of return on investments in nitrogen.

The maximum percentage recovery also decreases rapidly if farmers do not give credits for nitrogen supplied by legumes, animal manures or other sources. It is noteworthy that recommendations sometimes do not include adjustments of fertilizer applications for crops following legumes and that many farmers make no adjustments even when substantial adjustments are recommended. It also is noteworthy that, just like fertilizer nitrogen, unused nitrogen from manures and legumes can decrease the quality of water supplies.

Overall, the foregoing discussion indicates that a substantial portion, probably more than half, of the fertilizer nitrogen applied to soils in the Corn Belt remains unaccounted for. It is likely that a substantial portion of this nitrogen may have found its way to groundwater supplies.

Early Losses of Fertilizer Nitrogen

Losses of fertilizer nitrogen that occur between the time of fertilization and the maturity of the first crop are of special interest because these losses may be easier to control than are losses of nitrogen that occur between cropping seasons. The most obvious method of controlling these losses is through more judicious selection of the time at which the fertilizers are applied. In the Corn Belt, the most common fertilization practice is to apply fertilizers immediately before crops are planted in the spring. The problem with this practice is that significant amounts of rainfall occur during the late spring and early summer, and this rainfall may result in significant losses of fertilizer nitrogen by leaching and denitrification. The following is a summary of results from recent ^{15}N-tracer studies that indicate that early losses of fertilizer nitrogen by leaching and denitrification in the Corn Belt may be greater than has been generally believed.

In the spring of 1982, ^{15}N-labeled urea was applied to triplicate microplots on 10 different soils extensively used for corn production in various parts of Iowa (Priebe et al., 1983). The urea was surface-applied and not mixed into the soil because the initial objective was to determine amounts of nitrogen lost by ammonia volatilization following surface applications of urea. Design of the study was based on the customary assumption that fertilizer nitrogen is not rapidly leached from Iowa surface soils. Analyses of soil samples collected during excavation of the microplots to a depth of one meter after six weeks showed recoveries of urea-derived nitrogen ranging from 100 percent to 35 percent (mean recovery for 10 sites was 65%). Because losses of fertilizer nitrogen did not show a significant relationship to soil pH, there is little evidence to support the idea that ammonia volatilization was the major mechanism for loss where large losses of nitrogen occurred. Because most of the urea-derived nitrogen found was as nitrate and because most was found below a depth of 25 cm, leaching and denitrification must be considered probable mechanisms of loss where large losses of fertilizer nitrogen occurred.

In the spring of 1983, ^{15}N-labeled urea was surface-applied to microplots on four soils (Priebe and Blackmer, 1984). Triplicate microplots from each soil were excavated at two-week intervals and the resulting soil samples were analyzed to determine percentage recovery of the labeled nitrogen. Similar results were obtained for each of the four soils. The results from one of these soils are summarized in Table 3, which shows that significant amounts of fertilizer nitrogen were lost. Although this soil was selected for study because its pH and buffering capacity were conducive for high losses of fertilizer nitrogen by ammonia volatilization, it is unlikely that ammonia volatilization was the major mechanism for nitrogen loss because most nitrogen was lost after the urea-derived nitrogen was converted to nitrate. It is not possible to determine the amounts of fertilizer nitrogen lost by denitrification. However, the distributions of nitrate found showed that fertilizer-derived nitrate moved rapidly downward to the lowest depths sampled.

It should be noted that the amounts of rainfall that occurred in the spring of 1983 were above average (see footnote of Table 3) and, therefore, that losses of fertilizer nitrogen were greater than could be expected for most years.   However, it is relevant that farmers who applied fertilizer nitrogen by this method in 1983 would have lost similar amounts of their fertilizer nitrogen.

Although the studies using ^{15}N-labeled urea suggest that substantial amounts of fertilizer nitrogen may be lost by leaching, interpretations of the results of these studies are complicated by reactions associated with ammonium. This problem was eliminated in studies using ^{15}N-labeled nitrate in 1984 and 1985.   ^{15}N-labeled nitrate was surface applied to six microplots at each of three dates at three locations in Iowa in 1984 and at one date at six locations in the spring of 1985 (Cerrato et al., 1985). At 2, 4, and 6 weeks after application, duplicate microplots at each location were destructively sampled to determine the amounts of fertilizer-derived nitrate at various depths. Soil samples were collected to a depth of 1.8 m in 10 cm increments.   The results of these studies showed that large amounts of rainfall resulted in substantial losses of nitrate from the soil profile. High recoveries of nitrate were observed when small amounts of rainfall occurred.   Overall, the studies with labeled nitrate confirm that surface-applied fertilizers may be much more susceptible to loss by leaching than has been generally recognized.

Table 3. Recovery of urea-derived ^{15}N in the surface meter of a calcareous overwash soil at various times after surface application of urea.[1]

Time of sampling	Urea-derived ^{15}N recovered			
	As ammonium	As nitrate	As other[2]	Total
weeks	— — — — — — — — — — — — % — — — — — — — — — — —			
2	51	6	46	103
4	1	41	29	71
6	2	22	19	43
8	2	4	21	26

[1]Amounts of rainfall during the first, second, third, and fourth two-week period were 0.5, 6.0, 22, and 4.7 cm, respectively.

[2]Amounts of nitrogen recovered as soil organic matter and fixed ammonium.

Although a reasonable discussion of the topic is beyond the scope of this paper, the results of these studies present strong evidence that nitrate moves through many Iowa soils by mechanisms that have not been recognized. The major difference is in the importance of preferential flow of water and nitrate though macropores (old root channels, worm holes, cracks, etc.) in the soil. The macropores act as "super highways" that permit water and nitrate to bypass most of the soil within the rooting zone and move rapidly out of this zone.

The consequence of preferential flow through macropores is a tendency for nitrate to disappear from the surface few centimeters of soils without accumulating in subsoils. In absence of knowledge about preferential flow through macropores, failure to detect nitrate in subsurface layers often is incorrectly interpreted as direct evidence that leaching did not occur. Such an interpretation makes denitrification and ammonia volatilization seem like probable mechanisms of nitrogen loss, even when there is no direct evidence that denitrification or ammonia volatilization occurred. It is quite probable that, because of this misinterpretation, the importance of nitrate leaching has been seriously underestimated in many studies.

The finding that early losses of fertilizer nitrogen by leaching may be more important than has been believed is not good news when discussing the effects of nitrogen fertilization practices on groundwater quality. However, if large amounts of nitrate are being leached from soils via macropores, recognition of this as an important mechanism of nitrogen loss is the first step toward finding ways to prevent these losses.

THE NEED FOR EXPANDED RESEARCH
AND EDUCATION PROGRAMS

At this point, it needs to be recognized that farmers use fertilizer nitrogen only because it is a powerful tool for increasing productivity and farm income. Using data from the long-term studies at Kanawha (see the discussion that accompanies Table 2) as an example, it can be shown that appropriate amounts of fertilizer nitrogen can almost triple the yields of corn grain when this crop is grown continuously on the same soil. When prices for corn and fertilizer that were common in the early 1980s are used, investments in the first 50 kg of N/ha for continuous corn in the long-term studies returned more than 10 fold in terms of increased crop yields. Investments in economically optimum amounts of fertilizer returned more than six fold. The importance of nitrogen fertilizers to agricultural productivity and the economy of states in the Corn Belt must be considered carefully when developing strategies for protecting groundwater.

The potential for high rates of economic return has made fertilizers seem relatively inexpensive. Therefore, losses of fertilizer nitrogen seem relatively unimportant. But farmers must pay for the fertilizers, and fertilizer nitrogen represents a major cost of production for many crops. It would be irrational for a farmer to apply more fertilizer than he believes to be necessary. And all evidence suggests that farmers would rapidly accept any proven and cost-effective technology that could be used as a substitute for inefficient fertilization practices. However, the need to develop and prove such technologies has not seemed sufficiently important to invest in research and education programs on a scale that is commensurate with the problem.

If current fertilization practices are considered to be a major threat to groundwater supplies, the most rational solution to this problem is to invest in research and education programs that show farmers how to use fertilizers more efficiently. Because of the amounts farmers are spending for nitrogen fertilizers and the amounts of fertilizer nitrogen being lost from soils, it is likely that farmers, as well as society, would benefit greatly from these programs even if groundwater problems are not considered.

LITERATURE CITED

Baker, J.L., K.L. Gabel, H.P. Johnson and J.J. Hanway. 1975. Nitrate, phosphorus and sulfate in subsurface drainage water. J. Environ. Qual. 4:406-412.

Cerrato, M.E., A.M. Blackmer and D.L. Priebe. 1985. Movement of ^{15}N-labeled nitrate in the rooting zone of Iowa soils. Agronomy Abstracts, p. 23.

Chichester, F.W. and S.J. Smith. 1978. Deposition of ^{15}N-labeled fertilizer nitrate applied during corn culture in field lysimeters. J. Environ. Qual. 7:227-233.

Crutzen, P.J. and D.H. Ehhalt. 1977. Effects of nitrogen fertilizers and combustion on the stratospheric ozone layer. Ambio 6:112-117.

Gast, R.G., W.W. Nelson and G.W. Randall. 1978. Nitrate accumulation in tile drainage following nitrogen applications to continuous corn. J. Environ. Qual. 7:258-261.

Hallberg, G.H., R.D. Libra, E.A. Bettis III and B.E. Hoyer. 1984. Hydrogeologic and water-quality investigations in the Big Spring basin, Clayton County, Iowa: 1983 Water-Year. Iowa Geological Survey, Report 84-4.

Hanway, J.J. and J.M. Laflen. 1974. Plant nutrient losses from tile-outlet terraces. J. Environ. Qual. 3:351-356.

Kitur, B.K., M.S. Smith, R.L. Blevins and W.W. Frye. 1984. Fate of ^{15}N-depleted ammonium nitrate applied to no-tillage and conventional tillage corn. Agron. J. 76:240-242.

Meisinger, J.J., V.A. Bandel, G. Stanford and J.O. Legg. 1985. Nitrogen utilization of corn under minimal tillage and moldboard plow tillage. I. Four-year results using labeled fertilizer on an Atlantic Coastal Plain Soil. Agron. J. 77:602-611.

Olson, R.V. 1980. Fate of tagged nitrogen fertilizer applied to irrigated corn. Soil Sci. Soc. Am. J. 44:514-517.

Pierre, W.H., L. Dumenil and J. Heneo. 1977. Relationship between corn yield, expressed as a percentage of maximum, and the N percentage in the grain. II. Diagnostic use. Agron. J. 69:221-226.

Priebe, D.L., A.M. Blackmer and J.M. Bremner. 1983. ^{15}N-tracer studies of the fate of surface-applied urea. Agronomy Abstracts, p. 159.

Priebe, D.L. and A.M. Blackmer. 1984. Recovery of surface-applied urea ^{15}N in the surface meter of soil at various times after application. Agronomy Abstracts, p. 217.

Sanchez, C.A. and A.M. Blackmer. 1985. Recovery of anhydrous ammonia-derived N by corn in Iowa. Agronomy Abstracts, p. 183.

Stanford, G. 1973. Rationale for optimum nitrogen fertilization for corn production. J. Environ. Qual. 2:159-166.

Webb, J.R. 1982a. Rotation-fertility experiment. Annual Progress Report of the Northern Research Center and Clarion-Webster Research Center. Iowa State University, Ames.

Webb, J.R. 1982b. Rotation-fertility experiment. Annual Progress Report to the Northwest Research Center. Iowa State University, Ames.

Webb, J.R. and K.W. Ross. 1982. Rotation-fertility experiment. Annual Progress Report of the Northeast Research Center. Iowa State University, Ames.

CHAPTER 6

IMPACTS OF CHEMIGATION
ON GROUNDWATER CONTAMINATION

James S. Schepers
Agricultural Research Service
U.S. Department of Agriculture
University of Nebraska-Lincoln
Lincoln, Nebraska 68583-0915

DeLynn R. Hay
Agricultural Engineering Department
Institute of Agriculture and Natural Resources
University of Nebraska-Lincoln
Lincoln, Nebraska 68583-0726

INTRODUCTION

Contamination of groundwater by agricultural chemicals as the result of chemigation is of concern to producers as well as the public. A portion of this concern is well founded because of what is known about agricultural and industrial contaminants already in the groundwater. Other aspects may be emotional, because contaminants may reside in the soil many years before being discovered. This concern is reinforced in a recent report by the U.S. Geological Survey (1985) indicating that areas with well water concentrations of nitrate-nitrogen exceeding 10 mg/l corresponded to areas with extensive irrigation development, regions with high livestock or animal densities, or cultivated areas over unique porous land forms. Since chemigation is a relatively new practice in the U.S., it is unlikely chemigation is the sole cause of this nitrate contamination. Such examples illustrate the close ties between cultural practices and nitrate leaching, and may serve as a warning regarding leaching of other chemicals. The fact that over 40 percent of the sprinkler irrigated area in the U.S. involves chemigation is reason for concern (Threadgill, 1985).

The earliest use of chemigation was for the application of fertilizer, primarily nitrogen, with the irrigation water to increase nitrogen-use efficiency. More recently, there has been increasing interest in and use of chemigation to apply herbicides, insecticides, fungicides and nematocides. In most cases, an existing irrigation system is used for the chemigation application; the system is not designed specifically for chemigation. Applying chemicals with an irrigation system usually accomplishes the intended application; however, water application rate and uniformity may not be compatible with time of application for maximum pesticide efficacy or fertilizer utilization. In contrast, well designed irrigation and chemigation systems can be more accurate and precise than other methods (Eisenhauer, 1985). Nevertheless, environmental concerns have arisen and can be grouped into two broad categories. Unique to chemigation is the possibility of groundwater contamination due to accidental backflow or siphoning of chemicals down the well which is a form of point source pollution. The other type of groundwater contamination that can result from any method of chemical application (including chemigation) is the nonpoint leaching of chemicals through the soil and into the aquifer. Each mode of contamination can cause serious problems, but the processes involved in each mode are unique as are the remedial actions required after contamination. Therefore, each type of contamination will be discussed independently.

WELL CONTAMINATION

Groundwater contamination due to backflow into the well after the pumping operation ceases can take at least two forms. Both are the result of an inoperative or ineffective backflow prevention device on the main water supply line and would occur only after an unexpected shutdown of the irrigation pumping plant while it is unattended. In the first case, the water-chemical mixture in the distribution system can flow back into the well and potentially mix with the groundwater. The extent of this type of contamination depends on a number of factors such as the type of chemical, volume of water mixed with the chemical, amount of chemical in that volume of water, depth of the aquifer, diameter of the well casing, and elapsed time before the accident is discovered. Safety equipment to prevent this type of groundwater contamination has been described by Hay and Eisenhauer (1985).

An example may help quantify the magnitude of contamination caused by allowing a pipeline full of chemical laden water to flow back down into the well. A typical 1294 ft (394 m) center pivot with an inside pipe diameter of 6.4 inches (163 mm) holds 2170 gal (8.21 m^3) of water. If the water application depth was 1.0 inch (25.4 mm) and fertilizer nitrogen (N) was being applied at a rate of 30 lb N/acre (33.6 kg N/ha), the pipeline would contain 2.39 lb (1.09 kg) nitrogen at any one time. The nitrogen concentration in the pipeline would then be 132 mg/l. Approximately 50 percent of the water in the pipeline could drain back down the well (16-inch [0.41-m] well casing with 30 m depth of aquifer) before contaminated water could replace the volume of water in the casing and possibly be forced out into the aquifer. This estimate may be an oversimplification of the processes involved because of differences in density of the liquids and location of the pump inlet relative to the well screen. Only a portion of the water in the pipeline would be expected to flow back down the well even with a faulty check valve because of the arched nature of the pipeline between towers and discharge from the sprinklers and pipeline drains, unless the center pivot is on a significant incline. The extent of groundwater contamination would depend on the type and depth of aquifer and recovery characteristics of the cone of depression. Starting the well immediately would remove much of the contaminated water and minimize contamination. If the contaminated water completely mixed in a 30 m radius with 30 m of saturated thickness (6 m equivalent thickness of water in the sand and gravel aquifer), the resulting nitrogen concentration would be 0.06 mg/l.

The second possible method of contamination can occur if the chemical injection device continues to operate after the irrigation pump stops or gravity causes the chemical in the supply tank to flow into the irrigation pipeline. In either case, the entire contents of the chemical supply tank could potentially flow into the well if the backflow prevention equipment on the main pipeline fails. A 28 percent nitrogen solution (w/w) would have a nitrogen concentration of approximately 350,000 mg/l. Failure of safety devices to prevent drainage of the entire tank of nitrogen fertilizer intended for a 130 acre (52.7 ha) field (3900 lb N applied at 30 lb N/acre) would contaminate the 30 m depth and assumed 30 m radius around the well to a concentration of 104 mg N/l. If the groundwater has a gradient, a plume will develop downgradient and increase the difficulty of recovery (Nielsen and Warner, 1986).

Had the chemical in the above example been a pesticide applied at a rate of 1 qt/acre in 0.25-inch depth of water (2.34 l/ha in 6.35 mm water), the concentration in the pipeline would have been 37 mg/l. This amount of pesticide will result in a concentration of 0.02 mg/l if allowed to mix in the aquifer to a radius of 30 m from the casing and within the 30 m saturated depth. Doubling the water application rate would decrease the pesticide concentration in the pipeline by 50 percent. If the entire quantity of pesticide for a typical center-pivot system in Nebraska (32.5 gal [123.7 l] for 130 acres [52.7 ha]) would empty into the well and completely mix with the 30-m radius by 30 m saturated depth, the resulting concentration would be 7.3 mg/l.

Recovery of chemicals considered to be soil adsorbing (most pesticides to some degree) and nonadsorbing (nitrate) from a con-taminated well depends on the cation exchange capacity of the materials in the aquifer. Many of the same principles apply to recovery of contaminated groundwater as apply to herbicide applica-tion rates. For example, the recommended herbicide application rate and the potential for leaching through the soil depends, in part, on the soil type and organic matter content. Nitrate anions are usually considered mobile in the soil and aquifer unless immobilized by microbial activity. In contrast, organic compounds are adsorbed to residues and soil particles, which can reduce the efficacy of the compound, increase the potential for degradation, and reduce the potential for leaching. In the event that any of these chemicals reach the groundwater where the potential for adsorption may be very low because of the somewhat inert characteristics of sand and gravel, recovery by pumping may follow similar trends for many water soluble chemicals. Adsorbed chemicals migrate more slowly in the aquifer, but because they must be desorbed before the well can be reclaimed, restoration by pumping may be difficult and require more time than for those chemicals that are not bound.

Movement of agricultural chemicals with the groundwater away from a well cannot be generalized because of differences in groundwater hydraulics and the adsorption-desorption capacity of aquifer materials is not well characterized. Sand and gravel aquifers are among the best in that they yield large volumes of water; however, the hydraulic conductivity is also large. Therefore, the potential for movement of contaminated groundwater is also great (Table 1). The sand and gravel aquifers may contain interbedded layers of fine-grained materials that will restrict movement of or adsorb contaminants. Spalding (1986) calculated average groundwater movement in the sand and gravel aquifer of the Platte River Valley near Grand

Table 1. Representative values of porosity, specific yield and hydraulic conductivity for selected geologic materials (Lawton, 1985).

Material	Porosity[1]	Specific Yield	Hydraulic Conductivity[2]
Clay	40-70	1-10	0-0.000094
Sand	25-50	10-30	0.005-0.14
Gravel	25-40	15-30	0.05-0.71
Sand and Gravel	20-35	15-25	0.009-0.24
Sandstone	5-30	5-15	0.000005-0.002
Shale	1-10	0.5-5	0-0.000005

[1]Values in percent by volume

[2]Values in cubic meters per second (m^3/sec)

Island, Nebraska, to be approximately 0.45 m per day. Even at that rate, groundwater associated with an undetected chemical contaminant would migrate about 3.0 m per week. The longer elapsed time between well contamination and discovery of the accident, the greater the difficulty to reclaim the aquifer.

Another consideration when reclaiming a contaminated well by pumping is disposal of the contaminated water. Over application of herbicides carried in the contaminated water could result in carry over to succeeding years and could restrict growth of some crops. Excessive applications of fertilizer nitrogen in the water may be harmful to the crop. Even if it was not harmful, it could be expected to leach toward the groundwater as has been shown with very high fertilizer nitrogen application rates (Schuman et al., 1975; Hergert, 1986). The fate of pesticides applied at excessive rates in an attempt to reclaim a contaminated well is largely unknown, but even at recommended application rates, a small portion of applied atrazine has been shown to leach to the groundwater under irrigated corn production (Wehtje et al., 1984). In most cases, the chemical concentration in the water pumped after a contamination event will be relatively low.

LEACHING

Groundwater contamination by agricultural chemicals through leaching may be less dramatic and harder to visualize than by back flow down a well, but in terms of chemical recovery, contamination by nonpoint leaching may present an almost insurmountable task. For this reason, every effort must be made to minimize groundwater contamination by leaching, regardless of the method of chemical application. Because leaching may be a relatively slow process and chemical dilution in the groundwater can be extensive, detection of contaminated groundwater may require many years, unless sophisticated sampling techniques are used. Although documented cases of groundwater contamination attributed to leaching of agricultural chemicals are not extensive, many more instances exist where fertilizer nitrogen and some pesticides have been detected moving downward through the soil (Spalding and Cady, 1986). Given enough time, these chemicals could assumably reach the groundwater. A point of major concern is the reservoir of chemicals that may be moving through the soil and have not yet been detected in the groundwater. Considering the likelihood of chemical leaching, the question to be addressed is no longer "Will it leach?" but rather "How soon, to what extent, and under what conditions?" Many things can happen to the variety of agricultural chemicals as they move through the soil, and each process or factor must be considered as society makes a concerted effort to minimize the risk of chemigation.

It is probably necessary to examine the potential for nitrate leaching apart from the other agricultural chemicals used in chemigation because there can be several sources of nitrogen contributing to the soil nitrate pool. In addition, fertigation is intended to directly promote plant growth while the objective of pesticide application is to indirectly enhance crop production through control of one or more crop pests. Unlike well contamination where things happen very quickly, microbial processes play a major role in leaching of all chemicals. Some chemicals may be degraded by soil microbes. Others may selectively inhibit one or more groups of organisms, while others are intended to control above-ground pests where efficacy is not a function of soil microbial activity. It should also be noted that some chemicals such as herbicides adversely affect microbial and macro faunal populations, but usually do not reduce crop yields. In another sense, soil microbes and climatic factors largely regulate the soil organic matter content, which in turn can have a major effect on the chemical adsorption-desorption process.

Soil organic matter is also a primary component of the nitrogen cycle in that carbon serves as the substrate for microbial activity. As such, microbial decomposition of carbon-rich organic residues can result in temporary immobilization of inorganic nitrogen, with the subsequent release of plant available nitrogen which is susceptible to leaching. Other nitrogen-rich residues such as manure can supply large amounts of plant nitrogen upon application to the land. These organic sources of nitrogen function similarly to slow release forms of nitrogen fertilizer; however, mineralized nitrogen is difficult to characterize and quantify, so its full benefit to the crop is seldom recognized. Failure to give proper credit for manure, mineralized soil nitrogen, and legumes when making fertilizer recommendations can result in overfertilization and increase the potential for nitrate leaching.

Conceptually, fertigation is the ideal way to apply small amounts of nitrogen fertilizer when needed to supplement other nitrogen sources. Timing of fertigation to attain maximum nitrogen use efficiency without impairing production is probably the most difficult task a producer faces. Gascho and Hook (1985) found small frequent nitrogen applications (fertigation) on sandy soils were superior in terms of corn yield compared to sidedress or dribble applications. Since nitrogen fertilizer is relatively inexpensive compared to the yield response, irrigated corn producers can usually justify applying enough nitrogen to produce near maximum yields. However, tagged nitrogen studies indicate reduced nitrogen uptake efficiency as application rates increase. Fertigation offers the possibility for reduced nitrogen application rates, increased yields, and reduced potential for leaching.

The environmental concern with any fertilizer program deals with the nitrogen that is not utilized by the crop. An equally difficult situation can arise when the coordination between fertigation and crop requirements is interrupted when rainfall reduces or eliminates the need for supplemental water (McIsaac et al., 1984). In such cases, it may be necessary to risk a nitrogen deficiency or irrigate to apply nitrogen fertilizer at the risk of leaching nitrates below the root zone. To reduce the risk of a nitrogen deficiency, fertigation practices usually insure adequate soil nitrogen for several weeks in the future. In terms of nitrate leaching, fertigation is usually no worse than sidedress applications and is more desirable than preplant fertilizer nitrogen applications (Hergert, 1985). There are still many aspects of fertigation that could improve fertilizer utilization and reduce nitrate leaching; however, these refinements also require a higher degree of management and perhaps more technical inputs (Schepers and Martin, 1986).

Computer simulation models may assist in evaluating nitrogen and water management practices. Primary advantages of such models are that they make it possible to evaluate and compare an infinite combination of production practices over a variety of soil types and climatic conditions. Process oriented models that predict nitrate leaching have been used to evaluate nitrogen management practices on sandy soils (Watts and Martin, 1981) as well as fine textured soils that have a large potential for nitrogen mineralization (Schepers and Martin, 1984). The real utility of such models may not be in predicting the absolute amount of water percolation or nitrate leaching under a site-specific condition, but as a tool to evaluate different cultural practices (Hubbard et al., 1985; Schepers et al., 1984). The result is usually an increased awareness for the processes involved in water movement and nitrogen management. Frequently, predictions also identify other factors like temperature and/or precipitation that can dominate certain processes and make cultural practices seem trivial. An example is where the combination of irrigation scheduling, soil testing and nitrogen management may minimize nitrate leaching below the root zone during the growing season if the fertilizer nitrogen applied the previous fall remained in place during the winter and spring (Schepers et al., 1983). The same cultural practices resulted in large nitrate leaching losses during the growing season of irrigated corn when above normal precipitation in the spring leached nitrate deep in the root zone before the crop was ever planted. In this example, the common practice of fall nitrogen fertilization was identified as being risky relative to spring, sidedress or fertigation methods of application.

Fertigation may be regarded by producers as an essential cultural practice for maintained productivity on shallow sandy textured soils. In addition, fertigation should minimize nitrate leaching to the groundwater and maximize fertilizer nitrogen use efficiency. These assumptions are probably only valid if a majority of the nitrogen is applied via fertigation and total nitrogen application does not exceed crop requirements. This point has been illustrated in the Nebraska Sandhills where a small amount of tagged nitrogen was applied to corn in irrigation water during late June, mid-July or early August. In this study, total fertilizer applications exceeded those necessary to reach maximum yields and less than 10 percent of the tagged nitrogen was taken up by the crop (unpublished data). Nitrogen not taken up by the crop would be available for incorporation into the microbial biomass or subject to leaching.

Other currently unpublished research involving isotopic nitrogen tracers in irrigation water indicates that fertigated nitrogen may be no more effectively utilized than preplant or sidedressed nitrogen, especially if a large portion of the total plant nitrogen uptake has occurred prior to fertigation. Once in the soil, nitrogen applied by fertigation should have a fate similar to other inorganic soil nitrogen.

The probability of leaching inorganic soil nitrogen, as with many pesticides, is largely dependent on the driving force of water and on soil texture. Fine textured soils are expected to infiltrate and percolate water more slowly than coarse textured soils. These guidelines are confounded as soils mature, clay particles migrate downward, and carbon cycling (crop residues and microbial biomass) results in better aggregation. The net effect may be a heterogenous soil in terms of water flow characteristics and leaching of nitrates and pesticides. Soil cracks, root channels, macro faunal cavities, and fracture plains all contribute to nonuniform water percolation and complicate efforts to measure leaching. For these reasons, prediction of nitrate and pesticide leaching is far from an exact science. The situation is complicated even further by the adsorption-desorption characteristics of the soil and chemical in question. Highly structured soils may increase percolation but reduce the potential for chemical leaching provided a large portion of the percolation takes place in the macropores, and the chemical is uniformly distributed in the surface soil and largely inaccessible to the macropores.

Chemigation could result in accelerated leaching if macropores are continuous to the soil surface and have not been extensively disrupted by tillage. Increased percolation rates caused by macropores may be short-lived if the subsoil texture limits percolation.

With these factors in mind, it should be apparent that broad spectrum statements regarding the impact of chemigation on groundwater quality should be taken with extreme caution. Rather, the fate of chemicals applied in irrigation water must be examined as a composite of processes and addressed on an individual soil, chemical, climate, and irrigation system basis.

LITERATURE CITED

Eisenhauer, D.E.   1985.   Irrigation system characteristics affecting chemigation. In: E.F. Vitzthum and D.R. Hay (Eds.), Proceedings of the Chemigation Safety Conference, Lincoln, NE, April 17-18, 1985. Cooperative Extension Service, Institute of Agriculture and Natural Resources, University of Nebraska, Lincoln, NE 68583, pp. 5-8.

Gascho, G.J. and J.E. Hook.   1985.   Development of a nitrogen fertigation program for corn grown on sandy soil. In: S.C. Sharad (Ed.), Proceedings of the Third National Symposium on Chemigation, Tifton, GA, August 21-23, 1985. University of Georgia, College of

Agriculture and Rural Development Center, Tifton, GA 31793, pp. 42-50.

Hay, D.R. and D.E. Eisenhauer. 1985. Anti-pollution protection when applying chemicals through the irrigation system. In: E.F. Vitzthum and D.R. Hay (Eds.), Proceedings of the Chemigation Safety Conference, Lincoln, NE, April 17-18, 1985. Cooperative Extension Service, Institute of Agriculture and Natural Resources, University of Nebraska, Lincoln, NE 68583, pp. 12-15.

Hergert, G.W. 1985. Fertilizer selection, efficacy, management and safety for chemigation. In: E.F. Vitzthum and D.R. Hay (Eds.), Proceedings of the Chemigation Safety Conference, Lincoln, NE, April 17-18, 1985. Cooperative Extension Service, Institute of Agriculture and Natural Resources, University of Nebraska, Lincoln, NE 68583, pp. 37-40.

Hergert, G.W. 1986. Nitrate leaching through sandy soil as affected by sprinkler irrigation management. J. Environ. Qual. 15:272-276.

Hubbard, R.K., W.G. Knisel and G.J. Gascho. 1985. A comparison of nitrate leaching losses under N applied conventionally or by fertigation. In: S.C. Sharad (Ed.), Proceeding of the Third National Symposium on Chemigation, Tifton, GA, August 21-23, 1985. University of Georgia, College of Agriculture and Rural Development Center, Tifton, GA 31793, pp. 51-57.

Lawton, D.R. 1985. Groundwater Hydraulics. In: E.F. Vitzthum and D.R. Hay (Eds.), Proceedings of the Chemigation Safety Conference, Lincoln, NE, April 17-18, 1985. Cooperative Extension Service, Institute of Agriculture and Natural Resources, University of Nebraska, Lincoln, NE 68583, pp. 22-25.

McIsaac, G.F., D.L. Martin and J.S. Schepers. 1984. Nitrate leaching in sandy soils. Amer. Soc. Agric. Eng., Paper No. 84-2610.

Nielsen, K.F. and J.W. Warner. 1986. Aquifer contamination potential from chemigation. Public perception of groundwater quality and the producers dilemma. Proceedings of the National Well Water Association Conference on Agricultural Impacts of Groundwater, Omaha, NE, August 11-13, 1986. (In press.)

Schepers, J.S., K.D. Frank and D.G. Watts. 1984. Influence of irrigation and N fertilization on groundwater quality. Proceedings of the International Association of Hydrological Sciences Symposium No. HS2, "Relation of Groundwater Quality," Hamburg and Koblenz, Germany, August 25-26, 1983.

Schepers, J.S. and D.L. Martin. 1984. Management practices to reduce nitrate leaching with irrigated corn. In: D.M. Nielsen and L. Aller (Eds.), Proceedings of the National Well Water Association Western Regional Conference on Groundwater Management, San Diego,

CA, October 23-26, 1983. Water Well Journal Publishing Co., 500 W. Wilson Rd., Worthington, OH 43085, pp. 85-91.

Schepers, J.S. and D. Martin. 1986. Public perception of groundwater quality and the producers dilemma. Proceedings of the National Well Water Association Conference on Agricultural Impacts on Groundwater, Omaha, NE, August 11-13, 1986. (In press.)

Schuman, G.E., T.M. McCalla, K.E. Saxton and H.T. Knox. 1975. Nitrate movement and its distribution in the soil profile of differentially fertilized corn watersheds. Soil Sci. Amer. Proc. 38:1192-1197.

Spalding, R.F. and R.E. Cady. 1986. Excursion from chemigation backflow. Proceedings of the U.S. Committee on Irrigation and Drainage, Regional Meeting, Denver, CO, September 11-12, 1986. (In press.)

Threadgill, E.D. 1985. Chemigation via sprinkler irrigation: Current status and future development. Applied Engineering in Agriculture 1:16-23.

U.S. Geological Survey. 1985. National Water Survey. 1984. USGS Water-Supply Paper 2275. 467 pp.

Watts, D.G. and D.L. Martin. 1981. Effects of water and nitrogen management on nitrate leaching loss from sands. Trans. Amer. Soc. Agric. Eng. 24:911-916.

Wehtje, G., L.N. Mielke, R.J.C. Leavitt and J.S. Schepers. 1984. Leaching of atrazine in the root zone of an alluvial soil in Nebraska. J. Environ. Qual. 13:507-513.

CHAPTER 7

ASSESSING ANIMAL WASTE SYSTEMS IMPACTS ON GROUNDWATER: OCCURRENCES AND POTENTIAL PROBLEMS

James N. Krider
Soil Conservation Service
U.S. Department of Agriculture
Washington, D.C. 20013

With the rush of concern about agrichemicals in groundwater, particularly commercial fertilizer constituents and pesticides, it is easy to overlook animal waste as a potential source of groundwater pollutants. While not as significant a source as some others when viewed on a regional basis, animal waste nonetheless poses a clear threat to groundwater quality. Today's concern over groundwater quality has renewed interest in better management of organic waste materials. Protection of the quality of groundwater can be the result of better management. Human and livestock health is dependent on good quality drinking water, much of which comes from groundwater sources.

INTRODUCTION

In assessing the impacts of animal waste on groundwater, it is essential to first define animal waste in terms of the pollutants involved. Knowledge about the details of animal waste (management) systems is also necessary in the assessment of potentials. Vulnerable locations in a system need to be identified and characterized in terms of probable pathways for pollutants. Once the types of pollutants and their pathways to groundwater are understood, it is possible to evaluate potential for creating impacts on groundwater that can cause an impaired use and/or a risk for the user. The potentials are developed by identifying and applying the needed assessment techniques at the desired assessment levels. Assessment levels relate primarily to geologic characteristics that affect the occurrence and availability of groundwater. Most groundwater can be categorized as that found under saturated conditions--shallow, local aquifers and deep, regional aquifers, and that found under unsaturated conditions--sometimes referred to as the vadose zone.

ASSESSING POTENTIAL PROBLEMS

Animal Waste

Physiologically, animal waste is material that is unused or unusable by the animal and is excreted. What is waste to the animal's system is often a food source for other organisms; hence, it can be a beneficial product. But, when the concentration of waste is greater than these other organisms can use, environmental pollution can occur.

In the context of groundwater contamination, only a few animal waste constituents create health related problems for livestock and humans. Although there are a large number of animal waste constituents that can cause groundwater contamination, most constituents occur at such low concentrations as to be of little concern. Some potentially contaminating constituents, such as aluminum, zinc, iron, and potassium are essential micronutrients for humans and livestock and are not ordinarily recognized as a concern in drinking water. Some of these elements that can be toxic at high concentrations are tied up in the soil and are not often found in groundwater.

Constituents found in groundwater that are of concern are the soluble forms of nitrogen and phosphorus, bacteria, and at a much lower occurrence, organic materials. The nitrate form of nitrogen is commonly recognized as hazardous to humans, mostly in the very young. Cases of nitrate poisoning in humans are well known.

Methemoglobinemia, or Blue Baby Syndrome, is a problem associated with nitrate-to-nitrite reduction in the intestines of infants, but is also a phenomenon that occurs in adults, but without significant health effects. There is, however, little documentation about the effects on livestock of water-borne nitrates. The National Academy of Sciences published valuable works on the health effects of nitrates and nitrites in water (NAS, 1974; 1977). It cited numerous studies relating variabilities in responses, and concluded that it was the differences in an individual's "resistance" that controlled changes in health. With the Environmental Protection Agency (1973), the NAS established recommended standards for maximum allowable concentrations of nitrates and nitrites in drinking water for livestock and poultry, and for humans.

Phosphorus is usually adsorbed in the soil profile; hence, it is not often found in the groundwater. Soluble forms will leach into groundwater in areas of coarse-grained soils. Phosphorus, as such, does not present a significant health risk. It can, however, accelerate the growth of certain blue-green algae toxic to livestock.

Soluble phosphorus can flow into surface water supplies by ground-water return flows.

Certain pathogenic organisms in animals pose health risks for humans. Some forms of bovine tuberculosis, strains of salmonella, and cholera are known to affect humans. Most of these organisms are filtered out in the soil, and except in areas of very permeable soil, are not found in groundwater. Inflow along well casings and through sink holes are possible routes for these organisms to enter groundwater.

Other than imparting an undesirable taste and color in drinking water, organic materials found in groundwater are not known to be dangerous to health. However, their presence suggests that other contaminants are flowing directly into groundwater. Animal waste could be the source in those cases because of its high content of organic material. There are multiple routes by which these types of organics move into groundwater. The most prevalent routes are by percolating through unconsolidated soil media, by direct entry through sink holes and other exposed recharge areas, and by seeping down-ward along well casings. Once there, contaminants in groundwater are made available for consumption through pumping of wells and groundwater return flow into surface water streams. It is then that they pose a potential threat to the user.

Animal Waste (Management) Systems

The Soil Conservation Service (SCS) emphasizes the need for total systems planning for animal waste management. The need for management is stressed. The expression "animal waste systems" alone does not capture the essence of the SCS objective. A system without management often has neither the needed components nor the desired results. Conservation plans developed and implemented for manure management should account for all sources of potential pollutants, and they should be stylized to recycle the nutrients, improve soil organic matter, and protect both groundwater and surface water quality.

System plans for the management of animal waste come in many arrangements. The amount and characteristics of the land for recycling of waste materials is usually the controlling factor for the plans. Where land areas and characteristics are adequate for the amount of manure resource, the desired approach is to collect and store the manure for later application to agricultural land. Where land is inadequate, plans are developed so that some manure con-stituents are used in other ways besides land application, and excesses are stored for long periods of time, years in some cases. Each of these systems has components that should be evaluated for their potential for contaminating groundwater.

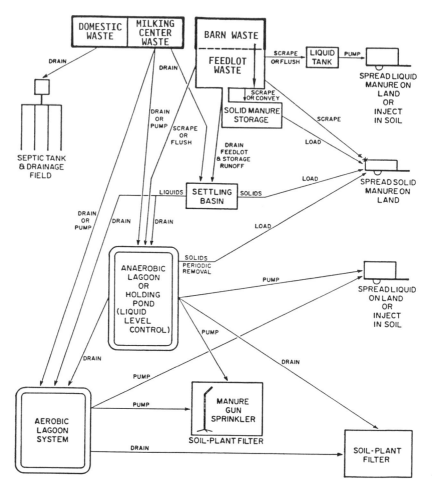

Figure 1.   Flow chart for dairy waste management system alternatives.

Figure 1 (WVU, 1975) schematically shows the numerous choices for structuring an animal waste management system. With regard to groundwater contamination potential, the major components are barnyards, storage facilities, treatment facilities and the land application activity.

If properly managed, barnyards and other open animal-holding areas that are paved do not present a significant threat to groundwater quality.   Also, there is little potential for groundwater contamination in the housing areas where animals are continuously housed such as in many poultry and swine operations. A potential for groundwater contamination exists under unpaved areas where animals are concentrated.

Earthen facilities used for the storage and treatment of manure and wastewater are the system components that present the greatest threat to groundwater quality. Debate continues about the adequacy of animal waste materials to seal the soil-waste interface in storage ponds and treatment lagoons. Field experience, and corroborating research results indicate that under certain circumstances, manure solids effectively seal soil surfaces, thereby reducing seepage to low levels in a few days to a few months (DeTar, 1977; Miller et al., 1985). Conversely, some operations and research indicate that there are certain combinations of soil conditions and waste characteristics that are conducive to unacceptably high rates of out-migration of liquids and pollutants for long periods of time (Barrington and Jutras, 1985; Phillips and Culley, 1985). SCS is planning to sponsor research that will provide planning and design guidance on combinations of waste and soil characteristics that are needed to protect groundwater quality.

Animal waste applied to the land at appropriate rates and during the proper times generally will not cause groundwater contamination problems. Threats to groundwater quality exist in those cases where excess amounts are applied to the land, mainly for disposal purposes, and in those cases where there are permeable soil conditions present on the application site. Ritter et al. (1986) showed very clearly how rapidly nitrates will leach through a permeable soil during periods of excessive rainfall.

Waste management techniques are more thoroughly described in SCS's Agricultural Waste Management Field Manual.

Groundwater

Groundwater is defined as that part of the subsurface water that is in the zone of saturation, including underground streams (American Geological Institute, 1980). Saturated zones can generally be categorized into shallow, local aquifers and deep, regional aquifers. The shallow aquifers are usually found within the upper 100 ft. of the earth's surface and are in unconsolidated materials. They are locally important, but usually do not extend over large areas--counties and other substate areas, for example. Movement of water and pollutants through them is controlled primarily by the uncemented soil materials overlying the deeper aquifer systems.

The deeper, more extensive groundwater zones have been studied extensively by the U.S. Geological Survey (USGS). USGS divided the country into groundwater regions (Figure 2) that are characterized mainly on the basis of the occurrence and availability of groundwater (1984). The features that distinguish the regional delineations are the nature of water-bearing openings, solubility of aquifer materials, water storage and transmission characteristics, and nature and location of recharge areas. All of the features are significant in terms of how they affect susceptibility to pollution, dispersion of pollutants, and rate of pollutant movement.

The unsaturated zones, known as vadose zones, located above the saturated groundwater zones are important in the total evaluation of subsurface water. These are the zones through which pollutants move and the zones that influence how far and at what rate the pollutants move. The unsaturated zones are the upper depths of soil profiles, the root zones for example. Unsaturated zones play a critical role in the successful implementation of animal waste management systems.

Assessment Levels

Two broad scope studies, national and regional, can be used to assess potentials for groundwater contamination in large geographical areas. National level studies are compilations of data about large regional areas cumulated to a national set of totals and averages. An example of this type study is the 1980 Appraisal for the Soil and Water Resources Conservation Act (USDA, 1981). These studies usually do not give enough information about specific impacts on a more comprehensible geographical area, but they are useful for the purpose of comparing overall water quality restoration achievements against cost. Groundwater regions established by USGS are invaluable in assessing potential contamination of national and regional groundwater.

Assessments are often conducted within political boundary lines such as those established by states and counties. They are useful for identifying pollution sources and implementing site-specific control strategies. Through these types of assessments, county governments can be effective in planning and implementing needed management activities. For example, Rock County, Wisconsin, in a cooperative effort with the Wisconsin Geological and Natural History Survey, established a groundwater protection plan (WGNHS, 1985). The plan identifies specific actions for minimizing pollution potentials. It presents a pollution risk assessment indexing procedure that can be of help in assessing potentials.

To many people, on-site assessments of animal waste operations are the most important. As an inseparable part of planning and implementing animal waste management systems, site conditions, characteristics, and features must be assessed. This assessment process should recognize the need to protect groundwater at the site being considered and the adjoining sites.

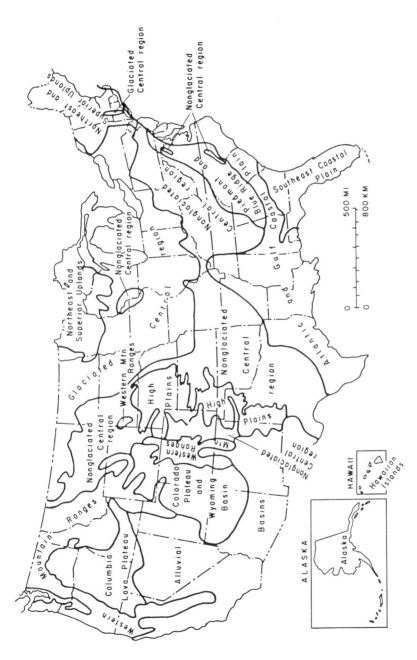

Figure 2. Groundwater regions of the United States (U.S. Geological Survey, 1984).

Assessment Techniques

Basically, two assessment techniques can be used to evaluate potential groundwater quality impacts. One is a surrogate technique that can be called presumptive, and the other is mathematical modeling. Another technique, groundwater monitoring, is a more direct way of assessing actual groundwater quality conditions.

The presumptive method of evaluating potential problems uses available information about concentrations of animals, effectiveness of animal management, soil and other site features, and groundwater zones. For a given area, these data are compiled and ranked according to their accumulated effects and the potential for groundwater contamination is established in terms such as high, medium, or low.

Mathematical models for groundwater analysis, the solutions to which are not highly evolved, provide the best technique for integrating a large number of variables in a continuous fashion. Researchers identify the physical processes and interrelationships, and convert them to mathematical expressions or models. Computer programs link the variables in proper sequence in the integration process. Results from the current generation of models provide insight into the effects on groundwater of different management options. The Agricultural Research Service's model GLEAMS (Knisel et al., 1986) is one that SCS is beginning to test.

The most widely used assessment technique is physical monitoring of groundwater quality conditions. Frequent monitoring is conducted to assess actual occurrences of groundwater contamination on an immediate basis. Less intensive monitoring activities are conducted to examine long term trends, and are useful in assessing effects of such features as land use.

Health Effects

Bacterial organisms occasionally get into groundwater. The most probable routes into groundwater for these organisms are through sinkholes and along well casings. Most of these organisms have passed through the intestinal tracts of warm-blooded animals. Some pose a threat to human and animal health. But because their survival outside the host animal is short, only a few hours in many cases, the threat is not great; however, the potential for infection is real and should not be dismissed.

Some of the more well known of the principal bacterial agents that cause intestinal diseases associated with drinking water are: (1) Salmonella group responsible for typhoid fever, paratyphoid fever, and intestinal fever; (2) Shigella group responsible for bacillary dysentery; (3) Vibrio cholerae responsible for cholera; and (4) Escherichia coli (or E. coli) responsible for gastroenteritis. In humans, gastroenteritis is by far the leading cause of waterborne infectious disease in the U.S. (NAS, 1977). Deficiencies in water treatment and groundwater contamination are responsible for most outbreaks (65%) and cases (63%). The consequences of an increasing prevalence of these organisms in livestock and their excreta has not been explored in any detail. Antibiotics in animal feeds, good sanitary conditions, and veterinary care probably prevent widespread problems.

Due to the ubiquitousness of nitrogen, it clearly presents the most significant threat to human and livestock health. The nitrate and nitrite forms of nitrogen are of the most concern. Nitrites are by far the most toxic to livestock. Fortunately, nitrite concentrations are usually low in natural waters. NAS (1974) reported that nitrite in water at a concentration of 200 mg/l decreased feed intake of turkeys, and caused decreased growth in chicks, but had no effect on the egg quality and production of laying hens. By contrast, nitrate at a concentration of 300 mg/l in water had no effect on chickens, pigs, and sheep, and at 450 mg/l had no effect on turkeys. At 667 mg/l in lambs and 1000 mg/l in sheep it produced significant changes in the oxygen carrying capacity of the blood in these animals.

The toxic effects on livestock of nitrates and nitrites in feed rations are much more severe. Silage or grazed vegetation that has accumulated nitrogen from any source, such as animal waste, irrigation water or commercial fertilizer, is a potential source of nitrogen-related health problems. For example, sheep that ingested 112 mg nitrate in their feed ration per kilogram of body weight experienced a 30 percent weight loss, and sheep fed 13 mg nitrite per kilogram of body weight died in 2 hours (NAS, 1977). In order for sheep to be subjected to a similar exposure in drinking water, concentrations of nitrate and nitrite would have to be approximately 670 mg/l and 78 mg/l, respectively. Concentrations at these levels in natural waters are uncommon, but they do occur in areas of large livestock populations (see section on Assessing Occurrences).

Recommended concentration limits have been set for nitrate and nitrite in drinking water for livestock and poultry by EPA (1973), and NAS (1974). They are 100 mg/l nitrate and 10 mg/l nitrite. These limits are considered to be far below the LD_{50} (LD_{50} = lethal dose that causes death in 50 percent of the animals involved) intakes of those elements.

Monogastric animals are more tolerant of high levels of nitrate than ruminants. For example, healthy adult humans seemingly can consume large quantities of nitrate in drinking water with little, if any effect. For human drinking water, however, the nitrate level is set at 10 mg/l as nitrogen based on effects in infants. This is considered by NAS (1977) to be a value that has little margin of safety for some infants.

Although speculative, there is potential for nitrate in water to act as a procarcinogen. Nitrate converts to nitrite which in turn reacts with amines in the body or in food to form N-nitroso compounds, more commonly known as nitrosamines. Most N-nitroso compounds studied were found to cause cancer in animals. There is no reason to expect that man is not also susceptible (NAS, 1977).

ASSESSING OCCURRENCES

An assessment of the occurrences of contaminants in groundwater from animal waste is complicated by the contribution from other sources of the same contaminants. In particular, there is no known way to distinguish nitrates that originate from animal manures from those of commercial fertilizer. Research work at the University of Rhode Island using nitrogen tracing techniques is investigating differentiation techniques.

Numerous studies of groundwater quality have determined the presence of nitrates, and in some there is evidence that animal waste has contributed. Where evidence regarding sources is not clear, suffice it to say that excesses of nitrate, whether from animal waste or other sources, will migrate into groundwater. Consequently, in many locations where there are concentrations of animals, there is likely to be elevated levels of nitrates in the groundwater. The concentration and persistence in the groundwater are the key concerns.

Case Studies

Delaware. To determine how rapidly nitrates can move through a coarse-grained soil profile, irrigated fertilizer plot work was conducted by Ritter et al. (1986). Before any fertilizer was applied, groundwater nitrates (as N) varied between 2.4 and 11.8 mg/l. Fertilizer at the rate of 300 lb/ac was applied in split applications beginning in 1984. In August, 1985, concentrations in monitoring wells (12 plots, 2 wells per plot) ranged from 8.9 to 65.7 mg/l. Following 11.8 inches of rain in late August, the groundwater nitrate levels in all wells at the 60-inch depth dropped to an average of about 12 mg/l, approximately the same as prior to any fertilizer applications. Ritter concluded that for maximum corn yields, nitrate leaching on sandy soils cannot be eliminated in humid regions.

Iowa. Groundwater studies in Iowa are well documented, especially in the karst topography area of the Big Springs Basin in northeastern Iowa. It has been described by Hallberg (1986) as a basin having severe groundwater quality problems. He showed that the average groundwater nitrate-N varies proportionally with the amount of fertilizer and manure nitrogen applied on the land. For example, in 1960, when manure nitrogen was applied on the basin's agricultural land at about 700 tons and commercial fertilizer nitrogen

at about 200 tons, the groundwater nitrate-N was about 3 mg/l.   In
1984, when manure nitrogen application was 1100 tons, and fertilizer
nitrogen 2700 tons, the groundwater nitrate-N averaged about 10
mg/l.

Since 1980, there have been 46 public water supply wells
throughout Iowa found to exceed the "maximum contaminant level" of
10 mg/l nitrate-N.   In some of these areas, 70 percent of the wells
sampled exceeded the nitrate contaminant level.   The nitrate-N
concentrations in some basins are between 35 and 40 mg/l.   Of all
sources of nitrogen applied to cropland, fertilizer nitrogen contributes
between 55 and 60 percent; manure nitrogen, between 10 and 15
percent; and legumes, rainfall, and natural soil nitrogen contribute
the remainder (Hallberg, 1986).   Threat to public health and the
economic loss of nutrients to the groundwater are the main concerns.

Pennsylvania.   In the first year (1985) of a nutrient management project in a sub-basin of the Conestoga Headwaters Rural Clean
Water Program, 71,310 pounds of manure nitrogen (total), and 23,160
pounds of fertilizer nitrogen (total) were applied (USDA, 1985).
Application rates of nitrogen varied between 136 and 828 lbs/ac.
This represented a 15 percent reduction in fertilizer applied nitrogen
(plus phosphorus) from the preceding year.   Fertilizer application
occurs mostly in a 30 day period in the spring.   Manure is applied
throughout the year and is dependent on climate and field conditions.
Domestic wells in the sub-basin showed nitrate-N concentrations
remaining fairly constant, three of those tested being near 1 mg/l,
three others ranging around 10 mg/l and a seventh showing a
decrease from about 30 mg/l in 1984 to just below 20 mg/l in the fall
of 1985.   It is expected that a more pronounced reduction in groundwater nitrate concentrations will occur from nutrient management in
the sub-basin.

California.   In the Chino Basin of Southern California, about
226,000 dairy animals reside on 25,000 acres of agricultural land.   The
area's groundwater quality has been extensively assessed.   It offers a
good opportunity to study the movement of groundwater contaminants
from an area of little to no contaminant input (mountainous areas to
the north), through a residential area, and then through the livestock
areas to the south (USDA, 1986).   Additionally, municipal wastewater
recharge plumes flow into the livestock area.   The nitrate (as NO_3)
concentrations at the upgradient edge of the livestock area are in the
30 to 40 mg/l range, and downgradient from the livestock area, 40 to
50 mg/l.   Groundwater nitrates upgradient from the livestock area
average 110 mg/l when influenced by the municipal wastewater plumes
(personal communication with Stan Moorhead, River Basin Planning
Staff, SCS, Davis, CA, Oct. 1986).

Of 34 groundwater samples taken in February, 1986, in the
vicinity of individual livestock operations, 13 exceeded 80 mg/l
nitrates as nitrate.   The highest single test was 345 mg/l.   This basin

is an example of how a mixture of intensive land use can cause some difficult groundwater quality problems.

WHAT NEEDS TO BE DONE?

A series of actions can be taken to help protect groundwater from contamination by animal waste constituents. Cooperative actions are needed by individuals, farmers and non-farmers alike, local, county, and state governments, soil and water conservation districts, and universities. Some of the actions apply equally to the management of commercial fertilizer. They are:

1. Identify areas that are vulnerable to groundwater contamination. The more susceptible areas are recharge areas, and areas where shallow soils overlie soluble bedrock (limestone). Make use of the available information on soils, geology, and groundwater in assessing groundwater contamination potentials.

2. Develop and implement groundwater quality management strategies. Risk assessment procedures such as developed in Wisconsin (Rock County) can be valuable.

3. Prepare totally integrated conservation plans for managing animal waste. Plans should account for all sources of on-farm animal wastes. Urge landowners to implement total plans.

4. To ensure that groundwater quality is protected, focus plans for animal waste management systems on the following critical functions and components:

    (a) Remove wastes from barnyards and other areas of animal concentrations and frequently convey them to waste storage or treatment facilities.

    (b) Prevent contaminants from flowing into wells by ensuring that the external areas around well casings are properly sealed and that wastes are kept the recommended distance from wells.

    (c) Ensure that earthen waste storage ponds and waste treatment lagoons do not leak contaminants into groundwater. Install "clay" or other type seals according to site needs when there is any doubt.

    (d) Apply proper rates of animal waste on land for crop utilization. Treat animal waste as a resource by balancing animal waste nutrient application amounts with nutrients in commercial fertilizer.

5. Support or implement research on organic waste management, especially as it relates to human and animal health.

6. Participate in the development of mathematical models that will help track the fate of animal waste type contaminants in the environment. They will be useful in determining the effects of different management decisions on groundwater quality.

CONCLUSIONS

It is clear that animal waste can contribute contaminants to groundwater. The health threat to humans and livestock from ingestion of water contaminated with animal waste should not be taken lightly. Nitrogen and bacterial organisms are the main constituents for which there should be concern. The potential for cancer in humans and in livestock from nitrates in water is an epidemiological area that needs study.

Commercial fertilizer contributes more nitrogen to groundwater than does animal waste, but this does not reduce the importance of proper management of manure nitrogen.

Certain segments of animal waste management systems are responsible for most of the contribution of contaminants to groundwater. These are areas of high animal concentrations such as barnyards, locations of manure storage ponds and treatment lagoons, and sites of manure applications on agricultural land. Each separately, or in combination, can be responsible.

Because medical care and human health in the United States is generally good, and animal health management is an integral part of livestock industry operations, the incidence of water-borne diseases is low. Although incidences are low, the number of cases per incident can often be quite high, numbering in the thousands. This is the signal that contamination of groundwater from animal wastes needs to be prevented.

LITERATURE CITED

American Geological Institute. 1980. Glossary of geology. American Geological Institute, Falls Church, Virginia.

Barrington, S.F., and F.J. Jutras. 1985. Selecting sites for earthen manure reservoirs. Proceedings of the 5th International Symposium of Agricultural Wastes. American Society of Agricultural Engineers, St. Joseph, MI.

DeTar, W.R. 1977. The concentration of liquid manure affects its infiltration into soil. Paper No. 77-2060, Annual meeting, American Society of Agricultural Engineers, St. Joseph, MI.

Hallberg, G.R. 1986. Nitrates in groundwater in Iowa. Proceedings, Nitrogen and Groundwater Conference, Iowa Fertilizer and Chemical Association, Ames, Iowa.

Knisel, W.G., R.A. Leonard, and D.A. Still. 1986. GLEAMS user manual, Version 1.8.52. ARS, Southeast Watershed Research Center, Tifton, GA.

Miller, M.H., J.B. Robinson, and R.W. Gillham. 1985. Self-sealing of earthen liquid manure storage ponds. Journal of Environmental Quality, Vol. 14, No.4, 1985.

NAS. 1974. Nutrients and toxic substances in water for livestock and poultry. Subcommittee on Nutrients and Toxic Elements in Water, National Academy of Sciences, Washington, D.C.

NAS. 1977. Drinking water and health. Safe Drinking Water Committee, National Academy of Sciences, Washington, D.C.

Phillips, P.A., and J.L.B. Culley. 1985. Groundwater nutrient con-centrations below small-scale earthen manure storages. Proceedings of the 5th International Symposium of Agricultural Wastes. American Society of Agricultural Engineers, St. Joseph, MI.

Ritter, W.F., F.J. Humenik, and R.W. Skaggs. 1986. Effect of irrigated agriculture on ground-water quality in the Appalachian and northeast states. Water Forum '86, American Society of Civil Engineers, New York, NY.

USDA. 1985. Conestoga headwaters, rural clean water program. 1985 Progress Report, Berks and Lancaster Counties, PA. U.S. Department of Agriculture (Agricultural Stabilization and Conservation Service).

USDA. 1986. Chino Basin land treatment study---detailed work outline. San Bernardino and Riverside Counties, CA. U.S. Depart-ment of Agriculture (Soil Conservation Service and Forest Service).

USEPA. 1973. Proposed criteria for water quality. U. S. Environ-mental Protection Agency, Washington, D.C.

USGS. 1984. Groundwater regions of the United States. USGS Water Supply Paper 2242, U.S. Geological Survey (Department of the Interior), Washington, D.C.

WVU. 1975. Livestock waste management guidelines for planning, designing, and managing livestock waste systems in West Virginia. WVU Pub. No. 517, West Virginia University, Morgantown, WV.

WGNHS. 1985. Groundwater protection principles and alternatives for Rock County, Wisconsin. WGNHS Special Report 8, Wisconsin Geological and Natural History Survey, Rock County Health Depart-ment, University of Wisconsin Extension, and Wisconsin Department of Natural Resources, Madison, WI.

CHAPTER 8

ENVIRONMENTAL IMPACTS OF
LARGE SCALE HOG PRODUCTION FACILITIES

Ruth Shaffer
Joe VanderMeulen
Legislative Science Office
Legislative Service Bureau
Lansing, Michigan 48909

INTRODUCTION

Hog production is a significant part of Michigan agriculture. As of December 1, 1984, Michigan ranked 11th nationally in hog inventory with 1.31 million hogs. The majority of the state's hogs are produced in the southwest and south central counties (Figure 1). In 1984, Cass, Allegan and Ottawa Counties had the most hogs (Michigan Agricultural Reporting Service, 1985).

Swine production offers many advantages to the farmer. A hog enterprise can be adapted to specialized or diversified farming. For example, hogs can be raised in large numbers or small numbers, in pastures or in confinement. Since hogs mature quickly (approximately 23 weeks according to Pond and Maner, 1984), there is a rapid return on the farmer's original investment. In addition, it is relatively easy for most farmers to get in or out of the hog production business. Finally, labor requirements are lower in hog production than for dairy or poultry production (Bundy et al., 1984).

Over the last 50 years, the nation's total agricultural production has been concentrated in fewer and larger facilities. Although a large number of small farms remain, they now account for only a small percentage of farm product sales and total land in agriculture (Brooks, 1984). These changes have been evident in swine production as well.

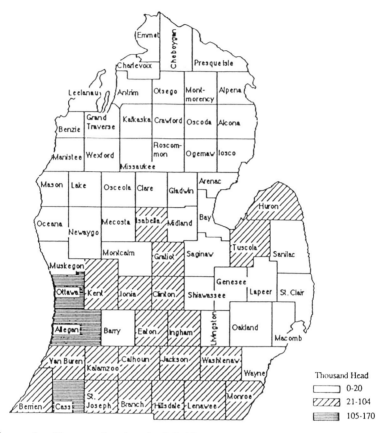

Figure 1. Hog production in Michigan's Lower Peninsula (population in thousand by county; 1985 Michigan Agricultural Statistics, for 1984).

In 1982, 73 percent of the Michigan swine producing farms had inventory of one to 99 head of hogs, 19 percent had 100-499 head and only 7 percent had greater than 500 head (Figure 2). However, there was a decrease in the number of small producers between 1978 and 1982, with an attendant increase in large producers. Changes have also taken place in the contribution of large producers to total state production. Between 1978 and 1982, the percentage of total production by producers with greater than 1,000 head of hogs increased from 27 to 40 percent (U.S. Department of Commerce, 1984). These trends have been seen nationally and reflect a shift towards production in larger units and increased specialization (Rhodes and Grimes, 1983).

One of the most significant concerns with any large scale livestock production facility is the management of animal wastes.

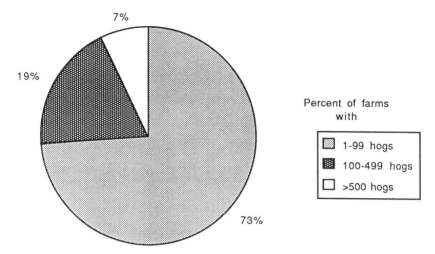

Figure 2. The 1982 Michigan inventory of hogs.

These wastes are the primary source of odor from swine production facilities (Miner and Barth, 1985) and pose significant risks to groundwater quality (OTA, 1984).   Over their lifetime, each hog produces about 0.5 to 1.5 gallons of manure a day.   Therefore, the size of the waste management problem depends on the number of hogs at the facility.

Usually regarded as less offensive when first excreted, manure that is allowed to undergo anaerobic or septic decomposition (i.e., without oxygen) is considered much more offensive.   Therefore, manure odors vary in intensity and character with the type of storage and handling facilities (described below) as well as the feed used and the animal's metabolism.

In addition to odor, animal wastes can contribute pollutants to surface water and groundwater if not properly managed.   Depending on its state of decomposition, manure contains significant amounts of chemical nutrients, as well as oxygen-demanding substances and bacterial organisms that degrade water quality.   If these wastes are spread on the land surface, rain and snow melt may wash the pollutants into local ponds, lakes, and streams.   Where these wastes are concentrated and allowed to seep or wash through the upper soil zones, these chemical substances may also degrade the groundwater (primarily through nitrate contamination).

As stated by Melvin et al., (1980), "Waste management in hog facilities has become a more important consideration as a result of larger production systems, scarcity of labor, and more stringent environmental restrictions." Many factors must be considered before deciding on the best waste management system for a particular producer. Further, proper system maintenance is important regardless of the facility size or configuration.

PRIMARY METHODS OF HOG PRODUCTION

There are three major systems used to produce hogs: pasture, semi-confinement (open feedlots) and roofed confinement facilities. A description of each system follows, together with a discussion of manure management approaches used for each type of production system.

Pasture

There are many advantages to the use of pastures for hog production. In general, less feed is needed. Also, where a legume cover crop is grown, protein supplements can be reduced. In addition, less labor and management skills are needed. In the open setting of a pasture, hogs tend to spread their own manure, odors are less intense, and hogs stay cleaner. This approach is particularly economical where pasture land unsuitable for cropping is used. However, the number of hogs per acre is limited by the kind of forage available, soils, weather, and the amount of supplemental feeding used.

Because the manure is left in the pasture, no handling equipment is needed. However, pasture sites must be away from steep slopes, streams and drainageways to limit runoff and potential surface water pollution. Proper site selection and rotation of pasture sites reduce the hazards of disease transmission. In addition, the number of pigs per acre should be reduced in humid weather (Melvin et al., 1980; Bundy et al., 1984).

Semi-Confinement Facilities

Open lot systems with semi-enclosed housing require a larger investment than the pasture method, but less than complete-confinement systems. Semi-confinement is advantageous where land costs are high and when large numbers of hogs are produced. If hogs are fed commercial feed, less land is needed than if the hogs forage for food in pastures. Thus, more hogs can be fed and kept in the smaller area in confinement systems. Because housing is provided in winter and shade in summer, there are less losses of young pigs from chilling or heat or from weed poisoning (Bundy et al., 1984). These facilities (Figure 3) can also be used more months of the year than pastures.

Waste management at open-lot facilities includes the need to remove manure from the open lot and housing units. The lots can be

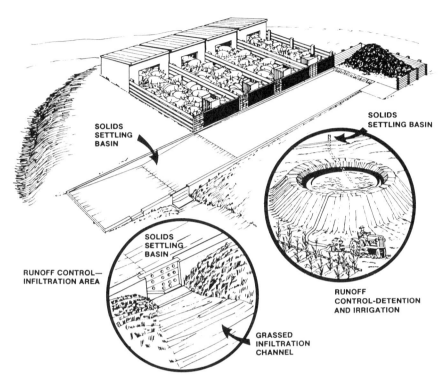

Figure 3.   Paved lot with shelter, solid manure handling.   Inserts show alternatives for runoff control (Melvin et al., 1980).

paved or unpaved and are sloped for drainage.   Usually, the solid manure is scraped from the lot using a tractor scoop and spreader, then taken directly to the field for spreading or stockpiled for later field spreading (Melvin et al., 1980).

Runoff control is a critical part of semi-confinement facility management.   Runoff from open feedlots contains high levels of pollutants from manure, spilled feed and other materials which should not enter streams.   Facilities to control runoff generally include a settling basin for solids separation, plus a holding pond for storage until final disposal on land or an infiltration area with a vegetative filter.   Adequate runoff systems must be designed for a particular location (i.e., topography, soil type, rainfall and snowmelt), lot size and configuration of the facility (Vanderholm and Nye, 1979).

Roofed Confinement

Nearly 80 percent of all hogs marketed in 1982 were raised in confinement systems, sometimes referred to as "hog hotels."   As

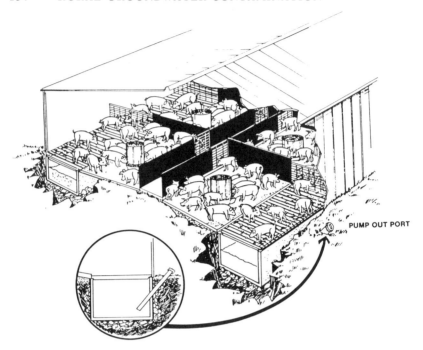

PUMP OUT PORT

Figure 4.   Confined, partially slotted floors, pit storage, liquid handling (Melvin et al., 1980).

mentioned previously, confinement facilities require less space than production in pastures.   Although they require a relatively large investment, these facilities are cost-efficient if production continues year-round with four or more farrowings (litters) per year.   However, hog production in confinement requires considerably more mechanization and better management skills than pasture facilities (Bundy et al., 1984).   For example, special care must be taken to control and prevent disease (Figure 4).

There are many options available for waste management at these facilities.   Removal of wastes from the buildings is the first step in a waste management system.   Flooring in confinement facilities is solid, slotted or a combination of both.   Absorbent material such as straw can be spread on solid floors to allow waste removal as a solid.   Slotted floors allow wastes to fall through.   This eliminates the cost in labor and machinery needed to remove manure, provides drier footing for the hogs and eliminates the need for bedding.   For these reasons slotted floors are increasingly used.

Removal of wastes from beneath the structures can be done in several ways.   In the past, deep pits under slotted floors were used to store manure for three to six months.   More recently, mechanized methods for waste removal have become popular.   For example, scraper systems use scrapers located in shallow pits under slotted floors. Flushing systems use water to remove wastes from open gutters at floor level or from beneath slotted floors (Melvin et al., 1980).

The second step in a waste management system involves storage and/or treatment of wastes before disposal.

Removing Wastes. Waste can be managed as a slurry (4 to 15 percent solids) or as a liquid (up to 4 percent solids). The form of waste is a result of the method used to remove the manure from the buildings. If a scraper system is used, the manure retains a semi-solid (slurry) form. Flushing systems dilute manure to a liquid. Using vacuum loading equipment, slurry can be removed from pits or tanks below the housing structures where deep-pit storage directly below slotted floors is used. Scraper systems automatically remove wastes from the housing facility. Liquid manure can be allowed to flow directly to lagoons by a drainage system or can be pumped to storage structures. Outside storage structures for manure can be tanks located below or above ground, excavated pits and lagoons, or lined pits and lagoons.

Wastes from hog production facilities can be treated using standard wastewater treatment technologies prior to discharge into a stream. However, the cost of this option for an individual facility is usually prohibitive (Melvin et al., 1980). An alternative is land application which is inexpensive and consistent with other farming practices. The nutrients in the wastes are used efficiently as fertilizer. Proper land application requires planning to control odors, prevent nitrogen overloading of soils which contributes to ground-water pollution, and limit runoff which contributes to surface water pollution.

Pretreating Wastes. Lagoons are used to pretreat hog manure when fertilizer value is less important than odor and pollution control and where land for waste application is limited. Oxidation ditches and aerated (aerobic) lagoons can be used to degrade organic solids, reducing odors and land requirements for spreading. However, mechanical aerators are used to supplement oxygen supplies adding energy and maintenance costs. Anaerobic lagoons offer low cost, low labor waste storage and treatment for large volumes of manure, but the potential for odor is great. Lagoons are not recommended where porous subsoils or a high water table exists unless a waterproof lining is installed to prevent infiltration and to protect groundwater quality. Additionally, lagoon design specifications are a critical factor in subsequent efficiency. In every case, there is a potential for odor problems and fly problems if the system is poorly managed (Sweeten et al., 1979; Bundy et al., 1984).

Field Application of Wastes. Irrigation equipment can be used to apply untreated or treated manure onto fields. Slurry can be irrigated onto the soil surface using a tank truck with pumping and agitation equipment. Untreated slurry retains maximum fertilizer value but odor is particularly intense. To reduce odor, slurry can be injected or knifed in under the soil surface. Liquid manure can be spray irrigated on fields using standard irrigation equipment if special solids-handling equipment and screens are used to prevent clogging (Jones et al., 1979; Melvin et al., 1980). Although fertilizer value of wastes is lowered by treatment in lagoons (especially aerobic lagoons), end products of this treatment including water and sludges retain some value and may be applied to land using irrigation approaches.

CONTROLLING OFF-SITE IMPACTS

Site Selection

Perhaps the single most important decision which must be made to minimize waste management problems is site selection. Soil characteristics such as soil type and depth together with the surface slope and resulting drainage characteristics must be taken into account to prevent runoff to surface waters and leaching to groundwater.

Proximity to residential areas is an important factor. Local zoning and land use plans are important criteria for site selection. The amount of land available is also important because some waste disposal systems such as lagoons or runoff control systems require large parcels of land.

Odor Control

In the past, site selection has been an effective means of limiting the disturbances caused by odors from agricultural activity. Location of the facility itself and of waste holding and treatment facilities (i.e., lagoons or runoff systems) away from residential areas can be very important. Good management practices such as timely disposal, proper animal carcass handling and spill control will help prevent some odor and pollution problems as well. As indicated above, design and construction of a facility affects the extent of odor production. In open feedlots, well-drained areas with properly designed and maintained runoff control systems appear best. In containment facilities, such options as slotted floors, flushing gutters and frequent pen scraping help control odors in the buildings. Covered manure storage tanks can also help reduce odors escaping from the facility.

Ideally, cropland that can readily assimilate the waste should be used for waste disposal. Proper techniques and timing of manure spreading can be very important to limit odor. Where possible, manure should be spread on windy, dry days, downwind from neighboring residences. Applications during cool times of the year are usually the least odorous (e.g., spring and fall). Immediate covering or injection into the soils also minimizes odors.

As mentioned earlier, waste treatment using lagoons is effective in reducing odor from wastes prior to land application. This treatment also reduces the organic content of manure. The amount of land needed for final disposal is reduced as well.

THE IMPACT OF SWINE MANURE MANAGEMENT ON WATER QUALITY

Most agronomists recognize that livestock manure is a valuable resource. When applied at proper rates and times, manure can increase soil fertility and improve soil tilth and water holding capacity. However, if manure is improperly stored, transported or disposed of, nutrients, organic matters, salts and pathogens can be carried by runoff to streams and lakes or leached to the groundwater. The decomposition of this excess organic matter can reduce dissolved oxygen content in surface waters. Similarly, nutrient overloads in surface waters can generate algal blooms that, on dying, remove dissolved oxygen. Where releases of animal wastes to surface water occur, rapid decreases in dissolved oxygen content and increased sediment loadings can destroy aquatic life and cause significant fish kills.

Manures are valuable for their plant nutrient content, as shown in Table 1. However, excessive or poorly timed applications of manure can overload soils with these nutrients. In such cases, nitrogen is a particularly troublesome pollutant. Through the processes of ammonification and nitrification, nitrogen can be converted to nitrate (NO_3) in the soil, becoming a highly mobile and potentially harmful pollutant. Although subject to rapid degradation at the soil surface, nitrate moves freely in many groundwater systems and is not easily removed or degraded.

Nitrate has been identified as the most common contaminant in groundwater. Further, it is becoming an increasingly widespread problem due to agricultural activities and human sewage disposal practices such as septic tanks and tile fields (Freeze and Cherry, 1979). High nitrate concentrations in groundwater can pose a serious health risk to humans and livestock. Infants under six months of age are susceptible to a potentially lethal blood disorder called methemoglobinemia, caused by large amounts of nitrates in drinking water. For this reason, the EPA and Michigan Department of Public Health have set a limit of 10 mg/l nitrate as nitrogen (NO_3-N) for public water systems (MDPH, 1983). Excessive nitrates can also cause nitrate poisoning of livestock. Therefore, efficient use of manure is important to ensure crop nutrient needs are met while protecting surface and groundwater quality.

Table 1. Estimated nutrient composition of swine manure at the time
of land application (Vitosh et al., 1986).

Manure Handling System	Dry Matter	Nitrogen		P_2O_5	K_2O
		NH_4^+	Total		
	percent	pounds/ton manure			
Solid:					
without bedding or litter	18	6	10	9	8
with bedding or litter	18	5	8	7	7
	percent	pounds/1000 gal. manure			
Liquid:					
liquid pit	4	26	36	27	22
lagoon*	1	3	4	2	4

*includes lot runoff water

Farmers can receive technical assistance for design of waste
management systems from the U.S. Soil Conservation Service. Plans
and specifications can be developed for specific farm sites and types
of operation. Proper management of the facility is also critical.
Maintenance of structures and regular emptying of storage pits or
lagoons will reduce the risk of overflow to surface waters. Protective measures such as clay liners in lagoons will reduce the risk of
leaching to groundwater.

Proper management practices during land application of animal
wastes can reduce the potential for groundwater contamination. The
appropriate amount of manure applied to agricultural land is dependent on several factors. For example, soil type is an important
factor in determining leaching potential (i.e., greater for sandy,
coarse-textured soils compared to fine-textured soils with higher clay
or organic matter content). The actual nitrogen needs are based on
the crop grown, the expected yield and previous management practices. Therefore, soil testing is important to determine fertilizer
requirements.

Nutrient composition of the particular manure should be
determined by laboratory analysis. Swine manure composition varies,
depending on the feed ration, manure management system and amount
of dilution by straw or water (Table 1). Some nitrogen is lost by
ammonia volatilization during handling and storage. Phosphorus and
potassium are not appreciably lost from confined manure systems.
However, nutrient losses by runoff and leaching from open feedlots
may be high (Vitosh et al., 1986).

Spray irrigation of manure in liquid form allows the mechanized application of wastes when crop size or ground conditions do not allow ground equipment to be used in the field. Extension recommendations for the determination of acceptable application quantities are the same for irrigation of manure as for solid manure application. The amount applied is dependent on nitrogen content of the effluent, existing nitrogen concentrations in the soils, the crop grown, and the yield goals. Because of the increased mobility of liquid manure, these applications can increase risks of nitrate leaching and groundwater contamination particularly in highly permeable and saturated soils. Special management techniques are necessary to protect groundwater. For example, applications can be split into two or three parts and applied at different times during the growing season to match times that crops have maximum moisture and nitrogen needs. The amount of liquid manure applied should be calculated into irrigation scheduling considering rainfall, soil water holding capacity and water loss through evaporation and plant transpiration. Irrigation scheduling can also prevent leaching caused by overapplication of the waste water (Jacobs, 1987).

The timing of applications of solid animal waste is also critical. Applications either early or late in the growing season can release pollutants through both surface runoff and leaching to groundwater even when applied in small quantities. In sandy soils, manure should be applied as near to planting as possible for plant uptake of nitrates. Crops should not be planted into raw manure, however, as salts in the manure can injure plants. Manure should be incorporated into soil within 24 hours after application to reduce nitrogen losses by ammonia volatilization and to reduce odor emissions. Fall application is recommended only to level, fine-textured soils. Manure should never be applied to frozen sloping land, since runoff to surface water is likely to occur (Vitosh et al., 1986).

REGULATORY CONTROL

As described above, wastes from large scale hog production facilities release significant amounts of odor and, if mismanaged, can cause surface water and groundwater pollution. Concerns over odor and water pollution problems generated by large scale animal production facilities has increased with more frequent use of large scale systems in recent years. The following sections provide a brief review of the current regulatory framework and state efforts to address these concerns.

Regulatory Framework for Water Resource Protection

The Federal Water Pollution Control Act of 1972 (FWPCA) as amended in 1977 is commonly called the Clean Water Act. A long and complicated statute, the Clean Water Act (CWA) provides for several major programs to control pollution in the waters of the United States. In some cases, the management of these programs can be transferred to the states under the supervision of the U.S. Environmental Protection Agency (EPA). In Michigan, the Department of Natural Resources (DNR) has assumed the primary administrative authority for these water pollution control programs under the Water Resources Commission Act (1929 PA 245).

Where damage to water quality is shown to be caused by a specific activity, existing state statutes can be invoked to end or modify the activity. Under Section 5 of the Water Resources Commission Act (1929 PA 245), the Water Resources Commission is empowered to take "all appropriate steps to prevent any pollution which is deemed by the Commission to be unreasonable and against public interest. . ." Section 6 of the Act also makes it "unlawful for any persons to directly or indirectly" discharge into the water of the state substances which are "or may become injurious to the public health, safety or welfare."

The Federal Clean Water Act provides for the establishment of effluent standards for point source discharges of waste (e.g., a pipe or well) to the nation's navigable waters. These standards represent the maximum amount of pollutants allowed in each discharge of wastewater. Under the Clean Water Act, every major point source discharge must obtain a National Pollution Discharge Elimination System (NPDES) permit that sets effluent limits or minimum standards of treatment to remove pollutants before discharge. In Michigan, the Water Resources Commission, with staff assistance from the Department of Natural Resources, has been empowered to issue either NPDES permits (under EPA's supervision) or state permits for discharges or potential discharges to all surface waters and groundwaters of the state.

The Federal Clean Water Act requires states to establish water quality standards for each lake or stream segment for review and approval by the EPA. Under the federal law, these standards must conform with area-wide management plans (called Regional Plans in Michigan) based on the uses of the specific water bodies. Water quality must not fall below the level necessary to protect it for the specified uses (e.g., for swimming or as a cold water fishery). Therefore, water quality standards may supersede effluent limits or treatment standards in the permitting process by requiring greater limitations in pollution discharges to protect water uses. In Michigan, these water quality standards are contained in Part 4, Administrative Rules of the Water Resources Commission Act.

In general, federal regulations require NPDES permits for discharges of wastewater from large scale "animal feeding operations." Apparently, there are no large scale hog production facilities (or hog hotels) in Michigan that meet both federal criteria and discharge wastewaters directly to surface water bodies (O'Neill, 1986). However, the Water Resources Commission Act (1929 PA 245) requires the

Michigan Water Resources Commission to "issue permits which will assure compliance with state standards to regulate municipal, industrial and commercial discharges or storage of any substance which may affect the quality of the waters of the state" (Section 5). Therefore, large scale animal production facilities using manure storage lagoons or applying wastes to the land, or both, have been subject to review by Department of Natural Resources and the Water Resources Commission in recent year.   Two such facilities ("hog hotels") have obtained storage and land application permits from the Water Resources Commission to date (O'Neill, 1986).   These permits have included criteria concerning lagoon construction, waste application rates and water quality monitoring.   A third hog production facility storage and waste application permit is now under consideration by the Water Resources Commission.

Because large scale animal production facilities usually accumulate wastes in holding structures or apply the waste to land, the Water Resources Commission has begun regulating these as potential "point" sources of pollution.   Pollutants released in a more general manner, such as surface runoff, are said to be "nonpoint" source pollutants. Although most types of farming have been identified with some amount of pollution carried in surface runoff, this nonpoint pollution has not been specifically addressed in regulations to date. However, new efforts to control nonpoint pollution from all agricultural activities are anticipated under the new Water Quality Rules of the Water Resources Commission Act (Part 4 Rules) enacted in 1986. In addition, the recently passed Congressional reauthorization of the Federal Clean Water Act contains a requirement (Section 319) that EPA and the states address nonpoint source pollution in their water pollution control planning and programs.

Regulatory Framework for Air Quality Protection

The primary regulatory mechanisms for air pollution control in Michigan are the Federal Clean Air Act (CAA) of 1970, as amended in 1977, and the Michigan Air Pollution Act (1965 PA 348).   Both acts allow regional governments to enact and enforce their own requirements as long as they are no less stringent.   Presently, Wayne, Muskegon and Kent Counties have their own air quality control ordinances, exercising some degree of control.   (Wayne County maintains total permitting control.)

The Clean Air Act requires the EPA to establish standards or limits on the amount of pollutants allowable in the breathable atmosphere as a whole (i.e., the ambient air) to protect public health and welfare. In addition, the EPA provides guidance to the states on technologies necessary to control the release of air pollutants and establishes performance standards for various industry groupings.

Odor Control Regulation.   Under the Michigan Air Pollution Act (1965 PA 348), the Air Pollution Control Commission (APCC) is responsible for establishing rules, air discharge permit requirements and addressing complaints against air polluters. Existing Air Pollution Control Commission rules require both a permit to install and a permit to operate "any process, fuel burning or refuse burning equipment. . .which may be a source of an air contaminant" (Rule 201). The term air contaminant means "a dust, fume, gas, mist, odor, smoke, vapor, or any combination thereof." Therefore, most dischargers of air pollutants are subject to review by the Air Pollution Control Commission and may be required to obtain permits. Further, Rule 901 prohibits any emission of an air contaminant in quantities that may be injurious to humans, plants, or animals or property or is an "unreasonable interference with the comfortable enjoyment of life and property."

Farm Practice Exemptions. Although several farm "processes" are known to emit odors, they are usually considered exempt from regulation by the Air Pollution Control Commission. As defined by the Part 1 Rules, the term "air pollution" does not include "those usual and ordinary animal odors associated with agricultural pursuits and located in a zoned agricultural area if the numbers of animals and methods of operation are in keeping with normal and traditional animal husbandry practices for the area." However, complaints over the odors emitted from large scale hog production facilities have been formally reviewed by the Air Pollution Control Commission. In September, 1986, the Air Pollution Control Commission refused to exempt a proposed hog production facility (confinement system) from the permit process stating that it is not "normal and traditional" farming practice. To date, no formal policy concerning large scale animal production facilities and permit requirements have been established by the Air Pollution Control Commission.

In terms of public response, odor is considered the most frequently perceived form of air pollution. Fifty percent or more of all citizen complaints to air pollution control agencies concern odor problems (Leonardos, 1984; Gnyp et al., 1985). However, there are no completely objective methods for assessing or measuring odor problems.

Although certain odorous chemicals are capable of causing harm to human beings, odors in themselves are not harmful. In general, strong odors can present problems of mental and physiological stress. Stresses identified in the literature include nausea, headache, loss of appetite, impaired breathing and some allergic reactions (Gnyp et al., 1985). However, there is no *definitive* data linking odors and health effects (Jarke, 1985). Therefore, the EPA has proposed no ambient air quality standards for odor as has been done with other pollutants under the Clean Air Act. Several states regulate odors through the control of source emissions in relation to some established threshold limit (Leonardos, 1984). In Michigan, odors are regulated on a nuisance basis under Rule 901 described above.

RECENT DEVELOPMENTS

Summary of Issues

Last year, several existing and proposed hog production facilities generated substantial controversy. These facilities include: the Sands, Calderone, and Pigs Unlimited as well as the proposed Tobe Strong facility. There are at least three areas of dispute which have been discussed by the Department of Natural Resources, the Department of Agriculture (MDA), the Department of Public Health (DPH), the Air Pollution Control Commission, the Water Resources Commission, and the Michigan Environmental Review Board (MERB).

The first issue concerns the classification of large scale animal production facilities as normal farming or something else. Neighborhood citizen groups have opposed these facilities due to potential groundwater pollution and odor. These groups have appeared before the Air Pollution Control Commission, the Water Resources Commission and Michigan Environmental Review Board on several occasions. In their view, large scale animal producers are not protected under the Michigan Right-to-Farm Act (1981 PA 93) because they are not using "generally accepted agricultural and management practices." Therefore, they believe these facilities should not be sheltered from "nuisance" suits under that act. For the same reasons, these citizen groups argue such facilities are not exempt from control under the Air Pollution Act.

Another issue concerns the legal and technical regulation of odor emissions from these facilities. The Air Pollution Control Commission apparently believes that at least one of the hog hotels is subject to the permitting procedure (the proposed Tobe Strong facility). However, no technical standards for odor control have been issued. Further, no policy has been defined as to which animal production facilities (e.g., size and type) will now fall under the scrutiny of the Air Pollution Control Commission.

A third issue concerns the consistent regulation of land application of animal wastes and animal waste storage facilities. Citizens have asked the Water Resources Commission to take action against animal production facilities responsible for water pollution, but not required to obtain a permit. However, no formal regulations have been promulgated to stipulate the size, type, or location of facilities required to obtain a discharge permit from the Water Resources Commission. Apparently, the Department of Natural Resources and Water Resources Commission have considered each large scale facility on an individual basis.

Several important developments concerning large scale hog production facilities have recently intensified the debate. The owners of Pigs Unlimited have withdrawn their application to the Water Resources Commission for a discharge permit, arguing that a permit is not required under the Water Resources Commission Act and Rules.

The Department of Natural Resources is seeking enforcement action. The owners of the Calderone and Sands facilities have been sued by neighborhood groups for causing a nuisance. The Department of Natural Resources and Water Resources Commission were included in these suits to require a review of the facilities' discharge permits. Also, the Michigan Farm Bureau and others have sued the Air Pollution Control Commission, Water Resources Commission and Department of Natural Resources on behalf of the proposed Tobe Strong facility. This case is currently under administrative review.

Departmental Responses

In response to these and related issues, the Directors of Department of Natural Resources and Department of Agriculture have proposed new initiatives to formulate appropriate policy responses. At their August, 1986, press conference, Department of Agriculture Director Paul Kindinger assumed responsibility for developing "best management practices" for confined feeding operations and animal waste management. Dr. Kindinger proposed that MDA assume the overall permitting and monitoring of these operations. Department of Natural Resources Director Gordon Guyer also initiated a special effort to address the questions concerning odor control and Air Pollution Control Commission permits as well as the Water Resources Commission permitting process.

In September, 1986, the Department of Agriculture formed the Animal Waste Resource Committee to address issues related to animal production facilities. According to recent reports, the committee meetings have involved about 60 participants. This group has been broken into several working units that include: (1) Management Practices, (2) Information and Education, and (3) Legislative Concerns. Apparently, these working units are formulating recommendations for action and policy development. The final report is expected on March 1, 1987 (O'Neill, 1986).

An internal Department of Natural Resources working group proceeded independently on a much faster track to answer the immediate concerns generated by the proposed Tobe Strong hog production facility. Recommendations concerning feasible odor control techniques were developed by an independent team of experts chaired by Department of Natural Resources Air Quality Specialist Dennis Armbruster. Their report was presented to the Air Pollution Control Commission and the Water Resources Commission in November and December, 1986, respectively.

LITERATURE CITED

Bundy, C.E., R.V. Diggins and V.W. Christensen. 1984. Hog Production. 5th Edition. Prentice-Hall, Inc.: Englewood Cliffs, NJ.

Brooks, N.L.   1984.   Minifarms-form business or rural residence. Agricultural Information Bulletin Number 480.   Economic Research Service, U.S. Department of Agriculture.

Freeze, R.A. and J.A. Cherry.   1979.   Groundwater.   Prentice-Hall, Inc.: Englewood Cliffs, NJ.

Gnyp, A.W., C.C. St. Pierre and E.M. Poostchi.   1985.   Assessing the impact of odorous emissions from municipal landfill sites on the surrounding communities.   Air Pollution Control Association (APCA), 78th Annual Meeting, Detroit, June 16-21.

Jacobs, L.   1987.   Department of Crop and Soil Sciences, Michigan State University, East Lansing, MI, Personal Communication, March 4.

Jarke, F.H.   1985.   Odorous emissions from waste disposal sites owner/operator concerns.   Waste Management Inc.   Air Pollution Control Association (APCA), 78th Annual Meeting, Detroit, June 16-21 (reprint).

Jones, D.D., R. Smith and R. George.   1979.   Flushing systems for hog buildings.   Extension Bulletin E-1339, Michigan State University, East Lansing, MI.

Leonardos, G.   1984.   Odor sampling and analysis.   In: S. Calvert and H.M. Englund (Eds.), Handbook of Air Pollution Technology.   John Wiley and Sons, Inc.: New York, NY, pp. 847-857.

Levi, D.R. and S.F. Matthews.   1979.   Legal guidelines for hog waste management.   Extension Bulletin E-1160, Michigan State University, East Lansing, MI.

Melvin, S.W., F.J. Humenik and R.K. White.   1980.   Hog waste management alternatives.   Extension Bulletin E-1399, Michigan State University, East Lansing, MI.

Michigan Agricultural Reporting Service.   1985.   Michigan Agricultural Statistics, p. 52.

MDPH.   1983.   Nitrate in Drinking Water: A Public Health Problem. Michigan Department of Public Health, Lansing, MI. Revised October.

Miner, J.R. and C.L. Barth.   1985.   Controlling odors from hog buildings.   Extension Bulletin E-1158, Michigan State University, East Lansing, MI.

O'Neill, D.   1986.   Groundwater Quality Division, Department of Natural Resources. Personal Communication, October 17.

OTA.   1984.   Protecting the Nation's Groundwater from Contamination.   Office of Technology Assessment, U.S. Congress, Washington, DC: publication OTA-O-233.

Pond, W.G. and J.H. Maner. 1984. Hog Production and Nutrition. AVI Publishing Company: Westport, CT, p. 121.

Rhodes, V.J. and G. Grimes. 1983. The structure of the U.S. pork industry. Extension Bulletin E-1676, Michigan State University, East Lansing, MI.

Sweeten, J.M., C.L. Barth, R.A. Hermanson and T. Loudon. 1979. Lagoon systems for hog waste treatment. Extension Bulletin E-1341, Michigan State University, East Lansing, MI.

U.S. Department of Commerce. 1982. Census of Agriculture. Washington, DC, p. 273.

Vanderholm, D.H. and J.C. Nye. 1979. Systems of runoff control. Extension Bulletin E-1132, Michigan State University, East Lansing, MI.

Vitosh, M.L., H.L. Pearson and E.D. Purkhiser. 1986. Livestock Manure Management for Efficient Crop Production and Water Quality Preservation. Michigan Cooperative Extension Service, Bulletin WQ12, Michigan State University, East Lansing, MI.

CHAPTER 9

GROUNDWATER CONTAMINATION FROM LANDFILLS, UNDERGROUND STORAGE TANKS, AND SEPTIC SYSTEMS

Gary Klepper
George Carpenter
Denise Gruben
Groundwater Quality Division
Michigan Department of Natural Resources
Lansing, Michigan 48909

INTRODUCTION

The purpose of this paper is to examine the known ground-water contamination sites in Michigan from the perspective of landfills, underground storage tanks (UST), and septic systems. The data used in this report are derived from the *Michigan Sites of Environmental Contamination Priority Lists: May 1986 for Fiscal Year 1987*, published by the Michigan Department of Natural Resources (1986). These lists are published annually as required by the Michigan Environmental Response Act (MERA), Act 307 of the Public Acts of 1982, as amended. Among its provisions, MERA requires that the Governor or his designee, the Michigan Department of Natural Resources (MDNR), annually identify and evaluate known sites of environmental contamination for the purpose of assigning priority for evaluation and response actions. According to MERA, a site of environmental contamination is a location where the release or potential for release of a hazardous substance has resulted or may result in injury to the environment or the public health, safety or welfare.

The data in this study are presented in terms of point of contaminant release, that is, the structure that has or may release human-originated hazardous contaminants. While many sites are currently known and listed on the MERA inventory, more will undoubtedly be discovered as currently unknown contaminants migrate toward potential detection locations. Known groundwater contamination sites are those locations where contamination has been detected in groundwater monitoring wells or potable water supply wells, while suspected sites are shown on the MERA List as potentially impacting groundwater.

The sites on the inventory have been subdivided into types, with some types more narrowly defined than others. For example, landfills and highway salt storage piles were expected to be significant contamination sources, and large numbers of these site types were evaluated early in the MERA program. Landfills are a general category while highway salt storage piles are quite specific. In contrast, septic system incidents have not been included in the MERA inventory because of their great number, their tendency to involve only sanitary waste with local impact, and the general lack of hazardous materials associated with them. Although septic systems aren't included in the MERA inventory, some localized private well contamination incidents (especially those involving nitrates) are suspected to result from septic system releases. These problems are better addressed through public health/water supply channels rather than MERA. Shallow private wells and careless backyard disposal of spent car oil and antifreeze also cloud the quality and recoverability of septic system data. For these reasons, the problem of septic system releases will not be addressed in this paper.

Some further clarification of the terms "landfills" and "underground storage tanks" (UST) as used in this study is necessary. "Landfills" can include some modern facilities licensed under Act 641 of the Public Acts of 1978 (primarily domestic waste) and Act 64 of the Public Acts of 1979 (hazardous waste), old landfills operated prior to 1980, open dumps, haphazard dumping, and indiscriminant surficial and subsurface waste disposal at the back of private, commercial or industrial property. In this study, "underground storage tanks" are separated into gasoline and chemical storage categories, with oil transport and storage and other petroleum categories also distinguished.

DATA BASE ANALYSIS

As of May, 1986, there were 1254 sites of environmental contamination in Michigan (Table 1) as identified by the MDNR (1986) in the annual MERA inventory. Since that date, 285 additional sites have been identified, bringing the total number of sites to 1539 in October, 1986. This contrasts with 1010 sites in December, 1984 (MDNR, 1985) and 814 sites in December, 1983 (MDNR, 1984). The pattern of these data indicate a 19 percent annual increase in the number of sites identified by the MERA program. Within the May, 1986, data, 1194 of the 1254 known or suspected sites involve groundwater contamination, of which 739, or 59 percent, are sites which are known to have affected groundwater, and 455 sites are suspected to have affected groundwater (Table 1). Of the 739 sites, 417 involved only groundwater contamination and not surface water or air contamination (Table 2).

Table 1.   Trend   in   discovery   of   MERA   contamination   sites   in Michigan.

Date	Number of MERA Sites	Known Groundwater Contamination	Percent Known Groundwater Contamination
December, 1979	a	268	--
July, 1982	a	441	--
December, 1983	814	b	--
December, 1984	1010	591	59
December, 1985	1254	739	59
November, 1986	1539	b	--

a - prior to annual MERA inventory
b - data not available

GEOGRAPHICAL DISTRIBUTION

The geographical distribution of all MERA sites on the May, 1986, inventory generally correlates with state population density (Figure 1).   Although every county in the state has at least one contamination incident, site density is clumped within counties with major industrial cities, particularly in the southern part of the state. Thus, the Muskegon, Grand Rapids, Kalamazoo, Lansing, and Detroit/ Ann Arbor areas have the greatest number and diversity of sites.

Table 2.   Known groundwater contamination sites due to landfills and underground storage tanks as of May, 1986.

	Known	%	Suspected
Groundwater contamination*	739	59	455
Only groundwater contamination	417	33	---
Numbers of groundwater contamination due to:			
Landfills	150	20	173
UST	<u>119</u>	<u>16</u>	<u>24</u>
TOTAL	269	36	197

*322 sites include surface water or air contamination in addition to groundwater contamination

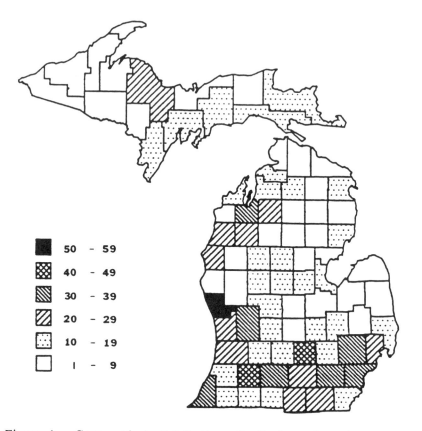

Figure 1.   Geographical distribution of all sites of environmental contamination on the May, 1986, MERA priority list.

Similarly, the geographical distribution of landfills and dumps which resulted in groundwater contamination (Figure 2) also correlates with counties containing the major population centers.   Many counties have only a few contamination sites due to landfills and dumps while 21 have none.   In all probability, historical development practices affect this distribution.   Prior to 1978, landfill development went relatively unchecked; indiscriminant dumping as well as siting landfills in wetlands, abandoned quarries and river floodplains was a common practice.

In contrast, underground storage tanks which caused groundwater contamination are not as numerous as landfills having the same effect.   The UST concentration near population centers is still pronounced due to the greater proportion of these tanks in urban areas (Figure 3).   Urban areas were defined as incorporated population centers of greater than 1000 people.   Leakage has generally been found to be linked to tank age and poor operation practices.

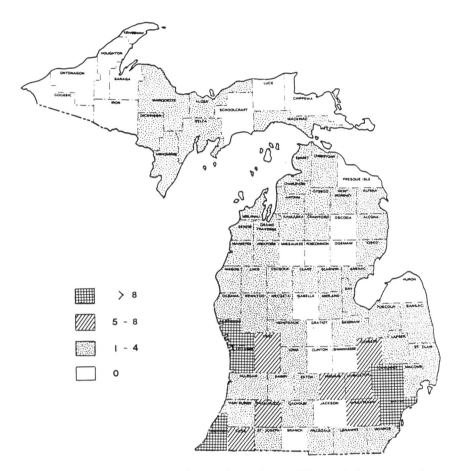

Figure 2.   Geographical distribution of landfills and dumps on the May, 1986, MERA priority list.

Currently, many problems are detected when tanks are removed from closed gas stations.   Because of a higher frequency of removal, leakage is often detected and arrested sooner in the urban areas.

When the point of release of the known groundwater contamination sites is examined (Table 2), it is clear that landfills and USTs are the largest group of points of release in the state (269 of 739 sites, or 36%).   Of the 269 sites, 150 are due to releases from landfills and dumps, while 119 are due to USTs.   An additional 197 of the 739 sites are suspected to have contaminated the groundwater, but contamination has not yet been detected.

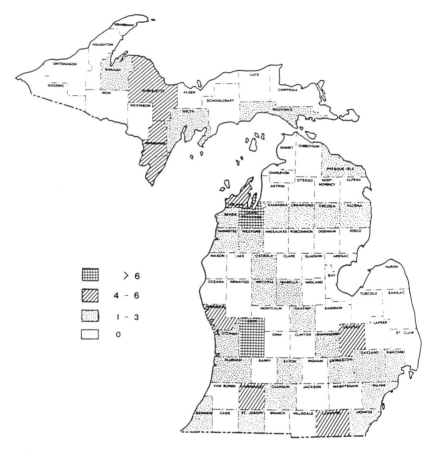

Figure 3. Geographical distribution of gasoline underground storage tanks on the May, 1986, MERA priority list.

Groundwater contamination data have been broken down into rural and urban areas. It is apparent that more contaminating landfills are found in rural than urban areas, but there is no real difference between the number of USTs in rural and urban areas (Table 3).

Figure 4 shows that there are more groundwater contamination sites in rural than in urban areas, although the differences are not pronounced. Landfills and dumps are located in the sparsely populated areas simply due to the larger amount of land needed and the siting practices mentioned earlier. When USTs are further separated into the chemical (CUST) and gasoline (GUST) categories, there still is no difference in number of tanks between rural and urban areas. Contamination due to these sources is likely to be discovered as a more widespread problem as the registration program discussed below is implemented.

Table 3.   Rural versus urban distribution of known groundwater contamination incidents due to landfills and USTs as of May, 1986.

	- - - RURAL - - -		- - - URBAN - - -	
	Number	Percent	Number	Percent
Landfills and dumps	85	12	65	9
UST	61	8	58	8
(gasoline UST)	53	7	49	7
(chemical UST)	8	1	9	1

In the May, 1986, data petroleum related sites other than leaking tanks (oil transport and petroleum sites) and salt storage sites accounted for a significant number of groundwater contamination incidents (Figure 4). Agricultural problems, also shown in the figure, include nitrate contamination, pesticide spills, and pesticide manufacture and are obviously more common in rural areas.

Many of the rural and urban characteristics described above for known release points are also true for suspected release points (Figure 5). Landfills, dumps, and both types of USTs comprise the majority of suspected or potential sites. Aside from landfills and salt storage sites, there are substantially fewer sites currently suspected to present problems. As mentioned before, the large number of landfills and salt storage sites reflects the MERA program emphasis and not necessarily the true rate of occurrence relative to other sites.

The possibility of general population exposure to all MERA contamination sites is greater in the urban areas simply because of the larger population near these sites. From the perspective of groundwater contamination, however, the correlation of exposure with population does not hold for two main reasons. First, the geographical distribution of sites which involve groundwater contamination is more uniform between urban and rural areas (Figures 2 and 3). Second, and more important, the rural areas are more dependent upon shallow groundwater for their potable water supplies. The larger population centers are supplied by municipal systems which, in the cases of Muskegon, Grand Rapids, Traverse City, Saginaw and the greater Detroit area, are offshore Great Lakes intakes. Their risk of exposure to groundwater contamination is therefore low. Other areas such as Kalamazoo, Battle Creek and Lansing depend upon deeper groundwater aquifers for potable water. The risk to these cities depends on how protected the aquifers are. In contrast, rural areas and private wells are the most vulnerable to exposure because every residence has its own shallow well. The number of sites where at least one residential well has been affected far outnumbers the

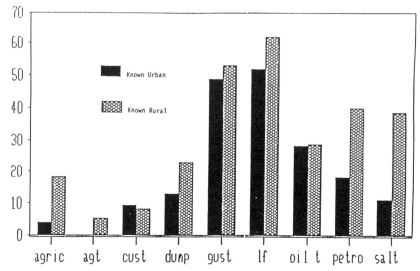

Figure 4. Frequency of urban versus rural contamination site release points which are known to have contaminated groundwater. (agric-agriculture, agt - above ground tanks, cust - chemical underground storage, gust - gasoline underground storage, lf - landfill, oil t - oil transport, petro - oil field operations)

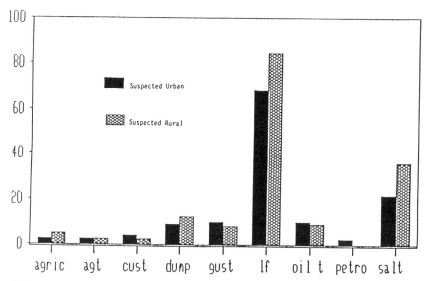

Figure 5. Frequency of urban versus rural contamination site release points which are suspected to have contaminated groundwater. (agric - agriculture, agt - above ground tanks, cust - chemical underground storage, gust - gasoline underground storage, lf - landfill, oil t - oil transport, petro - oil field operations)

Table 4.   The number of known groundwater contamination sites where at least one potable water supply well has been affected.

Affected Wells	Number	Percent
Municipal	38	5
Residential	<u>184</u>	<u>25</u>
TOTAL	222	30

sites where municipal wells have been affected (Table 4).   This illustrates that rural areas are actually at greater risk than urban areas.

CONTAMINATION SERIOUSNESS

When the relative risk score and rank of sites on the MERA priority list (MDNR, 1986) are examined, landfills and USTs do not appear to be consistently more higher ranking than other contamination sites. For example, only ten landfills and three USTs are ranked among the 50 most contaminated sites in the state. Four of these sites (two landfills and two USTs) are among the ten most contaminated sites, but this proportion is essentially the same as the occurrence of these site types throughout the list (39%).

The contaminants associated with the major point of release categories shown in Figures 2 and 3, are presented in Table 5. Landfills and dumps release the most diverse types of contaminants, followed by above ground tanks and the two UST categories. Contamination due to oil transport and other petroleum incidents resembles the GUST category. The relative toxicity posed by the various contaminants indicates that the risks posed by the landfill and UST release point categories are generally greater than other release point categories. Most of the chemicals listed are found on the Michigan Critical Materials Register (MDNR, 1980) and, depending on their concentrations, may pose substantial risk to the public.

The hazard of many of these chemicals in the groundwater may be greater than if they were found in surface water, air, or near the soil surface (Guthrie and Perry, 1980). Microbial degradation and chemical degradation via mechanisms such as photolysis do not occur in groundwater, and dilution is slower due to low flow rates underground. Underground retention is also enhanced by absorption and adsorption in the carrying soils. For these reasons, contaminant persistence is greater in groundwater. These factors impact cleanup problems because groundwater purging and soils excavation can be costly and time consuming.

Table 5.   Major contaminants associated with known groundwater contamination incidents excluding urban heavy manufacturing incidents.

Contaminant	LF	DMP	GUST	AGR	CUST	AGT	OIL	PET	SALT
Benzene	X		X			X	X	X	
Xylenes	X		X			X	X	X	
Toluenes	X		X		X	X	X	X	
Ethylbenzenes			X			X	X	X	
Cadmium	X								
Chromium	X	X							
Copper	X								
Lead	X		X						
Nickel	X								
Zinc	X								
Cyanide	X	X							
Arsenic	X	X							
Phenols	X	X				X			
Dichloroethanes	X	X			X	X			
Trichloroethanes	X	X			X	X			
Trichloroethylene	X	X			X	X			
Tetrachloroethylene	X	X							
Naphthalenes	X								
Chloroform	X								
Hexachlorobenzene	X	X							
PCBs	X								
Phthalates	X								
Paint Residues	X	X							
Nitrate				X					
Pesticides				X					
Salt/brine									X

LF - Landfill, DMP - Dump, GUST - Gasoline Underground Storage, CUST - Chemical Underground Storage, AGR - Agriculture, AGT- Above Ground Tanks, OIL - Oil Transport, PET - Oil Field Operations, SALT - Salt Storage

Correlations of release point category with contaminant concentration in the groundwater are not possible within this data base.   Concentration depends more on age of the release point, duration of release, chemical characteristics, and proximity of the detection location (monitoring well or residential well) to the release point. This information is not included in the MERA data system.

MANAGEMENT AND RESPONSE TO KNOWN SITES

Allocation of state funds for management and response at currently known sites can be made for three purposes under MERA: interim response, evaluation, and site cleanup. Interim response can be authorized to provide bottled water in the short term and well replacement in the long term when potable water supplies have been impacted. In addition, certain site stabilization activities such as elimination of fire and explosion hazard, repacking barrels and removing them to a secure area, and fencing to restrict site access can be accomplished.

Site evaluation and cleanup are part of a multiple step process of discovery, preliminary data collection and risk assessment scoring using the Michigan Site Assessment System (MDNR, 1983). Following these activities, a site is placed on a priority list for evaluation and interim response (the data source of this paper). Funds are allocated for determination of extent of contamination and design of the cleanup efforts. Once a final remedy is agreed upon, cleanup funds are allocated and final site cleanup begins. Similar processes are followed under the Federal Superfund.

If a responsible party has been identified during the discovery and site description stages, every effort is made to obtain responsible party funded responses due to the large number of sites in the state and the need to accomplish cleanups before limited state monies are allocated. This is often successful and accounts for much of the site response in the state. Litigation can be pursued against reluctant responsible parties. State and Federal Superfund money is available for the worst sites where the public health is in danger and no responsible party has been identified. Only through a combination of responsible party, MERA (State) and Superfund (Federal) action, can sites be addressed.

FUTURE CONSIDERATIONS

It is difficult to predict how many landfill and UST sites remain to be discovered and what frequency of failure may occur at currently safe locations. Because of the emphasis that older landfills have received in the MERA inventory program, the majority of problem landfill sites are probably known. In addition, open, indiscriminant dumping should be much less of a problem than in the past. Substantial improvement in regulatory tools such as passage of Acts 641 and 64 should result in fewer problems developing in the future. Encouragement of alternatives to landfills such as recycling and waste incineration for energy will also help alleviate future problems.

The extent of underground storage tank problems are only now becoming known. Federal legislation, Hazardous and Solid Waste Amendments of 1984, has added a comprehensive regulatory program for petroleum (including gasoline and crude oil) USTs. These amendments are to the Resource Conservation and Recovery Act (RCRA) of 1976. UST regulations developed under this program include notification/registration requirements for existing tanks and construction/installation standards for new tanks. This program is being implemented by the MDNR. Contamination site discovery is proceeding as the UST notification/registration requirements are being met and inactive or leaking tanks are being excavated. Because it has been estimated that there may be 200,000 such tanks in Michigan, and up to 30 percent may be leaking, it is anticipated that many new sites may results from the program. These efforts may stop contamination from many of these tanks before it becomes widespread and requires remedial action. Furthermore, the construction/installation standards should control future problems from new tanks.

CONCLUSION

Numerous contamination sites have been discovered in Michigan. Many of these have resulted in groundwater contamination, and a significant proportion have been found to result from landfills and underground storage tanks. More groundwater contamination incidents have occurred in rural areas rather than in urban areas. Rural areas appear to be at greater risk due to dependence on groundwater for potable water supplies and the tendency for private wells to be completed at shallower depths than municipal supply systems. Although larger in population, urban areas appear to be at less risk from contaminated groundwater.

Current state and federal regulations and improved public and business awareness should help reduce the number of contamination sites in the future. Responsible party, State and Federal responses to known contamination incidents and prevention of future groundwater contamination are critical to protecting the nearly one-half of Michigan's population which depends on groundwater for its potable water.

LITERATURE CITED

Guthrie, F.E. and J.J. Perry (Eds.) 1980. Introduction to Environmental Toxicology. New York: Elsevier North Holland, Inc. 484 pp.

MDNR. 1980. Critical Materials Register, 1980: Amended October 1, 1986. Michigan Department of Natural Resources, Publication Number 4833-5324. 66 pp.

MDNR. 1983. Site Assessment System (SAS) for the Michigan Priority Ranking System Under the Michigan Environmental Response Act (Act 307, P.A. 1982). Michigan Department of Natural Resources, Site Assessment Unit, Groundwater Quality Division. 91 pp.

MDNR.   1984.   Michigan Sites of Environmental Contamination
Priority List: February, 1984.   Michigan Department of Natural
Resources, Site Assessment Unit, Groundwater Quality Division.   43
pp.

MDNR.   1985.   Michigan Sites of Environmental Contamination
Priority List: February, 1985.   Michigan Department of Natural
Resources, Site Assessment Unit, Groundwater Quality Division.   185
pp.

MDNR.   1986.   Michigan Sites of Environmental Contamination
Priority Lists: May 1986 for Fiscal Year 1987.   Michigan Department
of Natural Resources, Site Assessment Unit, Groundwater Quality
Division.   219 pp.

CHAPTER 10

ABATEMENT OF NITRATE POLLUTION IN GROUNDWATER AND SURFACE RUNOFF FROM CROPLAND USING LEGUME COVER CROPS WITH NO-TILL CORN

M. Scott Smith
Wilbur W. Frye
Steven J. Corak
Department of Agronomy
University of Kentucky
Lexington, Kentucky 40506

Jac J. Vargo
Department of Agronomy
University of Mississippi
Mississippi State, Mississippi 39762

INTRODUCTION

The objectives of this research project are summarized as follows:

1. To compare the leaching of nitrate in soil treated with nitrogen fertilizer and soil with a legume cover crop as the source of nitrogen.

2. To compare fertilized and cover crop soil with regard to soil structure, porosity, and infiltration rates.

3. (a) To measure potential rates of soil water consumption under a living cover crop in the spring, under a killed cover crop in early summer, and in soil without a cover crop.

   (b) To relate the resulting approximate soil water budgets to the potential for leaching and surface runoff.

Unanticipated variability in the systems has made it impossible to address these questions in full. In this report, problems will be discussed, other research that has been conducted and provides information related to these objectives will be reviewed, and descriptions will be made on the experimental system which will make it possible to accomplish these objectives in the near future. Complete results of these studies will be made available in future publications.

Background

There exists a critical local and national need for developing and testing crop and soil management systems which will maintain agricultural productivity at a high level while reducing detrimental effects on soil and water resources. Nonpoint source pollution of ground and surface waters resulting from water and materials moving through and over agricultural soils continues to be a significant problem in some regions, including parts of Kentucky. Generally, about one-half the nonpoint source water pollution in Kentucky can be attributed to agricultural activities (Wolf, 1984).

Of major concern in surface water pollution are soil sediments, pesticides, and fertilizer nutrients. Nitrate and phosphate have been singled out as major causes of accelerated degradation of lake and stream waters. Nitrate and certain pesticides are potentially hazardous to groundwater quality. The amount of fertilizer nitrogen applied by Kentucky farmers increased by about 26 percent from 1976 to 1981 (Wolf, 1984). The same trend has continued nationwide since World War II. As fertilizer nitrogen use increases, so does the potential for water pollution from cropland.

In an effort to determine ways to alleviate the potential hazards to water quality of nonpoint source pollution from cropland, nitrate leaching is being investigated as well as soil and plant characteristics related to runoff and erosion in a novel cropping system employing legume winter cover crops.

Legume Cover Crop Systems

Three important features of modern corn production are: (a) large amounts of commercial nitrogen fertilizer are used, (b) corn requires large amounts of water for high yields, and (c) high rates of soil erosion may occur with conventional tillage on certain soils. No-till corn production with a mulch cover is usually advantageous over conventional till, with regard to the latter two features (Frye et al., 1983). However, these benefits are dependent to a large extent upon a winter cover crop that can be chemically killed in the spring to provide the mulch. Small grain crops have traditionally served in that role.

One of the greatest challenges in producing no-till corn is managing the nitrogen. Fertilizer nitrogen may be lost by leaching and denitrification and immobilized by high C:N ratio organic matter to a greater extent in the no-till system than under conventional tillage (Bremner, 1965; Rice and Smith, 1984; Thomas et al., 1973). In some cases, these effects have resulted in higher rates of nitrogen fertilizer being recommended for no-till. In doing so, the potential for nitrate pollution of water resources and production costs are increased because nitrogen fertilizer represents one of the greatest costs in corn production.

If a legume winter cover crop were used to form the mulch for no-till corn, it would provide the same benefits as small grains and provide some biologically fixed nitrogen as well. Additionally, immobilization of available nitrogen that usually accompanies decomposition of a small grain mulch would not likely occur with a legume mulch (Rice and Smith, 1984). The results would be greater use-efficiency of nitrogen and less commercial nitrogen fertilizer needed.

Research in Kentucky and other states (Ebelhar et al., 1984; Flannery, 1981; Mitchell and Teel, 1977) has shown that a legume winter cover crop adapted to the climate of the area can provide a substantial amount of biologically fixed nitrogen to a subsequent no-till corn crop. Based on yield responses, estimated amounts equivalent to from 80 to 125 kg/ha fertilizer nitrogen have been commonly reported. In Kentucky, Delaware, and New Jersey, hairy vetch has performed best; but crimson clover has been the most prolific producer of nitrogen in Georgia and other areas of the Southeast (personal communication with W.L. Hargrove, University of Georgia).

In these studies, hairy vetch usually produces about 3 to 5 Mg/ha dry matter above ground each winter and spring. This contains about 120 to 200 kg/ha nitrogen. When killed in the spring, it forms an excellent mulch cover that would be expected to increase water infiltration, decrease runoff, decrease evaporation losses, increase soil organic matter, and help provide protection from soil erosion.

Work in Kentucky by Utomo (1983) showed that a hairy vetch cover crop, compared to corn residue cover, resulted in a significant increase in soil organic matter in the 0 to 7.5 cm soil depth. There was greater soil water content during the first two months of the corn growing season after the hairy vetch was killed. However, a preliminary study by Ebelhar et al., (1984) in that same field in 1980, showed soil water contents under hairy vetch cover crop and corn residue at corn planting (May 16) to be 16.6 and 18.0 percent, respectively, in the 0 to 7.5 cm depth. In the 7.5 to 15 and 15 to 30 cm depths, soil water values were 17.1 and 16.7 percent, respectively, under hairy vetch and were 20.3 and 22.3 percent, respectively, under corn residue.

Studies in progress suggest that legume cover crops act as slow-release nitrogen fertilizer. Legumes, after being killed at the time no-till corn is planted, slowly decompose, requiring from 60 to 90 days or longer to mineralize most of the nitrogen they contain (J.J. Varco, Department of Agronomy, University of Kentucky, unpublished data). In another study, Huntington et al., (1985) observed that the release of legume nitrogen to plant available forms occurred slowly but continuously as the subsequent corn crop grew. Based on these observations, it seems plausible that during the early growing season, when conditions for leaching are most favorable, less nitrogen would be lost from cover crop systems because less is present as nitrate compared to nitrogen fertilized soil.

One interesting result of previous trials with legume cover crops has been that even with recommended rates of nitrogen fertilizer, corn yields are greater with cover crops than without them (Ebelhar et al., 1984). This may be due to the mulch providing water conservation during later summer, better weed control, or a number of other factors (Frye et al., 1985). A possible beneficial component which has been examined in this project is enhanced soil structure with legume cover crops. This could result from more extensive rooting or from greater soil organic matter additions by the legumes. Casual observations support the probability of improved structure, but it has to be quantitatively documented.

Although the focus in most of these studies has been on agronomic considerations, particularly nitrogen uptake and yield of crops, the importance of considering the environmental impact of new or modified cropping systems is also recognized. In this work the question of how legume cover crops influence nitrate transport from soils to ground and surface waters, with emphasis on nitrate leaching is addressed.

In legume cover crop systems, it is hypothesized that nitrate leaching from soils to groundwater will be minimized because nitrogen is released slowly from the decomposing proteins of the killed legume cover crop. In contrast to inorganic nitrogen fertilizer, legume nitrogen is not in a leachable form early in the season when soils are wet and leaching potential is greatest.

EXPERIMENTAL PROCEDURES

Experimental Sites

Studies were conducted on a Maury silt loam soil (Typic Paleudalf, fine-silty mixed, mesic). The Maury is a deep, well-drained soil formed in residuum of phosphatic limestone. Slopes at this site range from 2 to 6 percent. The first set of experimental plots was established in 1977 (and will be referred to as the 1977 plots). This experiment involves growing corn under two tillage systems, plowed and no-till, with varying commercial fertilizer nitrogen rates and different winter cover crop treatments. The cover crop treatments of interest for this study were: no cover crop, rye cover crop, and hairy vetch cover crop.

The second set of experimental plots was established nearby in 1985 (referred to as the 1985 plots). This experiment included nitrogen fertilizer rates of 0, 85, 170 and 255 kg N/ha. Cover crop treatments were: no cover crop grown, hairy vetch grown but clipped (5 cm height) and removed, hairy vetch grown and left, and hairy vetch supplemented with the clippings from the second treatment. Cover crops were seeded in early autumn and killed with herbicide as corn was planted. Corn for all treatments was planted by no-till methods (without soil disturbance).

Infiltration Rate

Two concentric cylinders were forced into the soil to a depth of approximately 25 cm forming double rings approximately 50 and 75 cm in diameter. Double rings were installed in the 1977 plots. Two suction probe lysimeters had been installed to depths of 30 and 90 cm in the center of each ring. Four replications of the two treatments annually receiving either 100 kg nitrogen fertilizer/ha without cover crops or 50 kg N/ha with hairy vetch cover were sampled. Steady state ponded infiltration rate was determined by flooding inner and outer rings to a depth of 5 to 10 cm with water, then periodically measuring the rate of fall in the water surface of the inner ring until this rate became constant. Suction was maintained on the suction probe lysimeters during the infiltration experiment, and water samples were collected and analyzed for nitrate.

Suction Probe Lysimeters

These consisted of ceramic cups sealed to the end of PVC pipes. A hole to 30 or 90 cm depth, equal in diameter to the pipe, was made in the soil and partially filled with a soil slurry; then the probe lysimeter was installed. The cap of the tube held two smaller tubing lines, one of which was used to collect soil solution from the ceramic cup, the other was used to apply a vacuum to the sampler. Samples were collected when it was possible during the 1985 growing season, i.e., following rainfall events or infiltration tests.

Drum Lysimeters

Lysimeters were also constructed from 55 gallon drums. A sloped bottom and an outlet tube were installed in the drums. Sixteen of these drums were placed in 2 rows in a trench constructed by backhoe. The drum tops were then 5 to 10 cm above the natural soil surface. A wooden frame was constructed between the two rows of drums in the trench allowing access to the outlet tubes and making it possible to backfill soil around the drums. Soil was placed in the drums so as to preserve the natural sequence of soil horizons. Construction was completed in the winter of 1985. In the spring of 1986, drums were amended either with 100 kg/ha nitrogen fertilizer or with an approximately equal quantity of nitrogen as hairy vetch. Half of the barrels received ^{15}N enriched fertilizer or vetch. Corn was planted in and around the drums.

Transformations of Legume and Fertilizer Nitrogen

The ^{15}N-labeled hairy vetch used in this study was grown under field conditions each spring in submerged 25 x 25 cm plastic pots with the bottoms removed and filled with one part vermiculite and two parts sand and buried to the rim in the soil. Enriched $K^{15}NO_3$ was added to the pots of actively growing vetch at a rate of 30 kg N/ha about every two weeks until 120 kg N/ha had been applied. The last application was made about two weeks before harvesting the vetch. One day prior to applying the treatments, all of the above-ground portion of the hairy vetch was removed, cut into approximately 5 cm sections and mixed thoroughly. The ^{15}N-labeled hairy vetch added to each core was determined on a fresh weight basis.

Soil cores (5 x 20 cm) were obtained from the 1977 plots prior to corn planting using a soil core sampler lined with a 5 x 24 cm plexiglass tube and driven into the ground by a slide-hammer. Each soil core contained within a plexiglass tube was trimmed at the bottom to 20 cm in length. The bottom of each tube was then enclosed with fiberglass screen fastened with duct tape.

To simulate plowing (conventional tillage) each soil core was pushed out of the tube and pulverized by hand, returned to the tube, and tamped until the length was again 20 cm. When hairy vetch was to be added to the conventional tillage soil, about 100 gm of soil was first placed in the tube and then the residue was thoroughly mixed with the remaining soil before adding it to the tube. For the no-tillage treatment, the residue was simply placed on the soil surface of each core.

Fertilizer nitrogen solution containing enriched $^{15}NH_4^{15}NO_3$ was applied to the corn residue treatment cores in 1984, and $(^{15}NH_4)_2SO_4$ was applied in 1985.

The prepared cores were returned to the appropriate plots in the field one day after corn planting and placed flush with the soil surface. Soil samples were taken from the plots each year at corn planting at 0 to 10 and 10 to 20 cm depths to obtain the residual amount of ammonium and nitrate.

In 1984, whole cores were removed 30, 50, 90, and 120 days after they were placed in the field. In 1985, cores were removed 15,

30, 45, 60, and 75 days after they were placed in the field. Cores removed from the field were stored frozen until processed and analyzed.

For no-tillage cores, the residue was removed from the soil surface, while for conventional tillage cores residue remains were meticulously removed from the soil and any soil adhering to the residue was carefully brushed off. Residue pieces larger than 1 mm were separated from the soil using a sieve. Residue which passed through the 1 mm sieve was separated from the soil by floating it in carbon tetrachloride.

A 20 g portion of moist soil was extracted by shaking for 1 hr with 200 ml $\underline{M}$ KCl. An aliquot of each sample extract was analyzed for ammonium and nitrate by the procedures of O'Brien and Fiore (1962) and Lowe and Gillespie (1975), respectively. Gravimetric water was determined on each soil sample to correct for field moisture content. To determine the ^{15}N atom percent of the inorganic nitrogen fraction, enough soil was extracted with 1 $\underline{M}$ KCl to obtain 0.5 mg of nitrogen. The extract was then steam distilled with MgO-Devarda's allow (Bremner, 1965), and the distillate was collected in 0.5 $\underline{M}$ HCl. Soil organic nitrogen and residue nitrogen content were determined using a micro-Kjeldahl method (Bremner, 1965). Prior to digestion, the inorganic nitrogen fraction was removed by extracting three times with 1 $\underline{M}$ KCl. The digest was steam distilled, and the nitrogen was collected as ammonium chloride for later determination of the ^{15}N atom percent.

The ^{15}N atom percent of the inorganic and organic soil nitrogen fractions and of the residue nitrogen were measured with a Consolidated Electrodynamics Corporation (CEC) 21-614 mass spectrometer using a freeze-layer technique whereby ammonium ions are converted to nitrogen gas with NaOBr (Wolf and Jackson, 1979).

DATA AND RESULTS

Infiltration Rate

It had been hypothesized that the growth of legume cover crops would increase soil organic matter content, improve soil structure and so increase water infiltration into soil. To test this hypothesis, an attempt was made to measure steady state ponded infiltration rates. The treatments chosen for comparison either had no winter cover crop or had grown a winter cover crop of hairy vetch for seven years. In both cases, corn was the summer grain crop. These were no-till soils; there was no plowing or disturbance of the soil.

Results for an infiltration test conducted during the corn growing season of 1985, after the vetch had been killed and was largely decomposed, are shown in Table 1. The mean infiltration rate

Table 1.   Steady state ponded infiltration rate of soil with and without hairy vetch cover crops.

Treatment	Block				Mean
	I	II	III	IV	
	————————cm/hr————————				
No vetch	15.5	3.8	3.0	6.0	7.1
With vetch	9.5	4.5	12.2	9.0	8.8

was, in fact, slightly greater for the soils which had grown winter legumes. However, inspection of the individual replications for these treatments makes it obvious that this difference is not statistically significant. This extent of spatial variability within treatments was not anticipated. Clearly, it will not be possible to obtain meaningful treatment comparisons from these data. Because the variability is apparently a property of the system, rather than the technique, it

will probably be impractical at this site to evaluate cover crop effects on soil structure and infiltration rates using any available technique.

Suction Probe Lysimeters

Similar problems were encountered with attempts to evaluate nitrate leaching by this approach. This technique has been widely used by others and generally has provided at least a relative indication of nitrate transport at depth in the soil profile. Total quantities of nitrate collected from these probes at the site are shown in Table 2. Once again, inspection of the variability among replications indicates that these results provide no meaningful comparison of treatment effects. Nitrate concentrations, as opposed to quantities, were also highly variable and provided no useful indication of a possible difference between treatments (data not shown). Volume of soil solution which could be collected also contributed to the variability of the data in Table 2. On many sampling dates several of the 30 cm probes gave no sample, in spite of recent precipitation. Failure to collect soil solution indicates that the soil surrounding the ceramic cup was too dry to release free water at the tension provided by the vacuum within the sampler.

Table 2. Total nitrate collected from suction probe lysimeters from June through August, 1985.

Depth (cm)	Treatment	Block I	II	III	IV	Mean
				— ug N —		
30	Vetch + 50 kg N	208	537	15,250	4,035	5,008
	No vetch + 100 kg N	4,944	0	7,094	0	3,009
90	Vetch + 50 kg N	379	172	1,120	21	423
	No vetch + 100 kg N	86	208	1,498	2,267	1,005

Nitrate Concentration In Deep Soil Samples

These measurements were made as one attempt to overcome the problem of variability documented above. Soil samples were collected periodically from the 1985 plots. At least eight cores per replicate plot were composited to minimize variability. Soil was sampled to a depth of 90 cm. (This is a relatively shallow soil in which most plant rooting and nutrient extraction occurs in the top 30 cm and is very limited below 90 cm).

Results are shown in Table 3. These data do provide an indication that nitrate is being transported through the profile. At the surface, concentrations were maximal on the first sampling date after fertilizer application. In the next depth increment concentrations peaked three to six weeks later, while concentrations continued to increase through the season at the greatest depth sampled. While variability was much less than observed with the previous techniques, it was still a problem. Coefficients of variation were typically 20 to 60 percent.

With few exceptions, significant differences between treatments were not commonly observed. In the surface 30 cm, concentrations of nitrate were initially greater in the plots without cover crops. This would be expected since more inorganic nitrogen was added to these plots, and the conversion of vetch nitrogen to nitrate would require some time. After the first month, surface nitrate was similar for the two treatments. At greater depths there were no consistent differences. Therefore, these results do not provide any evidence that nitrate leaching is affected by cover crop treatment.

Table 3. Distribution of nitrate in soil with hairy vetch plus 85 kg N/ha fertilizer or with 170 kg N/ha fertilizer only.

Depth (cm)	Treatment	Nitrate Concentration by Date					
		———— 1985 ————				— 1986 —	
		5/30	6/20	7/8	12/23	3/25	4/25
		———— mg N/kg dry soil —————					
0-30	Vetch + 85 kg/ha N	32.9	27.7	11.2	5.1	7.9	12.9
		(7.3)*	(5.6)	(3.6)	(0.6)	(2.0)	(5.6)
	170 kg/ha N only	44.7	39.1	12.4	3.7	8.6	11.5
		(12.5)	(2.5)	(2.8)	(1.0)	(1.8)	(4.2)
30-60	Vetch + 85 kg/ha N	3.4	9.8	4.8	3.9	4.4	11.4
		(1.4)	(2.3)	(2.8)	(0.9)	(1.0)	(2.3)
	170 kg/ha N only	3.8	6.0	6.8	3.7	4.8	6.7
		(2.9)	(5.1)	(2.7)	(1.4)	(0.8)	(2.8)
60-90	Vetch + 85 kg/ha N	0.5	0.8	3.5	4.1	4.8	8.6
		(0.3)	(0.7)	(1.3)	(0.9)	(1.7)	(1.3)
	170 kg/ha N only	1.4	1.8	3.7	3.1	3.4	8.3
		(1.2)	(1.3)	(2.1)	(1.1)	(0.3)	(2.8)

*Standard deviations are given in parentheses below the mean values.

Drum Lysimeters

It became apparent during the first year of this project that the methods described above would not be adequate to satisfy the objectives. To solve this problem, construction of contained lysimeters large enough to grow plants in was initiated late in the project. Of course, time was inadequate to complete measurements before termination of the project. It was anticipated that the useful life of these lysimeters will be three to five years, and data will continue to be collected throughout this period. Final results will be made available in the future.

Early results indicate that this approach will be useful in studying leaching in the cropping systems of interest. Reproducible, consistent differences have been observed in volume of water leached and total nitrogen leached. Water volumes have been significantly greater in hairy vetch treated lysimeters (Table 4). This can be attributed to smaller evaporative water loss in the vetch treated barrels due to the well-documented mulch effect.

Table 4. Volume of leachate collected from drum lysimeters treated with fertilizer or hairy vetch.

Treatment	Volume of Leachate by Date - 1986			
	5/30	6/4	6/9	6/16
	— — — — — — — ml/lysimeter — — — — — — —			
Fertilizer	1669	599	6	0
Hairy vetch	1535	1008	228	124

Since water movement through the profile has been greater in vetch treated barrels, it is not surprising that total nitrogen transport has also been greater (Table 5). Nitrate concentrations have not been significantly different (data not shown).

These early results suggest that leaching losses will be greater in systems with vetch. However, it is important to appreciate the limited nature of data collected to this time, which will be considered in the conclusions.

Transformations of Legume and Fertilizer Nitrogen

Losses from the soil of nitrogen added as fertilizer versus hairy vetch can be estimated from these experiments. It should be pointed out that gaseous losses by microbial denitrification cannot be distinguished from leaching losses to groundwater using this technique. Also, labeled nitrogen remaining in the soil has not been

Table 5. Nitrate leached from drum lysimeters treated with fertilizer or with hairy vetch.

Treatment	Total Nitrogen Leached by Date - 1986			
	5/30	6/4	6/9	6/16
	— — — — — — — mg NO_3^--N/lysimeter — — — — —			
Fertilizer	8.6	4.1	<0.1	0
Hairy vetch	9.8	6.6	1.2	0.6

Table 6. Effects of tillage on the recovery of nitrogen from ^{15}N-labeled $(NH_4)_2SO_4$ and hairy vetch as soil inorganic nitrogen in 1985.

Nitrogen Source	Tillage	Sampling Day				
		15	18	45	60	75
		———————%———————				
$(NH_4)_2SO_4$	No-Till	57	18	22	14	9
	Conventional	78	38	44	32	8
Hairy vetch	No-Till	12	4	7	8	13
	Conventional	47	15	26	23	17

measured below the depth of the cylinder. Thirdly, these experimental systems do not contain growing plants, in contrast to the drum lysimeters, and so are somewhat unrealistic.

As expected, more of the labeled nitrogen was found in the inorganic pool for fertilizer treatments than for vetch treatments (Table 6), at least during the first 60 days. More inorganic labeled nitrogen was present in the plowed compared to the no-till systems. Inorganic nitrogen would be susceptible to leaching loss, in contrast to organic forms. This indicates a greater potential for nitrate leaching of fertilizer nitrogen than of vetch nitrogen.

A greater percentage of added labeled nitrogen was immobilized in the soil organic fraction for vetch relative to fertilizer (Table 7). This was most apparent after 75 days in the soil. In general, plowing of the soil increased tie up in the immobile, nonleaching organic fraction. These results also suggest a reduced potential for leaching in vetch systems.

Table 7. Effects of nitrogen source and tillage on nitrogen immobilization in 1985.

Nitrogen Source	Tillage System	Sampling Day	
		15	75
		— — % — —	
Fertilizer	No-Till	15	15
	Conventional	22	18
Hairy vetch	No-Till	11	29
	Conventional	38	31

Table 8.   Effects of nitrogen source and tillage on total ^{15}N recovered in 1985.

Nitrogen Source	Tillage System	Sampling Day	
		15	75
		— — % — —	
Fertilizer	No-Till	71	22
	Conventional	97	23
Hairy vetch	No-Till	77	64
	Conventional	100	53

Table 8 shows total recovery of added nitrogen that is the sum of inorganic nitrogen, soil organic nitrogen and nitrogen remaining in residues.   The nitrogen not recovered is presumed to be lost by leaching or denitrification.   Losses were considerably greater with fertilizer than with vetch after 75 days.   Little nitrogen was lost during the first 15 days, and there was no remarkable effect of nitrogen source.   Plowing of the soil did increase recovery, or reduce losses.

In this well-drained soil, nitrogen leaching is likely to be a more important mechanism of nitrogen loss than denitrification. These results indicate that leaching loss was greater when fertilizer was the nitrogen source compared to vetch as the nitrogen source.

CONCLUSIONS

The methods which were to be used to accomplish the objectives have been shown to be inadequate at this site.   The primary reason is the extreme spatial variability.   This makes it impractical to determine treatment differences with any feasible number of samples. This variability suggests that water and nitrate flux from the soil is heterogeneous in space.   Previous research in this department does, in fact, indicate extensive macropore flow in this soil.   Simply stated, much of the water and the solutes in it flow down a few holes, cracks or channels through the soil.   Under these conditions, it might be expected that suction probe lysimeters, which collect water from a small soil zone, would fail to provide a representative sample.   In the extreme case, water would not flow in the sampling zone and no solution would be collected.   This occurred frequently in these experiments.

Another difficulty with these approaches is that they provide single point measurements, in both a temporal and spatial sense. Therefore, they do not directly assess flux of water or nitrate.

Enclosed lysimeter systems resolve these difficulties primarily by allowing collection of all flow from a defined area but also by sampling a larger volume of soil. An additional advantage is that by employing ^{15}N labeled amendments, it will be possible to distinguish soil-derived from fertilizer-derived nitrate. Disadvantages of these systems are that soil disturbance is required initially and that construction is costly and time consuming. With the time and resources available, during this project construction of 16 drum lysimeters was completed. Since then, additional data have been collected, and it is probable that this approach will work. Data collection will continue for at least two more years.

Early results suggested that nitrate leaching was greater with vetch amended soil than with fertilizer amended soil. However, it is not expected that this relationship will remain constant throughout the experiment. Because the vetch organic material reduced evaporation during late spring and early summer, more water moved through the profile. It is anticipated that differences in evaporative water loss will be minimal as the corn canopy closes and the mulch decomposes. Furthermore, it is predicted that growth of the cover crop during later fall and early spring will deplete soil water and soil nitrate, greatly reducing the potential for nitrate leaching at that time of year.

A study of the decomposition and transformations of vetch nitrogen using ^{15}N labeled plant material was successfully completed. The results indicate that vetch nitrogen, in comparison to fertilizer nitrogen, leads to lower concentrations of soil inorganic nitrogen and greater immobilization of added nitrogen in soil organic matter. This would reduce the potential for nitrate leaching. After two or three months in soil, total losses from a fertilizer nitrogen source were greater than from a vetch nitrogen source. This could be a result of greater nitrate leaching or greater gaseous losses through denitrification. On this soil, leaching is a more plausible loss mechanism.

Soil management and cropping practices will certainly have an effect on the quantity of nitrate lost by leaching from agricultural soils. This will have a significant impact on nitrate loading in ground and surface waters. In this project, the investigation of the potential for legume winter cover crops to reduce nitrate leaching has begun. At this time, the results do not provide a clear and definite answer. It is to be expected that these effects may be dependent on seasonal, climatic and soil factors. Continued observations using the lysimeters already constructed will help to reveal these relationships and lead to the development of feasible agricultural systems which minimize adverse effects on water quality.

ACKNOWLEDGMENT

The work upon which this report is based was supported in part by funds provided by the U.S. Department of the Interior, Washington, DC, as authorized by the Water Resources Research Act of 1984, Public Law 98-242.

LITERATURE CITED

Bremner, J.M. 1965. Inorganic forms of nitrogen. Methods of Soil Analysis. Agronomy 9:1179-1237.

Doran, J.W. 1980. Soil microbial and biochemical changes associated with reduced tillage. Soil Sci. Soc. Am. J. 44:765-771.

Ebelhar, S.A., W.W. Frye and R.L. Blevins. 1984. Nitrogen from legume cover crops for no-tillage corn. Agron. J. 76:51-55.

Flannery, R.L. 1981. Conventional vs. no-tillage corn silage production. Better Crops. LXVI (Summer-Fall):3-6. Phosphate-Potash Institute, Atlanta, GA.

Frye, W.W., J.H. Herbek and R.L. Blevins. 1983. Legume cover crops in production of no-tillage corn. In: W. Lockeretz (Ed.), Environmentally Sound Agriculture. New York: Praeger Publishers.

Frye, W.W., W.G. Smith and R.J. Williams. 1985. Economics of winter cover crops as a source of nitrogen for no-till corn. J. Soil Water Conser. 40:246-249.

Huntington, T.G., J.H. Grove and W.W. Frye. 1985. Release and recovery of nitrogen from winter annual cover crops in no-till corn production. Commun. in Soil Sci. Plant Anal. 16:193-211.

Lowe, R.H. and M.C. Gillespie. 1975. An Escherichia coli strain for use in nitrate analysis. J. Agric. Food Chem. 23:783-785.

Mitchell, W.H. and M.R. Teel. 1977. Winter-annual cover crops for no-tillage corn production. Agron. J. 69:569-573.

Nelson, D.W. and L.F. Sommers. 1973. Determination of total nitrogen in plant material. Agron. J. 65:109-111.

O'Brien, J.E. and J. Fiore. 1962. Ammonium determination by automatic analysis. Wastes Engineering 33:352-353.

Rice, C.W. and M.S. Smith. 1984. Short-term immobilization of fertilizer nitrogen at the surface of no-till and plowed soil. Soil Sci. Soc. Am. J. 48:295-297.

Thomas, G.W., R.L. Blevins, R.E. Phillips and M.A. McMahon. 1973. Effect of a killed sod mulch on nitrate movement and corn yield. Agron. J. 64:736-739.

Utomo, M. 1983. Effect of legume cover crops on soil nitrogen, soil temperature and soil moisture in no-tillage corn. Unpublished M.S. Thesis. Department of Agronomy, University of Kentucky, Lexington, KY 40546-0091.

Volk, R.J. and W.A. Jackson. 1979. Preparing nitrogen gas for nitrogen-15 analysis. Anal. Chem. 51:463.

Wolf, K. 1984. Nutrients: General description, sources, and impact on the environment in Kentucky. Volume II. Kentucky Water Quality Management Program. Kentucky Natural Resources and Environmental Protection Cabinet, Division of Water. Frankfort, KY.

PART III

ASSESSMENT AND MODELING

CHAPTER 11

USING CROP MODELS AS A DECISION SUPPORT SYSTEM
TO REDUCE NITRATE LEACHING

Joe T. Ritchie
Department of Crop and Soil Sciences
Michigan State University
East Lansing, Michigan 48824

Decision making in today's agricultural environment is a volatile process, mainly because of the intensity of influence of many factors. Prices for products and inputs (i.e. fertilizers, chemicals, and fuels) have greatly increased in the past two decades. Concern for the environment is increasing rapidly. The changing relative values of national currencies strongly affect export and import commodity prices. People function to a greater extent in a world economy where there is more global interdependence among all national economies. These economic realities often discourage the implementation of environmentally sound practices that either increase the cost of farm products or the management time required by producers.

The volatility of these factors affecting yield, environmental quality and price variations means that it is important for farmers to have tools available that can assist them in making appropriate cost and environmentally effective decisions. Almost every decision a farmer has to make is surrounded by natural and economic uncertainties, mainly weather and prices. The facts available to assist the farmer in making such decisions have come mostly from results of experimental trials. Relatively few field trials have experimental designs which allow response estimates for conditions beyond the experimental site and year because of soil spatial variability and weather variability. Even when experimental trial data are available, using or applying them is less than satisfactory. For example, consider a simple crop yield response to nitrogen (N) fertilizer as usually fitted from experimental trials by regression techniques (Figure 1).

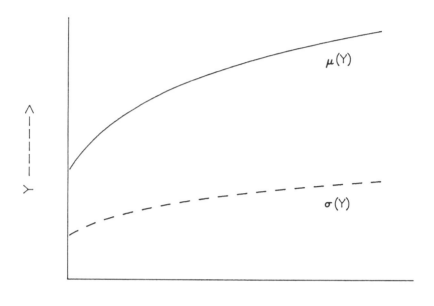

Figure 1. Response of final crop yield (Y) to nitrogen (N) fertilizer application showing the mean (μ(Y)) and standard deviation (σ(Y)) information needed in risk assessment.

For calculating response curves similar to those in Figure 1, it is assumed that any interacting factors are known and held constant for each nitrogen rate, but the shape and position of the response curve will vary from year to year as weather sequences during the crop growth stages vary. Realistically, only a few years of data can be used in estimating the curve.

Given the fitted curve for mean yield, as in Figure 1, it is simple to conduct an economic analysis to determine the optimum level of N, given the price of output and the cost of N. The fitted curve, however, provides only the mean response. Any calculated optimum could only apply to a farmer who takes no account of risk.

Farmers are not risk neutral. Rather they attempt to attain some maximum degree of utility, as expressed through different goals, such as (1) maximum profit or cash flow; (2) minimum variance of yield; (3) maximum stability of income; (4) minimum environmental degradation; (5) some mix of these goals.

Because yield response or nitrate loss functions are established from experimental data, they often provide little assistance to farmers in making decisions about farm management strategies that would maximize the utility of response functions. This is because experiments usually cannot be carried out over a wide enough range of weather sequences.

If weather does influence the results from experimental treatments, yield probability distributions about the mean response curve have to be established. From that information, standard deviations can be calculated, as shown in the lower curve of Figure 1. Data required to generate the probabilities of response usually cannot be obtained directly from field experiments because of the time involved, but they can be generated for any soil type with validated crop simulation models using many years of weather data.

For comparing alternative agrotechnology packages, it is necessary to determine the frequency or probability distribution of outcomes. Examples of agrotechnology packages affecting crop yield, nutrient or pesticide losses and production costs are (1) alternative N, P, or K levels and dates of application; (2) alternative cultivars; (3) alternative irrigation methods or amounts; (4) alternative sowing dates; (5) alternative sowing rates or plant populations; (6) alternative pest control methods; and (7) any combination of the above.

CROP MODELS FOR DECISION MAKING

Several research groups have developed crop models to evaluate how the environment affects crop growth, the soil water balance and the nitrogen balance in the soil and plant. Legg (1981) provides a comparative summary of some of the models. An interdisciplinary team of my colleagues have developed user-oriented models for several grain crops. The most widely tested of these are CERES-Wheat (Ritchie and Otter 1985) and CERES-Maize (Jones et al. 1986). The models are designed to predict the yield of crops sown anywhere at any time and to be a useful tool for decision making by farmers, researchers and governments.

The CERES models are multi-purpose simulation models that can be used for within-year crop management decisions, multi-year risk analysis for strategic planning, large-area yield forecasting, and definition of research needs. The models use readily available weather and soil input; are written in a familiar and widely used computer language; require minimal computation time; and can be adapted for use in both mainframes and minicomputers. Most input data are available from or can be readily estimated from routinely collected daily weather data, standard soil characterization data, and other data provided in the model documentation. The computer program is written in FORTRAN; versions that run on 1986 microcomputers such as IBM-PC-AT require about 20 seconds for running time to compute an annual crop production cycle.

One of the somewhat unique features of the CERES models is the emphasis on duration of crop growth stages as influenced by genetic differences in crop maturity type and in sensitivity to photoperiod, vernalization and temperature. Duration is emphasized because it is the major factor enabling crops to be selected for regions with different lengths of growing seasons. The CERES models do not depend on information regarding known yield potentials in a region. Rather, they calculate plant growth and development as affected by temperature, light and degrees of stress, and they partition assimilates into different organs in the plants at different stages of growth. The economic yield is not determined until the final growth stage is finished. Management is limited in present versions to selection of the genotype; time, density and depth of sowing; and the time and amount of irrigation and nitrogen fertilization.

Other crop models developed but less tested that are similar to CERES-Maize and CERES-Wheat include rice, sorghum, millet, barley and potatoes. Models of soybeans (SOYGRO) (Wilkerson et al., 1983), peanuts and field beans that are similar in the level of detail and require the same soil and weather inputs as the CERES models have been developed by a team of scientists at the University of Florida.

Field experiments are needed to check the validity of simulation models to ensure that the models can provide a suitable set of response data from which realistic decisions can be made. The data needed for running such models includes daily weather data; soil properties from which to calculate water and nutrient balances, initial soil water and nutrient conditions; and basic management information (planting date, plant population, variety, row spacing, fertilization, and irrigation). The minimum set of data needed to test the model includes yield and yield components (i.e. head numbers, grain numbers, and grain weight), intermediate products of yield (i.e. above ground dry matter and plant nutrient concentration) taken several times during the season, dates of major phenological events and general observations about pest damage and other conditions that may influence crop growth. Testing the water and nitrogen balance requires measurements of the soil water and nitrogen content several times and at several depths during the season, and evaluating the nitrogen in the plant several times.

If nitrogen leaching is considered to be a major factor to be tested for model validation, sampling of the soil may not provide satisfactory results because of soil and plant variability within a field. In such cases, an alternative is to install natural lysimeters that cover a fairly large area. Such lysimeters will intercept all of the water and chemical flow that occurs below the root zone. A diagram of such an installation is shown in Figure 2. This design is similar to ones reported by Brown et al. (1974) and Kissel et al. (1974). Four of these were recently installed in sandy soil regions of Michigan where nitrogen leaching is thought to be a contributing source of groundwater contamination. These are used to validate model calculations of nitrogen leaching from various nitrogen application strategies.

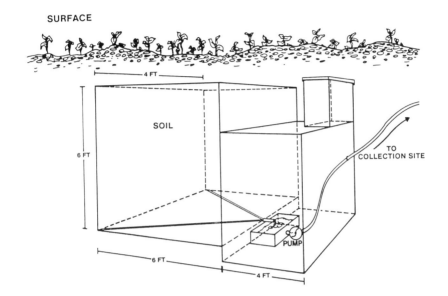

Figure 2. Lysimeters such as this intercept water and chemical flow that occur below the root zone.

STOCHASTIC WEATHER GENERATOR

An important need in running crop simulation models over a long time period is good weather data.   Some locations have no weather data available.   The complete set of temperature, radiation, and precipitation are usually not available from the same location for long periods nor are they collected with the same spatial resolution. The extensive climatic data collected by the national and interna- tional meteorological organizations is available, but expensive and time consuming to work with.   What is needed are generated weather data that have the statistical characteristics of the actual location.

Richardson (1981), Nicks (1974), and Stern et al. (1982) describe practical methods for generating weather data.   Woolhiser et al. (1985) have developed a user-friendly microcomputer program that uses the generator information for daily weather simulation.   Coupling such generators to simulation models provide a flexible and powerful tool for rapidly examining alternative management strategies for decision making.   Richardson's (1981) stochastic weather generator model (WGEN) provides values for daily precipitation, maximum and minimum temperature, and solar radiation.   The model requires approximately 50 stochastic parameters derived from known weather data to generate a realistic weather sequence for a location. Richardson and Wright (1984) have published model parameters required to generate weather sequences for the 48 contiguous states in the United States.

The occurrence of rain on a given day has a major influence on temperature and solar radiation for the day. The model (WGEN) generates precipitation for a given day independently of other variables. Maximum and minimum temperatures and solar radiation are generated according to the occurrence of a wet or dry day. In addition, the model preserves the dependence in time, the correlation between variables, and the seasonal characteristics in actual data for a particular location.

RISK ASSESSMENT

Methods for determining optimal strategies under risky or uncertain conditions have been proposed (Anderson 1973, 1974; Anderson et al., 1977). These strategies are used more often in economic research rather than in agronomic or environmental research. Dowling and Smith (1976) used a stochastic water balance model and risk analysis procedures to determine the best time for establishing pastures in Australia. Smith and Harris (1981) and Stapper (1984) used risk analysis to define optimal sowing times and maturity types for wheat grown in the Middle East. Boggess and Amerling (1983) and Boggess et al. (1983) utilized risk assessment techniques to evaluate irrigation investment and management decisions in humid regions. Godwin and Vleck (1985) coupled CERES-Wheat with a weather generator to simulate nitrogen yield responses and leaching losses at contrasting locations in the world.

EXAMPLE APPLICATION: WATER MANAGEMENT

To demonstrate the use of crop models to assess yield variations as influenced by soil and water relationships, a 30-year yield sequence was simulated using CERES-Maize and weather data from St. Joseph County, Michigan, USA. The weather patterns in St. Joseph County are typical of many humid and subhumid regions where rainfall during some part or all of the growing season is often insufficient for maximum crop productivity. Three soil properties were assumed--shallow, medium and deep--each having rooting depths of 50, 100 and 150 cm and available water capacity (AWC) of 6.5, 13 and 19.5 cm, respectively. All other management factors were held constant and nitrogen was assumed to be nonlimiting. Results from the simulation are shown in Figure 3, which shows that weather and soil both strongly influence yield.

In years 13 and 26 (Figure 3), the difference between the shallow and deep soils was quite small. In these types of years, a favorable soil-water environment exists even in shallow soils because rains are frequent throughout the growing season. However, years 6 and 29 in the sequence had almost total yield failures for the shallow

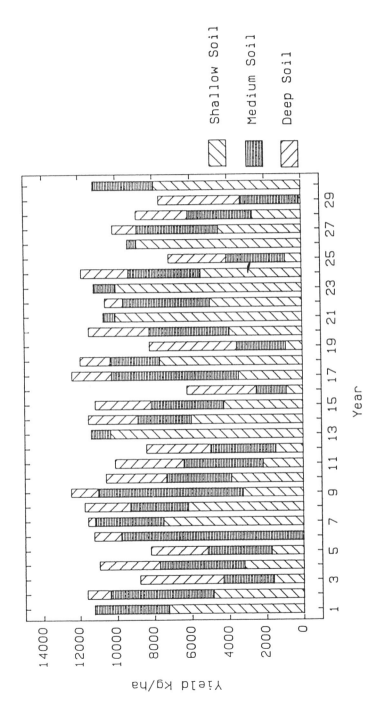

Figure 3. Simulated maize yields over a 30-year weather sequence (from St. Joseph County, Michigan, USA) on a field with three soil depths not irrigated. (The deep soil yields not shown are equal to the medium soils.)

soils, while the deeper soils provided relatively good yields.   In environments like this, spatial variations in AWC become very important in affecting the productivity of the whole field.

In regions where water is available to supplement rainfall, yields usually increase with irrigation. If the strategy is to minimize the amount of water added through irrigation, the potential AWC will be important in the decision making process regarding when and how much to irrigate. Using the same weather and management information as used for simulating the information in Figure 3, the model was altered to simulate an irrigation of 15 mm each time the soil water in the root zone reached a critical threshold value.   The amount of irrigation required for the season was then determined (Figure 4).

Although 15 mm is considerably less than the deficit below the drained upper limit soil water content, maintaining a relatively large soil water deficit allows the soil to store more precipitation should a rain follow an irrigation.   It also decreases the possibility of leaching. In the summer, 15 mm is equivalent to two to four days of evaporation; therefore, with this strategy, applications have to be frequent during rainless periods.   Martin et al. (1985) demonstrated that about 40 percent less irrigation water was needed with an application of 6 mm of irrigation per application, for this type of soil and weather when compared to 32 mm per application, the latter being a more common practice for farmers in the area. The smaller application leaves a larger soil water deficit in the event of a rain soon after irrigation.

Using the 15 mm irrigation strategy, simulated yields for the three soil types were practically the same every year.   The yields from year to year varied between 10,000 and 12,500 kg/ha, 90 percent of the time.

Because of soil spatial variability, it is possible to have soils similar to those depicted in Figures 3 and 4 contained within a relatively small field.   Most large field irrigation systems, such as center pivot sprinklers or linear move sprinklers, uniformly irrigate fields.   The critical management problem then becomes one of deciding which soil properties to use as a basis for making decisions concerning irrigation. If irrigation is applied whenever plants in the shallow soils are stressed, the plants in the medium and deep soils would receive more water than necessary. If irrigation is applied only when plants in the medium or deep soils are stressed, the plants in the shallow soils would invariably be stressed more than the other plants. For example, in year 1 of the sequence (Figure 4), 105 mm of water would be required if the shallow soil was used to indicate when to irrigate.   For this strategy, then, 45 and 77 mm of water more than was needed would be applied to the medium and deep soils, respectively.   In the simulation, yields for these three irrigation amounts were 11,200 kg/ha.   For the entire 30-year sequence and using the shallow soil as the reference point for irrigation, the

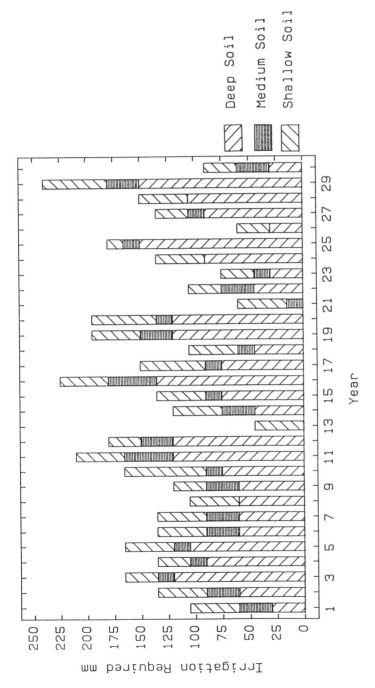

Figure 4. Simulated amounts of irrigation water, applied in increments of 15 mm, required for optimal plant growth over 30-year weather sequence (from St. Joseph County, Michigan, USA) on a maize field with three soil depths.

average percent of excess water applied would be 30 and 45 mm for the medium and deep soils, respectively.

An irrigation strategy that would conserve water and decrease leaching, but provide an environment so that soil water does not limit yield is desirable. A strategy is needed such that each part of the field would be irrigated according to its individual needs. This strategy would require knowledge of the soil spatial variability and a computer controlled irrigation system that responds to geographical information regarding soil variability. Once the spatial variation of AWC was known, each outlet along a moving irrigation line could be turned off or on, depending on the water needs of the different soils. Thus a minimum amount of water would be applied to achieve maximum yields for the entire field.

EXAMPLE APPLICATION: NITROGEN LOSS EVALUATION

Because of the concern regarding nitrate leaching into groundwater, it will be important to evaluate both yield and nitrate loss with models. The goal is to minimize leaching while maintaining reasonable profitability. A fortunate fact from research is that nitrate and ammonium concentrations in soil solution can be relatively low for optimum nitrogen uptake. It is also quite well established that most nitrogen uptake from soil by many grain crops occurs during the first half of the growing season while the leaves and stem are actively growing. Thus most of the nitrogen harvested in the grain is derived from nitrogen stored in the plant before grain filling occurs. It should be possible, therefore, to "spoon-feed" plants during vegetative growth with a sufficient supply of nitrogen fertilizer to maintain the optimum concentration in soil for unlimited nitrogen uptake. This strategy would approximately match the supply of and demand for nitrogen, leaving only a small margin for leaching.

The problem with "spoon-feeding" strategies is the cost in time and money for the frequent applications. If irrigation is practiced, application through the irrigation system is possible if the system spreads the water uniformly throughout the field. However, the time for applying the fertilizer may not coincide with the time when water is needed, thus a minimum amount of irrigation would be required to apply the fertilizer. For rainfed crops, ground or aerial application would be required. Also the rainfall necessary to transport the nitrogen to the root zone would be needed. If rain does not come, plants would become deficient in nitrogen and some fertilizer could be lost by volatilization. Thus with either rainfed or irrigation production, some risk is involved; the extent of risk is related to soil properties and weather uncertainty. Crop models can be of great assistance in minimizing this risk by planning management strategies that account for variations in soil and weather.

The CERES-Maize was used to simulate yields of corn and the nitrogen balance for a 30-year weather sequence for a loamy sand

Table 1.   The average yield and nitrate leaching simulated for six nitrogen fertilization strategies at the Kellogg Biological Station in Michigan.

Application	Total N (kg/ha)	Yield (kg/ha)	Leaching (kg/ha)
Split	300	10,900	75
Preplant	250	10,800	41
Split	200	10,800	22
Split	180	10,700	18
Preplant	175	10,500	23
No Addition	0	900	3

similar to some soils at the Kellogg Biological Station near Hickory Corners, Michigan.   This crop was assumed to be irrigated and six nitrogen fertilization strategies were compared.   The strategies ranged from no fertilization to 300 kg/ha, with some fertilizer applied pre-plant and some split applications.   Table 1 contains the average yields and average nitrogen leaching loss for the six strategies.   The split applications consisted of a preplant application and three others applied at fixed intervals during vegetative growth.

The result given in Table 1 demonstrate how fertilizer not used in producing grain is often lost by leaching with the soil and weather conditions of this simulation.   It should also be clear from these results, which agree qualitatively with observations made at KBS, that any combination of fertilizer amounts and application practices could be evaluated.

The primary decision criterion regarding leaching amounts to be allowed is difficult to establish, but future policy to protect ground-water from nitrate contamination may require that nitrate losses not exceed prescribed maximum level of 10 mg nitrogen per liter. Validated models, coupled with good weather and soil information, will be valuable in making such assessments.

CONCLUSION

Computer models, when coupled with generated or long-term weather and soils data, are valuable tools providing insight into many aspects of decision making for crop production systems.   With long-term weather data and a crop model, year-to-year variations in yield can be quantified and thus used in making risk assessment.   For the assessments to be correct and credible, however, crop models must be validated in the region where the analysis is needed.

Crop models of the type described herein have limitations for broad applicability. They can only account for variations in the factors included in the models. Potentially important factors not included in the model are assumed to be nonlimiting. The major factors not considered in the models described in this paper include pests and nutrient deficiencies other than nitrogen. Models of these factors, however, can be coupled with crop models, as was done with nitrogen in the CERES models, therefore resulting in even more valuable models.

It is believed that decision support systems that include crop models can and must play an increasing role in agricultural decision making. For this to happen, however, interdisciplinary teams will be needed to provide the necessary input information for running and validating models and for building new components into models. There is also a need to make models user-friendly to meet the needs of farmers and policy makers in various regions of the world.

Because of the increasing competition for natural resources, models should play an important role in helping to evaluate methods to enhance the quality of water.

ACKNOWLEDGMENTS

I would like to acknowledge the assistance of Sharlene Rhines in editing this manuscript and Edward Martin and Jim Jenkins for assisting with the analyses.

LITERATURE CITED

Anderson, J.R.   1973.   Sparse data, climatic variability and yield uncertainty in response analysis. Amer. J. Agric. Econ. 5:77.

Anderson, J.R.   1974.   Risk efficiency in the interpretation of agricultural production research. Rev. Marketing Agric. Econ. 42:131.

Anderson, J.R., J.L. Dillion, and J.B. Hardaker. 1977. Agricultural Decision Analysis. Iowa State University Press, Ames, Iowa, USA.

Boggess, W.G. and C.B. Amerling.   1983.   A bioeconomic simulation analysis of irrigation investments.   Southern J. Agric. Econ. Dec:85-91.

Boggess, W.G., G.D. Lynne, J.W. Jones, and D.P. Swaney. 1983. Risk-return assessment of irrigation decisions in humid regions. Southern J. Agric. Econ. July:135-143.

Brown, K.W., C.J. Gerard, B.W. Hipp and J.T. Ritchie. 1974. A procedure for placing large undisturbed monoliths in lysimeters. Soil Sci. Soc. Am. Proc. 38:981-983.

Dowling, P.M. and R.C.G. Smith. 1976. Use of a soil moisture model and risk analysis to predict the optimum time for the aerial sowing of pastures on the Northern Tablelands of New South Wales. Aust. J. Exp. Agric. Anim. Husb. 16:871.

Godwin, D.C. and P.L.G. Vlek. 1985. Simulation of nitrogen dynamics in wheat cropping systems. In: W. Day and R.K. Arkin (Eds.) Wheat Growth and Modeling. Plenum Press, New York, U.S.A.

Jones, C.A., J.T. Ritchie, J.R. Kiniry, and D.C. Godwin. 1986. Subroutine Structure. pp. 49-111. In: C.A. Jones and J.R. Kiniry (Eds.). CERES-Maize: A Simulation Model of Maize Growth and Development. Texas A & M University Press, College Station, Texas, U.S.A.

Kissel, D.E., J.T. Ritchie and E. Burnett. 1974. Nitrate and chloride leaching in a swelling clay soil measured by an undisturbed drainage lysimeter. J. Environ. Qual. 3:401-404.

Legg, B.J. 1981. Aerial environment and crop growth. In: D.A. Rose and D.P. Charles-Edwards (eds), Mathematics and Plant Physiology. Academic Press, London.

Martin, E.C., J.T. Ritchie and T.L. Loudon. 1985. Use of the CERES-Maize model to evaluate irrigation strategies. pp. 342-350. In: Advances in Evapotranspiration, Proc. Nat'l Conf. Dec. 16-17, 1985. Am. Soc. Agric. Engr., St. Joseph, Michigan.

Nicks, A.D. 1974. Stochastic generation of the occurrence, pattern and location of maximum amount of daily rainfall. In: Proceedings of Symposium on Statistical Hydrology. Misc. Pub. No. 1275. USDA. Washington, D.C.

Richardson, C.W. 1981. Stochastic simulation of daily precipitation, temperature, and solar radiation. Water Resources Res. 17:182-190.

Richardson, C.W. and D.A. Wright. 1984. WGEN: A Model for Generating Daily Weather Variables. U. S. Dept. of Agric., Agric. Research Serv., ARS-8.

Ritchie, J.T. and S. Otter. 1985. Description and performance of CERES-Wheat: A user-oriented wheat yield model. pp. 159-175, USDA-ARS. ARS-38.

Smith, R.C.G., and H.C. Harris. 1981. Environmental resources and restraints to agricultural production in a Mediterranean-type environment. Plant and Soil 58:31-57.

Stern, R.D., M.D. Dennet, and I.C. Dale. 1982. Methods for analyzing daily rainfall measurements to give useful agronomic results. II. A modeling approach, Exp. Agric. 18:237.

Stapper, M. 1984. The use of a simulation model for the prediction of wheat cultivar response to agroclimatic factors in semi-arid regions. Ph.D. Thesis, University of New England, Armidale, Australia.

Wilkerson, G.G., J.W. Jones, K.J. Boote, K.I. Ingram, and J.W. Mishoe. 1983. Modeling soybean growth for crop management. Trans. ASAE, 26(1), 63-73.

Woolhiser, D.A., C.L. Hanson, and C.W. Richardson. 1985. Microcomputer program for daily weather simulation. Proceedings of the Specialty Conference on Hydraulics and Hydrology in the Small Computer Age. HY Div.-ASCE, Lake Buena Vista, Florida, USA, Amer. Soc. of Civil Eng. p. 1154-1159.

CHAPTER 12

SOFTWARE FOR TEACHING PRINCIPLES OF CHEMICAL MOVEMENT IN SOIL

Arthur G. Hornsby
Soil Science Department
University of Florida
Gainesville, Florida 32611

INTRODUCTION

Farmers are faced with decisions on how to best manage agrichemicals each year to optimize return on their investment of capital and time. Improper management can result in chemical movement below the root zone with a loss of the chemicals from the production system and the possibility of groundwater contamination. Recent documentation of the occurrence of pesticides in groundwater throughout the United States (US EPA, 1986) substantiates that loss of agricultural chemicals by leaching, even though applied correctly, can become a serious concern.

Understanding the fate of agrichemicals in soil systems requires an understanding of the interaction of simultaneously occurring processes which affect chemical behavior in soil in response to various environmental conditions. While these interactions may be complex, the individual processes have been well studied and mathematical models developed to represent them. In recent years, research and screening models have been developed to simulate pesticide movement in soils. However, these models require considerable input data which are not generally readily available.

Farmers, county extension agents, state and federal action agency personnel, and the general public need to understand the behavior of chemicals in order to be responsible stewards of resources. To gain such an understanding, simple, uncomplicated methods are needed to provide insight into the processes which control movement of chemicals and to permit rational management decisions of pesticide use. A software package is described herein which can be used to better understand the behavior of pesticides in soil.

DESCRIPTION OF THE SOFTWARE

Chemical Movement In Soil, CMIS, (Nofziger and Hornsby, 1985) is a software package written to illustrate the influence of soil properties, chemical properties, plant rooting depth, precipitation, and evapotranspiration upon the movement and persistence of surface applied chemicals (pesticides) in well drained soils. This software is meant to serve as a teaching tool rather than a management tool since several assumptions were made to simplify the model and reduce data input requirements. The computational procedure is that of Rao et al. (1976) which assumes uniform soil properties and piston flow of water.

The procedure is as follows: Let d_i represent the depth of the solute pulse i days after the chemical was applied to the soil surface. Let I_i and ET_i represent the amount of water (precipitation or irrigation) infiltrating the soil surface and the potential evapotranspiration, respectively, on day i. The depth of the solute at the beginning of day i+1 is given by

$$d_{i+1} = d_i + q_i (R \ \Theta_{FC}) \qquad \text{if } q_i > 0$$

or   (1)

$$d_{i+1} = d_i \qquad \text{if } q_i \leq 0$$

where q_i is the amount of water passing the depth d_i, Θ_{FC} is the soil water content on a volume basis at "field capacity," and R is the retardation factor for the chemical in the soil. Assuming a linear and reversible adsorption model, the retardation factor R is given by

$$R = 1 + (BD^* K_d / \Theta_{FC}) \qquad (2)$$

where BD is the soil bulk density, K_d is the linear sorption coefficient or the partition coefficient of the chemical in this soil. The partition coefficient is given by

$$K_d = K_{OC} \ OC \qquad (3)$$

where K_{OC} is the linear sorption coefficient normalized by the organic carbon content (OC) of the soil. The organic carbon content can be obtained from the organic matter content by multiplying by 0.58. (Note that the use of K_{OC} as defined in equation 3 is applicable only to nonionic, nonpolar organic solutes). The majority of computations in the model is directed toward the determination of q_i in equation 1 from known values of I_i and ET_i. This process is described below.

Consider a soil with the solute at depth d_i. Due to evapotranspiration, the soil water content in the root zone may be less than Θ_{FC} (soil water deficit). When an infiltration event occurs, some water is needed to increase the soil water content above the solute pulse to Θ_{FC} (replenish the soil water deficit). The excess water (if any) contributes to downward movement of the solute pulse. That is,

$$q_i = I_i - swd \qquad (4)$$

where swd is the soil water deficit above the depth of the solute pulse, d_i. The soil water deficit is given by

$$swd = [\Theta_{FC} - \Theta_a] d_i \qquad \text{if } d_i < d_{root}$$

or $\qquad\qquad\qquad\qquad\qquad\qquad\qquad\qquad\qquad (5)$

$$swd = [\Theta_{FC} - \Theta_a] d_{root} \qquad \text{if } d_i \geq d_{root}$$

where d_{root} is the depth of the root zone and Θ_a is the average volumetric soil water content above d_i if $d_i < d_{root}$ or above d_{root} if $d_i > d_{root}$. If q_i in equation 4 is greater than zero, Θ_a is increased to Θ_{FC}. If q_i is less than or equal to zero, the solute depth does not change ($d_{i+1} = d_i$). Instead, the infiltrating water just increases the water content Θ_a as given by

$$\Theta_a = \Theta_a + I_i/d_i \qquad \text{if } d_i < d_{root}$$

or $\qquad\qquad\qquad\qquad\qquad\qquad\qquad\qquad\qquad (6)$

$$\Theta_a = \Theta_a + I_i/d_{root} \qquad \text{if } d_i \geq d_{root}$$

To deal with evapotranspiration in the model, the average volumetric water content, Θ_a, defined above is calculated for each day. In addition, during the time in which the solute pulse is in the root zone (i.e., $d_i < d_{root}$), a second average water content, Θ_b, is calculated for the soil between the solute depth and the maximum rooting depth. If $\Theta_a = \Theta_b$, both water contents decrease together to meet the evapotranspiration demand. In this case

$$\Theta_a = \Theta_a - ET_i/d_{root}$$

and $\qquad\qquad\qquad\qquad\qquad\qquad\qquad\qquad\qquad (7)$

$$\Theta_b = \Theta_b - ET_i/d_{root}$$

If Θ_a is greater than Θ_b, then Θ_a is decreased to meet the ET demand until $\Theta_a = \Theta_b$ at which point the remaining ET is removed uniformly from the entire root zone. The water contents in the root zone are not permitted to decrease below the water content corresponding to the "permanent wilting point" of the soil.

ASSUMPTIONS IN THE MODEL

In order to simplify the computations and to reduce the data input requirements for the software, the following assumptions are used:

1.   The soil is homogeneous. This assumption is reasonable for a model intended for instructional purposes. This model will not precisely describe movement in soils containing layers with different organic carbon contents, textures, or pore size distributions.

2.   All soil water residing in pore spaces participates in the transport process. Soil water initially present in the profile is completely displaced ahead of water entering at the soil surface. Rao et al. (1976) present data from different researchers which indicate that these assumptions are valid for many soils. If they are not valid and a portion of the soil water is bypassed during flow, this model would tend to underestimate the depth of the chemical pulse.

3.   Water entering the soil redistributes instantaneously to "field capacity" water content. This assumption is approached in coarse textured soils. If the water redistributes more slowly as in fine textured soils, the depths predicted here would need to be associated with an elapsed time a few days later than that specified.

4.   Water is removed by evapotranspiration from the wettest part of the root zone first. When the water content in the root zone is uniform, water is lost uniformly at all depths. The validity of this assumption will depend upon the root distribution in the soil. It will not be strictly valid for many situations. However, it is a reasonable assumption for an instructional model.

5.   Upward movement of water does not occur anywhere in the soil profile. Water is lost from the root zone by evapotranspiration, but soil water in the root zone is not replenished from below. This assumption seems reasonable for homogeneous, well drained soils.

6.   The chemical pulse is considered to be of infinitely small thickness and is taken as a point. Chemical pulses moving in soil tend to become dispersed. This dispersion depends upon several factors, of which water flux and pore size distribution have major roles. Since neither is used in the computational scheme or is not available a priori, representation in the model is not practical. The consequence of this assumption is that the model predicts the peak of the pulse rather than the leading edge. This means that some chemical may be in the

soil below the predicted peak due to dispersion. Therefore, the model will underpredict the depth of movement.

7. The adsorption process can be described by a linear, reversible equilibrium model. If the sorption coefficient is described by a nonlinear isotherm, the partition coefficient decreases with increasing solution concentration of the chemical. Thus, the depth to which the chemical will be leached will depend upon the concentration. This aspect is probably not significant for the concentration range of interest in most agricultural applications (Rao and Davidson, 1980). When adsorption equilibrium is not instantaneous, the chemical will be leached to a greater depth than predicted here. Irreversible sorption would result in less leaching.

8. The half-life for biological degradation is a constant with time and soil depth. Degradation rate coefficients are dependent upon several environmental factors, primarily temperature and soil water content. Also, with decreasing microbial activity at greater soil depths, the degradation rate coefficient may decrease with depth. Sufficient data are not available to formulate mathematical relationships to describe these effects. For instructional purposes, this additional complication was not considered important, but it would need to be modeled for actual prediction of pesticide fate under field conditions.

USING THE MODEL

The software is menu driven with a main menu (see Figure 1) to select between data handling routines and simulation of chemical movement. The option choices are explicit and self-explanatory. Option A sets up a sequence of choices to be made by the user to calculate chemical movement using data on soil and chemical parameters, precipitation, and evapotranspiration which have previously been entered into appropriate files. Options B, C, D, and E are data handling routines to enter, modify, or print data files for soil, chemical, rainfall (and irrigation), and evapotranspiration, respectively. These choices are made in sub-menus of the appropriate option.

Outputs can be presented in graphical or tabular form. The output options menu is shown in Figure 2. A graphical output resulting from a simulation using Option A is presented in Figure 3. The chemical of choice is nemacur, the soil is Tavares fine sand, and the rooting depth is 30 inches. The upper panel displays the rainfall record as individual events for the simulation period selected, and the lower panel displays the depth distribution of the chemical with time. Outputs can be presented in graphical or tabular forms. Table 1 presents the tabular form of the output for the same simulation.

Table 1. Example of tabular output.

Simulation of Chemical Movement in Soil

Chemical Data:
Common Name	: FENAMIPHOS
Trade Name	: NEMACUR
Partition Coefficient (ml/g OC)	: 171
Half-Life (days)	: 10

Soil Data:
Soil Name	: TAVARES FINE SAND
Soil Identifier	: S27-8-(1-6)
Percent Organic Carbon	: 0.09
Water Content at -0.1 bar (% by vol.)	: 8.2
Water Content at -15 bars (% by vol.)	: 0.9
Bulk Density (g/cc)	: 1.55

Root Depth: 30 inches

Rainfall File	: LOCAL83.R
Evapotranspiration File	: LOCAL83.ET

Starting Date	: 4 - 2 - 83
Stopping Date	: 6 - 10 - 83

Total Rainfall	: 17.74 inches
Total Evapotranspiration	: 12.15 inches
Potential Evapotranspiration	: 15.12 inches

Month Day Year	Rainfall	Solute Depth	Relative Mass	Elapsed Time
	-----	inches -----		Days
4 - 3 - 83	0.28	0.9	0.93	1
4 - 8 - 83	1.07	4.1	0.66	6
4 - 9 - 83	2.56	12.0	0.62	7
4 - 10 - 83	0.16	12.1	0.57	8
4 - 15 - 83	2.56	18.5	0.41	13
4 - 16 - 83	0.25	19.1	0.38	14
4 - 19 - 83	0.61	19.9	0.31	17
4 - 23 - 83	1.50	22.8	0.23	21
5 - 4 - 83	0.26	22.8	0.11	32
5 - 14 - 83	0.60	22.8	0.05	42
5 - 17 - 83	1.35	22.8	0.04	45
5 - 24 - 83	2.43	25.2	0.03	52
6 - 5 - 83	0.39	25.2	0.01	64
6 - 7 - 83	2.05	26.1	0.01	66
Chemical Movement Below Root Zone				
6 - 8 - 83	1.67	30.7	0.01	67

CHEMICAL MOVEMENT IN SOILS
by
D. L. Nofziger and A. G. Hornsby
Copyright 1984

Options:

    A. Calculate Chemical Movement in Soil
    B. Enter, Modify, or Print Soil Data File
    C. Enter, Modify, or Print Chemical Data File
    D. Enter, Modify, or Print Rainfall Data File
    E. Enter, Modify, or Print Evapotranspiration File
    F. Display File Directory
    Q. Quit. Terminate Program and Return to DOS

Desired Option ? _

Figure 1. Screen image of main menu.

Output Options:

    S. Output Table to Screen
    P. Output Table to Printer
    F. Output Table to File

    G. Output Graphs for NEMACUR
    N. Output Graphs for NEMACUR and
       for a Non-Adsorbed Chemical such as Nitrate

    M. Return to Main Menu

Desired Option ? _

Figure 2. Output options menu.

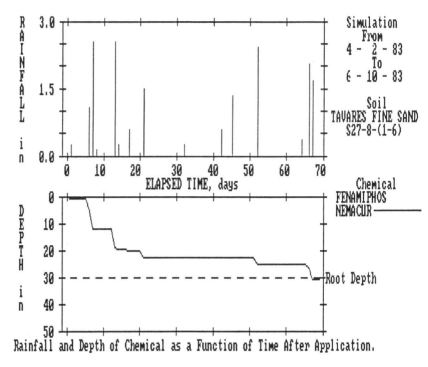

Figure 3. Example of graphical output to screen.

LITERATURE CITED

Nofziger, D.L. and A.G. Hornsby. 1985. Chemical Movement in Soil: IBM PC User's Guide. Circular 654, Florida Cooperative Extension Service. Institute of Food and Agricultural Sciences. University of Florida, Gainesville, 61 pp.

Rao, P.S.C., J.M. Davidson and L.C. Hammond. 1976. Estimation of nonreactive solute front locations in soils. In: Proceedings of the Hazardous Wastes Research Symposium, EPA-600/9-76-015, Tucson, AZ, pp. 235-341.

Rao, P.S.C. and J.M. Davidson. 1980. Estimation of pesticide retention and transformation parameters required in nonpoint pollution models. In: M.R. Overcash and J.M. Davidson (Eds.), Environmental Impact of Nonpoint Source Pollution. Ann Arbor Science Publishing Inc., Ann Arbor, MI, pp. 23-67.

US EPA. 1986. Pesticides in Ground Water: Background Document. U.S. Environmental Protection Agency, Office of Groundwater Protection, Washington, DC, 72 pp.

CHAPTER 13

GROUNDWATER MONITORING: AN OVERVIEW FROM FIELD DRILLING TO LABORATORY ANALYSIS

Charles S. Annett
Edward E. Everett
Keck Consulting Services
1099 West Grand River
Williamston, Michigan 48895

INTRODUCTION

Increased awareness concerning the extent and significance of groundwater contamination has prompted the development of reliable techniques to accurately monitor the quality of groundwater. The monitoring process involves many phases ranging from the initial drilling process to the final analysis of a water or soil sample. Each phase of the operation is subject to errors which might result in either the collection of a nonrepresentative sample or contamination of the sample. This paper provides an overview of the monitoring process and discusses available options and precautions for each phase. Hopefully, this will provide guidelines and a level of awareness which will promote the ability to institute and maintain an effective groundwater monitoring program.

DRILLING TECHNIQUES FOR MONITOR WELL INSTALLATION

The design of a groundwater monitoring system depends on the geological characteristics of the site and the specifications for pumping of the wells. These same factors dictate the type or method of drilling that can be used for the installation of the wells. The more common water well drilling equipment (i.e., cable tool, mud/air rotary and jetting rigs) can be effectively used for installing monitor wells; however, the most common and frequently specified technique in Michigan is the hollow-stem auger drilling.

Limitations of auger drilling include:

1. depth capability limited to 200 feet or less, depending on the geology;

2. cannot penetrate bedrock;

3. cobbles, boulders and very tight clays may inhibit drilling; and

4. well diameter is generally limited to 4 inches or less for hollow stem augers.

Most soil sampling and groundwater monitoring work is done at depths of less than 150 feet; thus, the hollow-stem auger drilling techniques are applicable in most situations. The greatest advantage of this technique is that no drilling fluid additives are required to complete the boreholes, which minimizes potential influences on the soil or groundwater samples collected.

Equipment

The hollow-stem auger consists of a core that may be from 2.25 inches in diameter up to 8.25 inches in diameter with continuous flighting wrapped the length of the core (refer to Figure 1). The outside diameter of the augers ranges from 2.75 inches to 10.75 inches. As the auger is advanced, soil cuttings are carried to the surface on the flighting. The drill bit, or cutter head contains cutting teeth for penetrating the ground. The bit is hollow to permit tools or materials to be placed through the end of the auger. The bit can be run open, with a center bit attached by rod to the drill rig, or with a knockout plug that remains in the auger until it is removed for well placement or sample collection.

Soil Sample Collection

There are several means of collecting soil samples from the drilling operation, and the method used for collection will be dictated by the end use of that sample. Cuttings from the flights can be used for logging the soil types, although with depth, the actual sample is a mixture of the formation penetrated during drilling. With care, good samples can be collected above saturation; however, below saturation, where the formation continues to flow into the well, the samples become less representative of specific zones. This type of sampling is completely suitable for most situations where specific samples are not critical for design purposes.

Undisturbed soil samples can be collected through the core of the auger by driving a sampling device beyond the end of the bit

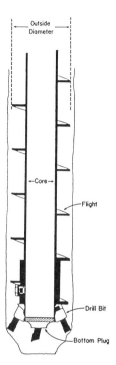

Figure 1. Hollow-stem auger.

(into formation that has not been drilled). Split-spoon samplers or Shelby tubes are most commonly used for collection of these samples. During drilling above the level of saturation, a core of sediment will form in the bottom of the auger. This portion of the undisturbed sample can be recognized and discarded from the actual undisturbed sample. Once saturation has been encountered, it is necessary to maintain a positive head of water in the augers to prevent heaving of the formation into the augers.

The split-spoon sampler consists of a drive shoe, split barrel and coupler to attach to rod. The sampler is lowered through the augers to the bit and then driven into the underlying formation with a 140 pound hammer. This type of sampling is used for penetration tests that allow determination of the bearing capacities of the formation. The sampler can be run with liners for retention of the sample for quality analyses or laboratory permeability testing. Without liners, the barrel is split and examination of the sample made. The sample would then be retained in a sample jar for lab analyses, if required.

Shelby tubes are thin-wall samplers that are pushed or lightly driven into the formation for sample collection. These are used for collection of samples that will be used for structural testing or hydraulic conductivity testing when a larger sample is required than obtainable from a split-spoon sampler. The samples to be collected must be cohesive since there is no means of retaining the sample other than wall friction. The tubes are more fragile than a split-spoon sampler and may not be usable in clay tills with gravel.

Water Sample Collection During Drilling

Determination of the vertical extent of groundwater contamination requires that samples be collected at various intervals through the thickness of the aquifer. This can be accomplished by two means. Well casing and screen can be set out the end of the auger and developed so that the sample from the selected interval can be obtained. The well must be pulled from the augers and the additional footage drilled between each sample. This can be time consuming and very difficult, considering the tendency for the saturated material to heave into the auger core and lock the casing in place, which necessitates the removal of the augers to free the pipe. An alternative to this is the "Keck Screened Auger Method" in which the lead auger is slotted and, using the core of the augers as the casing, can be developed and sampled at the selected intervals.

During drilling with the screened auger, the core of the auger can be checked for the presence of water. Once sufficient saturated thickness has been penetrated, the core and screened section can be developed using air-lift techniques or a surge block to produce clear water. Sampling equipment can then be used for proper evacuation of the core volume of water and collection of the water samples. Analyses of these samples for key parameters immediately after sampling allows determination of the proper well setting. Additionally, the full vertical extent of the contamination is defined.

Monitor Well Installation

If the intent of the drilling program is for setting monitor wells or the holes are drilled with the screened auger, the bit is sealed with a plug and there are no cuttings inside the augers. The well is then set by pulling the augers back to the desired depth or just pulling them back a foot if the well is to be set at the base of the borehole, then dropping the casing and screen through the plug. A positive head has to be maintained in the augers when the plug is knocked out to prevent sand lock. The formation will naturally collapse around the well screen during withdrawal of the augers. If a sand pack is required, it can be set by washing it in through the auger core as they are slowly pulled from the hole. Grouting is accomplished by injecting through the core of the augers as the augers are pulled from the hole.

Grouting

The purpose of grouting the annular space between the casing and the borehole wall is to assure that the borehole is less permeable than the surrounding soil. Additionally, it may be desirable to provide support for the well with the use of a cement grout.

Grouting materials are usually bentonite, cement or a mixture of the two. Bentonite can be used as a grout in powder or granular forms. Generally, this type of seal would be used above the water table. As wetting of the bentonite occurs, the bentonite swells and becomes less permeable. Bentonite can also be used as a slurry and injected through the auger core. A thick slurry is formed at mixtures exceeding one pound bentonite per gallon. This will provide an adequate grout for most applications; however, to assure even greater protection, after mixing the grout, fine granular bentonite can be added to the mix and injected in the hole. This additional bentonite gives the grout additional hydration capability.

Neat cement (without sand or gravel matrix) is frequently used as a grout material. It does provide a firm base for support of the well; however, it has no shrink and swell characteristics which, under certain conditions could result in some leakage through the annular space. This type of grout is used frequently in the water well industry and is used all the time in the oil fields without causing problems.

A mixture of cement and bentonite is recommended for many projects. This mixture provides both support and flexibility. The proper mixing procedure is to mix five to six pounds of bentonite in five to six gallons of water. To that mixture, add a sack of portland cement (94 pounds) and add water as needed to make the mixture pumpable (not to exceed 7.5 to 8 gallons of water total).

Equipment Decontamination

The prevention of cross-contamination between samples or boreholes by using equipment that has been contaminated is critical for maintaining the integrity of the sampling effort and the monitoring system. With analytical techniques that permit detection of contaminants at the low part per billion range, it is very easy to introduce contaminants that were not previously present.

When collecting undisturbed soil samples or water samples from a borehole, the equipment should be thoroughly cleaned between each use. The most widely accepted method for decontamination of these tools is by scrubbing with a solution of trisodium phosphate (TSP) and thoroughly rinsing the equipment. In order to assure that the decontamination procedures are correct, frequent samples should be taken from the rinse water and analyzed for the parameters of concern. Acetone, a solvent that was frequently recommended for decontamination in the past, was found to be difficult to remove from the equipment. Therefore, it should not be used except under extreme conditions where it is necessary to cut through greases. Thorough cleaning should follow any use of acetone or similar compound.

Drilling equipment (the rig, augers, drill rod and any tools used in the operation) should be thoroughly decontaminated between boreholes. The recommended method is to clean with high-pressure, hot-water washes, making sure that all dirt is removed from the equipment.

Well materials (unless purchased precleaned and wrapped) should be decontaminated before used. Metal casing will typically have grease or cutting oils left on it. Casing that is hauled on an open truck can also pick up contaminants from the road; therefore, the best cleaning location is on site.

The personnel operating the drilling equipment must take care not to handle fuel cans, grease or oils before handling tools or materials. Different sets of gloves for the various rig operations are highly recommended.

Conclusion

The critical nature of the information collected through drilling programs related to groundwater contamination dictates that knowledgeable, conscientious contractors be used for this type of drilling. Both water well drillers and drilling firms specializing in this type of drilling have the capability of completing the work appropriately. Care must be taken to assure that the contractor is aware of the protocols and the potential hazards related to doing this type of work.

The proper operation of drilling equipment is both a skill and an art that deserves greater praise than is generally given to the driller. The data the geologist or engineer collects from a drilling operation is no better than the drillers' ability to "read" the borehole and follow very strict protocols.

SAMPLING AND ANALYSES

Decisions concerning groundwater contamination are ultimately based on water quality data. The purpose of the drilling, well placement, well development, sampling, sample transport, preservation, and the analytical determinations is to provide accurate information on the water quality of the aquifer. All the steps in the process are critical, and they all point to the same goal--accurate data. In rural groundwater contamination situations, the major focus frequently concerns parameters whose natural, background concentrations are in the ppm (mg/l) range. For example, nitrate is generally only measured to the nearest hundredth of a mg/l, and the recommended maximum contaminant level (RMCL) for nitrate in drinking water is 10 ppm. Unfortunately, it is becoming apparent that groundwater contamination in rural areas frequently involves organic contaminants whose RMCLs are in the low ppb (ug/l) range, or zero.

It is important to realize the effectiveness of the modern analytical laboratory and the significance this holds for groundwater contamination decision making. It is difficult to conceptualize either a ppm or a ppb. A liter of solvent which is diluted with a million liters of groundwater may still exceed drinking water standards by orders of magnitude. Whether the primary concern is fertilizer leachate or organic contaminants, it is essential to maintain an understanding that the concerns frequently involve very small quantities. The release of only a small volume of substance may result in the contamination of a large aquifer system, and equally important, the careless collection of a sample of groundwater may result in the contamination of the sample. Both situations have the potential of causing improper assessments and decisions. As a result, extensive sampling protocols have been developed to insure precise drilling, collection and analytical techniques.

Sampling

Realistically, it is impossible to raise a sample of groundwater from the aquifer, preserve it, bottle it, transport it, probably store it before analysis and expect the analytical results to exactly represent the chemical composition of the sample. However, with great care in all phases of the operation, it can come close. Specific protocols and sophisticated equipment have been developed for the purpose of maintaining the integrity of the sample. Several of these options will be discussed here. One essential element of sampling is the removal of the influence of the prior drilling operation and the well casing. Well development must be conducted to remove fine residues from the drilling operation and to stimulate the formation of a natural gravel pack around the screen. Additionally, before sampling, it is necessary to remove the standing water in the casing which has been in extended contact with the metal or PVC side walls. This is accomplished by removing three to five volumes of standing water from the well before sampling.

Any sampling process utilized to collect a groundwater sample will influence the chemical composition of the sample. Adsorption onto surfaces, loss of volatiles into the air through agitation and changing pressures, oxidation and/or reduction--all are possible agents of influence which must be minimized. Two factors need to be stressed--the proper evacuation of the well and the utilization of clean equipment. Well evacuation may be performed by numerous devices ranging from bailers to sophisticated pumps. Teflon or stainless steel bailers have the advantage of being cost effective and easy to maintain. Disadvantages include the increased time required to perform the necessary evacuations and sampling plus the consideration of a fatigue factor. If it is necessary to sample numerous wells, or if the wells contain considerable standing water, fatigue may result in a progressive tendency to inadequately evacuate the wells. This progression may result in an artificially produced trend in groundwater quality data.

For the arm weary there are numerous submersible/electrical pumps which facilitate the evacuation and sampling process. These pumps have the advantage of being time-effective and encourage the complete evacuation process. The major disadvantage is the increased cost, usually in the thousands of dollars.

Both types of sampling devices involve a consideration of the materials utilized. Frequently, both bailers and pumps are made of relatively inert materials such as stainless steel or teflon. These materials minimize the process of adsorption and reaction. The operator should be aware of the materials of construction and their potential impact on the sample. This consideration extends to all sampling equipment which comes in contact with the groundwater, including drop lines, and static water level measuring devices. It is recommended that gloves are worn throughout the process to eliminate hand contact with the sample.

Sample Containers and Transport

Upon obtaining the sample, it is essential to contain it and transport it to the analytical laboratory. Laboratory technicians are understandably upset when they are assigned the precise analysis of samples collected in pickle jars. It is absolutely essential to begin the process with clean containers. It is recommended that when collecting samples which will ultimately be delivered to an analytical laboratory for analysis, the containers should be supplied by the laboratory. The laboratory has the facilities to insure complete cleaning of all containers. Furthermore, the containers supplied by the laboratory should be nonreactive and an appropriate size for the desired analyses. A small additional cost may be required for this service, but this is usually minor compared to the total analytical charge and insignificant compared to the costs of repeating the whole process due to unclean sample containers.

Following their collection, the samples must be preserved and transported to the laboratory. It is recommended that the samples should always be placed on ice in coolers, out of the sunlight, immediately following their collection. Cooling inhibits chemical reaction and the growth of organisms in the sample. This is an easy, effective preservation method. Additional preservatives are dependent upon the specific analyses required. Both EPA documents and analytical laboratories provide recommendations for preservation.

Because different analytical laboratories utilize different techniques, it is recommended that the laboratory providing the analytical services specify the preservation technique and provide the required reagents.

Concerning transport, it is recommended that the samples should always be delivered to the laboratory as soon as possible. If organic analyses are to be conducted, it is advisable to include an additional sample from a known uncontaminated source with the sample collection. It is possible for samples which have been stored in vehicles, garages, or facilities which contain organic vapors to become contaminated during the storage/delivery process. The inclusion of the blank sample would permit the detection of this contamination.

Analysis

The analytical process is a critical factor which should be given careful consideration. It would be unfortunate to waste all the effort expended in collecting a representative sample by allowing poor analytical techniques. Consideration should be given to the quality control of either the individual or the laboratory performing the determinations. Different levels of quality control are available within and between laboratories. These factors should be clearly understood before any analyses are conducted. Also, a broad range of prices exist between laboratories for the same services. Price alone should not be the determinant. Inexpensive, inaccurate analytical results are frequently not cost-effective in the long run.

OTHER ALTERNATIVES

In examining the overall process of utilizing monitor wells to assess groundwater quality, it becomes obvious that the total system is rigorous with numerous potential problems. It would be helpful, and cost effective, if remote sensing of the aquifer were possible. In special situations involving volatile organic compounds, it is possible to achieve some understanding of groundwater quality through analysis of the vapors present in the pore spaces of soils in the unsaturated zone.

When volatile organics are released through spills, leaking tanks or improper disposal, a portion of the substance may be transported to the water table. Here it mixes with the groundwater and migrates in the direction of groundwater flow. As it migrates, because of its chemical nature, a portion of the substance vaporizes into the pores of the overlying unsaturated soils. The presence of these vapors makes possible an indirect detection of the contaminated groundwater.

Soil-gas detection involves the collection of gas samples from the pore spaces of the soils followed by analysis using gas chromatography. The collection of the sample is the most variable factor because the mobility of the vapors in the soils is influenced by several variables, including the soil moisture content, and barometric pressure. These variables must be monitored to determine their influence on the sampling/detection process.

The advantage of soil-gas analyses is that probes may be placed only three to five feet deep to withdraw a sample of the soil pore gases. Therefore, twenty or more sites may be examined in one day. These analyses provide an important preliminary indication of the quality of groundwater and may be used to guide the placement of monitor wells which directly examine the groundwater. This combination minimizes the total number of monitor wells and provides a cost-effective investigative approach.

PART IV

DRINKING WATER STANDARDS, HEALTH IMPLICATIONS, AND RISK CONSIDERATIONS

CHAPTER 14

DRINKING WATER STANDARDS:
THEIR DERIVATION AND MEANING

Craig Vogt
Joseph Cotruvo
Office of Drinking Water
Environmental Protection Agency
Washington, D.C. 20460

INTRODUCTION

National Primary Drinking Water Regulations are currently being developed by the U.S. Environmental Protection Agency (EPA) for over 80 contaminants in drinking water under the Safe Drinking Water Act. Nonenforceable health goals, i.e., Maximum Contaminant Level Goals (MCLGs) are to be set at the no adverse effect level; and the enforceable standards, i.e., Maximum Contaminant Levels (MCLs) are to be set as close to the MCLGs as feasible. Feasibility includes the availability/performance/cost of treatment technologies, the availability/capability of analytical methods and other factors. This paper presents the fundamental concepts and approaches used by EPA in setting MCLGs and MCLs.

BACKGROUND

The Safe Drinking Water Act (SDWA) requires the EPA to establish primary drinking water regulations which: (1) apply to public water systems; (2) specify contaminants which, in the judgment of the Administrator, may have any adverse effect on the health of persons; (3) specify for each contaminant (a) MCLGs and (b) either (i) MCLs or (ii) treatment techniques. A treatment technique requirement would only be set if "it is not economically or technologically feasible" to ascertain the level of a contaminant in drinking water.

MCLGs are nonenforceable health goals and are to be set at a level at which, in the Administrator's judgment, "no known or

Table 1.   Contaminants required to be regulated under the SDWA of 1986.

VOLATILE ORGANIC CHEMICALS

trichloroethylene	benzene
tetrachloroethylene	chlorobenzene
carbon tetrachloride	dichlorobenzene
1,1,1-trichloroethane	trichlorobenzene
1,2-dichloroethane	1,1-dichloroethylene
vinyl chloride	trans-1,2-dichloroethylene
methylene chloride	cis-1,2-dichloroethylene

MICROBIOLOGY AND TURBIDITY

total coliforms	viruses	standard plate count
Giardia lamblia	turbidity	*Legionella*

INORGANIC CHEMICALS

arsenic	silver	vanadium
barium	fluoride	sodium
cadmium	aluminum	nickel
chromium	antimony	zinc
lead	molybdenum	thallium
mercury	asbestos	beryllium
nitrate	sulfate	cyanide
selenium	copper	

ORGANIC CHEMICALS

endrin	1,1,2-trichloroethane
lindane	vydate
methoxychlor	simazine
toxaphene	PAHs
2,4-D	PCBs
2,4,5-TP	atrazine
aldicarb	phthalates
chlordane	acrylamide
dalapon	dibromochloropropane (DBCP)
diquat	1,2-dichloropropane
endothall	pentachlorophenol
glyphosate	pichloram
carbofuran	dinoseb
alachlor	ethylene dibromide (EDB)
epichlorohydrin	dibromomethane
toluene	xylene
adipates	hexachlorocyclopentadiene
2,3,7,8-TCDD (Dioxin)	

RADIONUCLIDES

Radium 226 and 228	Gross alpha particle activity
radon	beta particle and photon
uranium	radioactivity

anticipated adverse effects on health of persons occur and which allow an adequate margin of safety." The House Report on the SDWA states that for carcinogens, the MCLGs should be set at zero.

MCLs must be set as close to the MCLGs as feasible. Feasible means "with the use of the best technology, treatment techniques and other means which the Administrator finds, after examination for efficacy under field conditions and not solely under laboratory conditions, are available (taking costs into consideration)." The SDWA requires regulation of 83 specific contaminants by June 19, 1989, as shown in Table 1. Deadlines set by the SDWA are shown in Table 2.

Primary drinking water regulations are also to include monitoring requirements. Specifically, regulations are to contain criteria and procedures to assure a supply of drinking water which dependably complies with MCLs including quality control and testing procedures.

DEVELOPMENT OF DRINKING WATER STANDARDS

Selection of Contaminants for Regulation

It is impossible to consider for regulation every chemical that may appear in drinking water and theoretically may adversely affect health in some remote circumstances. What is needed is some priority of contaminants for regulation so that a reasonable number of contaminants of sufficient concern can be addressed in regulations that will advance the goals of the SDWA and provide definitive guidance to address potential human health effects of exposure to hazardous materials in drinking water. The most relevant criteria for selection of contaminants are: (1) the potential human health risk, and (2) the occurrence or potential for occurrence in drinking water.

Table 2. Summary of deadlines for standards under SDWA of 1986.

What	When
9 MCLGs and MCLs/Monitoring	June 19, 1987
Public Notice Revisions	September 19, 1987
Filtration Criteria	December 19, 1987
Monitoring for Unregulated Contaminants	December 19, 1987
List of Contaminants	January 1, 1988
40 MCLGs and MCLs/Monitoring	June 19, 1988
34 MCLGs and MCLs/Monitoring	June 19, 1989
Disinfection Treatment	June 19, 1989
25 MCLGs and MCLs/Monitoring	January 1, 1991

A set of selection criteria have been developed which essentially expand the two primary factors listed above. Use of a specific formula to apply selection criteria is not believed to be appropriate because of the many variables associated with contaminants in drinking water. For each contaminant, the essential factors in the analysis are as follows:

Are there sufficient health effects data upon which to make a judgment on the potential health effects of human exposure?

Are there potential adverse human health effects from exposure to the contaminant via ingestion?

Does the contaminant occur in drinking water?

Has the contaminant been detected in significant frequencies and in a widespread manner?

If data are limited on the frequency and nature of the contaminant, is there a significant potential of drinking water contamination?

Factors considered in the analysis of potential occurrence include the following:

1. Occurrence in drinking water other than community water supplies;

2. Present in direct or indirect additives;

3. Present in ambient surface water or groundwater;

4. Present in liquid or solid waste;

5. Mobile to surface water (runoff) or groundwater (leaching);

6. Widespread dispersive use patterns; and

7. Production rates.

Determination of MCLGs

MCLGs are to be set at a level at which "no known or anticipated adverse effects on the health of persons occur and which allow an adequate margin of safety.

Noncarcinogens. For toxic agents not considered to have carcinogenic potential, "no-effect" levels for chronic/lifetime periods of exposure, including a margin of safety, are referred to commonly as Acceptable Daily Intakes (ADI). These ADIs are considered to be exposure levels estimated to be without significant risk to humans when received daily over a lifetime.

The intent of a toxicological analysis is to identify the highest no-observed-adverse-effect-level (NOAEL) based upon assessment of available human or animal data (usually from animal experiments). To determine the ADI for regulatory purposes, the NOAEL is divided by (an) appropriate "uncertainty" or "safety" factor(s). This process accommodates for the extrapolation of animal data to the human, for the existence of weak or insufficient data and for individual differences in human sensitivity to toxic agents, among other factors.

ADIs traditionally are reported in mg/kg/day but for MCLGs purposes, the "no effect" level needs to be measurable in terms of drinking water quality, i.e., mg/liter. An adjustment of the ADI to mg/liter is accomplished by factoring in an assumed weight of the consumer and the assumed amount of drinking water consumed per day. The "no effect level" in mg/l has been termed the Adjusted ADI (AADI) and is calculated as follows:

$$AADI = \frac{(NOAEL\ in\ mg/kg/day)\ (70\ kg)}{(UF)\ (2\ liters/day)}$$

Where: NOAEL is No-Observed-Adverse-Effect-Level.
70 kg is the weight of an adult.
2 liters/day is the assumed amount of water consumed by an adult per day.
UF is the uncertainty factor (usually 10, 100 or 1000).

The National Academy of Sciences recommended an approach for use of uncertainty factors when estimating ADIs for contaminants in drinking water (NAS, 1977). The NAS outlines are as follows:

An uncertainty factor of 10 used when good acute or chronic human exposure data are available and supported by acute or chronic data in other species.

An uncertainty factor of 100 used when good acute or chronic data are available for one species, but human data are not.

An uncertainty factor of 1000 used when acute or chronic data in all species are limited or incomplete.

Other uncertainty factors can be used to account for other variations in the available data.

To determine the MCLG, the contribution from other sources of exposure, including air and food, are taken into account. When sufficient data are available on the relative contribution of other sources, the MCLG is determined as follows:

MCLG = (AADI) - (contribution from food) - (contribution from air)

This calculation assures that the total exposure from drinking water, food and air does not exceed the ADI. However, comprehensive data are usually not available on exposures from air and food. In these cases, the MCLG is determined as follows:

MCLG = (AADI) (percentage of drinking water contribution)

The percentage of drinking water contribution often used is a 20 percent contribution for organic chemicals.

Carcinogens and equivocal evidence of carcinogenicity. Several groups of scientists have attempted to classify chemicals on the basis of available evidence for carcinogenicity. These include the International Agency for Research on Cancer (IARC), the National Academy of Sciences Safe Drinking Water Committee and EPA via its recently published risk assessment guidelines for carcinogenicity (Anon, 1986).

The IARC is responsible for a program on the Evaluation of Carcinogenic Risk of Chemicals to Humans, which involves the preparation and publication of monographs providing a qualitative assessment of the carcinogenic potential of individual chemicals and complex mixtures. The assessments are made by independent, international working groups of experts in cancer research. The program has existed since 1971 and has evaluated over 585 chemicals to date. Criteria used for evaluating carcinogenic risk to humans were first established in 1971 and were used by the IARC for the preparation of the first 16 volumes of the monographs. These criteria consisted of the terms "sufficient evidence" and "limited evidence" of carcinogenicity, referring to the amount of evidence available and not to the potency of the carcinogenic effect.

The EPA has published guidelines for carcinogen risk assessment which contain a classification system for chemicals using the degree of evidence of carcinogenicity (Anon, 1986). The categorization scheme places chemicals into five groups:

Group A: Human carcinogen (sufficient evidence from epidemiological studies)

Group B: Probable human carcinogen

Group B1:   At least limited evidence of carcinogenicity to humans

Group B2:   Usually a combination of sufficient evidence in animals and inadequate data in humans

Group C: Possible human carcinogen (limited evidence of carcinogenicity in animals in the absence of human data)

Group D: Not classified (inadequate animal evidence of carcinogenicity)

Group E: No evidence of carcinogenicity for humans (no evidence of carcinogenicity in at least two adequate animal tests in different species or in both epidemiological and animal studies)

The National Academy of Sciences (NAS, 1977) classified chemicals in four categories based upon the strength of the experimental evidence. These categories are: human carcinogens, suspected human carcinogens, animal carcinogens, and suspected animal carcinogens.

EPA has evaluated these three approaches, and a three-category approach based upon strength of evidence of carcinogenicity will be used to set the MCLGs. Category I includes those chemicals which have sufficient human or animal evidence of carcinogenicity to warrant their regulation as probable human carcinogens. The MCLGs for Category I chemicals will be proposed at zero. Category II includes those substances for which some limited inconclusive evidence of carcinogenicity exists from animal data. These will not be regulated as human carcinogens. However, MCLGs will reflect the fact that some possible evidence of carcinogenicity in animals exists. Thus, they will be treated more conservatively than Category III substances. Category III includes substances with inadequate or no evidence of carcinogenicity. MCLGs will be calculated based upon AADIs. These categories are summarized below:

Category I - Strong evidence of carcinogenicity
    EPA Group A or Group B
    IARC Group 1, 2A or 2B

Category II - Equivocal evidence of carcinogenicity
    EPA Group C
    IARC Group 3

Category III - Inadequate or no evidence of carcinogenicity
                in animals
    EPA Group D or E
    IARC Group 3

The method for determining the MCLGs for Category II chemicals is more complex than for the other categories. To be placed in Category II, chemicals are not considered to be probable carcinogens via ingestion, although some data are available that cause concern. Thus, these substances should be treated more conservatively than Category III "noncarcinogens," yet less conservatively than

Category I chemicals. Two options are available for setting the MCLGs for Category II chemicals; the first option involves basing the MCLG upon the AADI. To account for the possible evidence of carcinogenicity, an additional factor would be applied (e.g., AADI divided by a factor of 10 or some other value). The second option involves basing the MCLGs upon a lifetime risk calculation in the range of 10^{-5} to 10^{-6} using a conservative method. This risk range is commonly considered to be protective and in the future, if additional data led to reconsideration of a chemical's carcinogenicity, the MCLG would still be set at a level that would represent an extremely low nominal risk. The first option, basing the MCLG upon the AADI, will be used if sufficient valid chronic toxicity data are available. If sufficient data are not available, the MCLG will be based upon a risk calculation.

Determination of MCLs

MCLs are to be set "as close to" the MCLGs "as is feasible." The term "feasible" means "feasible with the use of the best technology, treatment techniques, and other means which the Administrator finds after examination for efficacy under field conditions and not solely under laboratory conditions and availability (taking costs into consideration)."

The general approach to setting MCLs is to determine feasibility of controlling contaminants. This requires an evaluation of (1) the availability and cost of analytical methods, (2) the availability and performance of technologies and other factors relative to feasibility and identifying those that are "best" and, (3) an assessment of the costs of the application of technologies to achieve various concentrations. Key factors in the analyses include the following:

Technical and economic availability of analytical methods: precision/accuracy of analytical methods that would be acceptable for accurate determination of compliance limits of analytical detection, laboratory capabilities, and costs of analytical techniques.

Concentrations attainable by application of best technology generally available.

Levels of contamination (i.e., concentrations) in drinking water supplies.

Feasibility/reliability of removing contaminants to specific concentrations.

Other feasibility factors relating to the "best" means of treatment such as air pollution and waste disposal and effects on other drinking water quality parameters.

Costs of treatment to achieve contaminant removal.

SUMMARY OF DRINKING WATER REGULATIONS

Drinking water regulations are being developed for the contaminants listed in Table 1 in several stages. Tables 3 through 7 present MCLGs and MCLs for contaminants that have been developed to date.

Table 3 summarizes the final MCLGs and proposed MCLs for volatile organic chemicals. Final MCLs are scheduled to be published in mid-1987. The MCLG and MCL for p-dichlorobenzene may be revised to zero and 5 ug/l, respectively, because of new toxicology data that possibly shows that it is a probable human carcinogen.

Tables 4, 5 and 6 present the MCLGs proposed on November 13, 1985, for inorganic chemicals (IOCs), synthetic organic chemicals (SOCs), and microbiological parameters, respectively. Final standards for fluoride were established on March 15, 1986. The MCLG and MCL are both 4 mg/l, and a secondary MCL (i.e., aesthetic effects) is 2 mg/l (nonenforceable).

Table 7 presents draft MCLGs for radionuclides, published in an Advance Notice of Proposed Rule-Making on September 30, 1986.

Table 3. Final MCLGs and proposed MCLs for volatile organic chemicals.

Volatile Organic Chemical	Final MCLG	Proposed MCL
benzene	zero	5 ug/l
vinyl chloride	zero	5 ug/l
carbon tetrachloride	zero	5 ug/l
1,2-dichloroethane	zero	5 ug/l
trichloroethylene	zero	5 ug/l
1,1-dichloroethylene	7 ug/l	7 ug/l
1,1,1-trichloroethane	200 ug/l	200 ug/l
p-dichlorobenzene	750 ug/l	750 ug/l

Table 4. Proposed MCLGs for inorganic chemicals.

Inorganic Chemical	Existing Interim MCL (mg/l)	Proposed MCLG (mg/l)
arsenic	0.05	0.050
asbestos	--	7.1 MFL*
barium	1.0	1.5
cadmium	0.010	0.005
chromium	0.05	0.12
copper	--	1.3
lead	0.05	0.020
mercury	0.002	0.003
nitrate	10.0	10.0
nitrite	--	1.0
selenium	0.01	0.045

* MFL = million fibers per liter

Table 5. Proposed MCLGs for synthetic organic chemicals.

Synthetic Organic Chemical	Existing NIPDWR* (mg/l)	Proposed MCLG (mg/l)
acrylamide	--	zero
alachlor	--	zero
aldicarb, aldicarb sulfone and aldicarb sulfoxide	--	0.009
carbofuran	--	0.036
chlordane	--	zero
cis-1,2-dichloroethylene	--	0.07
DBCP	--	zero
1,2-dichloropropane	--	0.006
o-dichlorobenzene	--	0.62
2,4-D	0.1	0.07
EDB	--	zero
epichlorohydrin	--	zero
ethylbenzene	--	0.68
heptachlor	--	zero
heptachlor epoxide	--	zero
Lindane	0.004	0.0002
methoxychlor	0.1	0.34
monochlorobenzene	--	0.06
PCBs	--	zero
pentachlorophenol	--	0.22
styrene	--	0.14
toluene	--	2.0
2,4,5-TP	0.01	0.052
toxaphene	0.005	zero
trans-1,2-dichloroethylene	--	0.07
xylene	--	0.44

*NIPDWR - National Interim Primary Drinking Water Regulations

Table 6. Proposed MCLGs for microbiological parameters.

Parameter	Existing Interim MCL	Proposed MCLG
total coliforms	1-4/100 ml	zero
turbidity	1-5 NTU	0.1 NTU*
Giardia	--	zero
viruses	--	zero

*Nephelometric Turbidity Unit.

Table 7. Draft MCLGs for radionuclides.

Parameters	Draft MCLG
radium -226	zero
radium -228	zero
gross alpha	zero
uranium	zero
radon	zero
gross beta and photon emitters	zero

The views and opinions stated in this article are those of the authors and not necessarily those of the U.S. EPA.

LITERATURE CITED

Anon. 1986. Federal Register 51 FR 33992.

NAS. 1977. Drinking Water and Health. National Academy of Sciences, Washington, DC, 939 pp.

CHAPTER 15

HEALTH IMPLICATIONS OF GROUNDWATER CONTAMINANTS

Michael A. Kamrin
Center for Environmental Toxicology
Michigan State University
East Lansing, Michigan 48824

If this paper was written one hundred and fifty years ago, very little could have been said on this topic. However, by the middle 1800s, a few clever physicians recognized the relationship between water consumption and serious health effects, especially communicable diseases. The first strong linkage was established between cholera and drinking water in London, England. Since then, it has become clear that many infectious agents are commonly transmitted in water. As a result of this knowledge, a variety of public health measures, such as disinfection of water, have become commonplace in developed countries. Although these steps have not been 100 percent successful and there still are occasional outbreaks of disease due to waterborne organisms in the United States, it is clear that this particular threat has been minimized.

Recently, the issue of health effects of water contaminants has returned to public attention; however, this time with respect to chemical rather than biological contaminants. The present situation is different from the previous one in more than just the source. Instead of a clearly defined disease, cholera, and an unknown source, there is clearly defined contamination and less well defined disease. Chemicals can be detected at minute levels, parts per trillion in many cases and even lower in some. (A part per trillion is equivalent to a pinch of salt in ten thousand tons of potato chips.) Possible health effects, however, vary tremendously from nonspecific symptoms such as headache and nausea to life-threatening illnesses such as liver damage and cancer. As a result, it has not been possible to link particular chemical contaminants to particular effects in most instances. This has led to the current situation where there is a great deal of concern but few answers. The focus of this chapter will be the sources of present difficulties and the prospects for solutions in the near future.

First, although the title of this chapter refers to groundwater contamination, many of the same principles also apply to chemical contamination of other parts of the environment. Indeed, as will be shown, the presence of multiple sources of chemicals in the environment adds to the complexity of ascribing particular effects to particular sources.

Before discussing particular health effects, some basic distinctions must be made. The first is between acute and chronic effects. Acute effects occur within minutes, hours or days while chronic effects appear only after weeks, months or years. The quality and quantity of evidence that can be gathered is quite different for each type of effect and, as a result, the confidence that can be placed in the conclusions that are reached is also quite different. This distinction is often overlooked in the overwhelming concern with a few specific health effects, namely cancer and birth defects.

Another important distinction is between an exposure assessment and toxicity assessment. The former is a measure of the extent and duration of exposure of an individual or population and the latter a measure of the extent and type of adverse effect associated with a particular level of exposure. Both measures are essential for assessing the possible link between health effects and environmental contaminants. Oftentimes, this is forgotten and conclusions are reached without any measures of exposure having been made. For example, dioxin is often referred to as the most toxic man-made chemical known, and this is taken to mean that it poses the greatest risk to society. This is not the case because the potential for exposure is very small.

How can exposure assessment be accomplished? There are three basic approaches: direct measures of the environment (groundwater in this case), indirect measures of the environment (e.g. sentinel animals), and direct examination of the population thought to be exposed. Direct measures of the environment are generally the first step and often provide the majority of usable information. Using this approach, the level of each contaminant can be carefully measured at one point in time. The results of this analysis generally reflect recent as well as present contamination since changes in the concentration of chemicals in groundwater are usually gradual.

This type of direct assessment is most useful when concern is focused on short-term effects since the symptoms show up shortly after exposure and before contaminant levels have had much chance to change. Use of direct measures is more complex when chronic effects are of interest. In this situation the exposure may have occurred over decades, and present measurements may not reflect either the type or amount of previous contamination. Currently, some researchers are trying to use direct measurements in conjunction with careful hydrogeological studies to project the measured current values backwards in time. This is potentially a powerful tool, but more research is needed before it can be utilized to its greatest effect.

Indirect measures of exposure also have their place. While direct measurements would be the methods of choice for short exposures, indirect ones are most useful in assessing chronic effects. In the case of groundwater contamination, this type of assessment would be part of the hydrogeological studies; however, in other cases it would represent a totally different approach. For example, in the case of surface water, measurements of fish living in the water or the sediments underlying the body of water would provide an independent measure of the amounts of various persistent chemicals which are and were present. Using the age and size of the fish, and information about how rapidly these organisms accumulate various chemicals, it is possible to estimate previous levels of contamination. Thus, indirect measurements can be important in exposure assessment but are not applicable in all situations.

The last type of exposure measure, examination of the population itself, would appear to be the most suitable. In determining if people are exposed, why not examine the people themselves. Unfortunately, it is usually not that simple. While measures of tissue or blood levels may be useful in cases of acute toxicity, they tend to diminish in utility as the time of initial exposure fades into the past. The acute situation represents the classical poisoning incident where the toxicant can be detected in the body and where enough knowledge exists to use the measured levels to calculate what the initial exposure must have been.

In the chronic situation, measured body levels cannot be used in the same way. There are a number of reasons for this. One is that the body usually reaches some type of balanced state where there is no longer any change in response to continued exposure. Another is that effects may persist even after exposure has ceased, and current levels may not reflect initial exposures. A third is that basic understanding of what happens to chemicals in the human body is lacking for many environmental contaminants. Last, and perhaps most important, there is no way to determine whether or not water was the source of the substances that are detected. Thus, direct examination of a population may provide information as to whether or not exposure has occurred but not the extent, duration or source of the exposure. (Present evidence suggests there is probably no unexposed population in the United States; all citizens contain minute amounts of a large variety of environmental contaminants.)

Overall, exposure assessment is not a very exact process when applied to the health effects of most concern--chronic toxicity. It is most often the limiting factor in trying to accurately appraise the connection between an environmental contaminant and toxicity. The assessment process is undergoing close examination and undoubtedly will be improved in the near future. However, because of inherent limitations in most cases, there will continue to be significant uncertainty. The few situations where greater confidence can be placed in this type of measure will be mentioned later.

Turning from exposure assessment to toxicity assessment, a new set of considerations must be addressed. Again, acute toxicity is easiest to deal with. Short-term studies with animals provide evidence as to which effects are linked with which chemicals and also give clues as to the levels at which these adverse effects occur. Often, some human experience is available as a result of accidental exposure, and this can serve as a source of supporting evidence in making a judgment. With these two types of evidence, it is usually possible to estimate what levels of a particular toxicant will lead to what type of acute adverse effect. Indeed, this approach is the basis for much of current regulation of toxic substances, especially with regard to occupational health. In this type of situation, standards are often set as the maximum air concentration allowable over a specific period of time.

Chronic toxicity is much more difficult to assess. At present, there are a variety of specific tests for particular types of adverse effects such as reproductive damage, behavioral effects, cancer, etc. It is not possible to examine all of these in this presentation, but a look at cancer assessment should reveal some of the problems inherent in long-term effect assessment and should also provide an examination of the effects which seem to be of most public concern.

In cancer assessment, it is not only the chronic nature of the disease but also the low incidence which is of concern, that causes difficulty. Currently, society has set a level of cancer of one person in 100,000 or one million as an acceptable incidence, so assessment measures must provide this sensitivity. In Michigan, the one in 100,000 criterion was adopted in establishing Rule 57 which is used in setting effluent limits for toxic substances. To perform low incidence cancer assessments, two types of evidence are utilized. One is based on experiments on animals, and the other is based on experience with humans.

Ideally, animal experiments would be performed with millions of animals exposed to environmentally relevant amounts of chemical. This large number is needed so that an increase in one cancer in a million could be detected. Unfortunately, there are neither the scientific nor the economic resources to carry out this type of study for the large number of chemicals of concern. Instead, investigations are performed on smaller numbers of animals, a few hundred, who have been exposed to very large amounts of the chemical in question. These large amounts are necessary to produce a high enough incidence of cancer to be detectable in this small population. Thus, the results of such studies indicate what levels of a chemical will cause cancer in a high proportion of the population.

How can this information be used to assess the level of chemical that will cause one cancer in a million animals or, more importantly, in a million humans? To accomplish this, a variety of assumptions must be made. The most important is how the animals will respond as the amount is lowered. Because there is no conclusive experimental evidence on this point, different mathematical models are used to predict the effect as exposure decreases and the level at which the cancer incidence decreases to one in a million. There are a variety of available models, and the one generally chosen is that which provides the greatest margin of safety; i.e., which overestimates rather than underestimates the potency of the chemical.

The other question, how human responses compare to those of rodents, has not really been addressed. It is generally assumed that rodents react identically to humans so that the results of the animal experiments can be translated directly to human toxicity assessment using the appropriate factors to account for differences in size and weight. Thus, the level which is calculated to produce a risk of one in a million rats is utilized directly as the basis for the number which describes the equivalent risk for humans.

The other type of evidence utilized in chronic toxicity assessment is human experience, better known as epidemiological evidence. In this type of study, human populations are carefully observed and possible associations between specific chemical exposures and particular health effects are investigated. Considering the previous discussions about exposure assessment, it should be clear that this is not an easy task. It is made even more difficult by the requirement of detecting very small changes in incidence, e.g., one extra cancer in a million people.

As a result, epidemiological assessments have only been useful in certain situations. One is exposure in the workplace, a place where levels are usually much above environmental ones and where the duration of exposure can be determined. Even there, a sizable increase in cancer incidence is needed before a connection can be established. The conclusion that asbestos causes lung cancer is based on this type of situation. An exception to the need for a high cancer incidence is the situation where the effect is unique so that even a few cases represent a considerable deviation from expectation. An example of this was the observation that a small number of vinyl chloride workers developed a rare form of liver cancer, angiosarcoma. However, even with these known carcinogens, the question of what happens at low exposures, e.g. common environmental ones, has not been answered.

In sum, the techniques available for assessment of chronic toxicity, especially carcinogenicity, provide rather clear evidence as to whether or not a particular chemical causes a particular effect in animals. However, they are not designed to provide quantitative assessments at environmentally relevant levels; and their applicability to human populations is not clear, especially as to the exact levels which give rise to a particular outcome. These uncertainties, taken together with the difficulties in exposure assessment, contribute to the current absence of definite conclusions about the relationship between most environmental exposures and chronic health effects.

The situation is even more complex than indicated above. In the previous discussion, each individual chemical was treated separately. However, most environmental exposures involve several chemicals simultaneously, and the relationship between effects due to a combination of chemicals and the effects of each one acting individually has not been determined. In general, the effects have been treated as additive, but the validity of this assumption has not been adequately tested.

How does this general discussion apply to groundwater contaminants? All of the uncertainties that were described are usually magnified in the groundwater contamination assessment. It is not possible to use many indirect environmental measures in groundwater; for example, fish do not live in this environment nor are sediments available. In addition, almost everything found in groundwater is also present in other environmental compartments so it is difficult to be sure human body burdens result from water contamination. Some common water contaminants such as gasoline contain many chemicals, and anyone who has consumed gasoline-tainted water has been exposed to a number of different toxicants. Further, the toxicity assessments performed in animals are usually performed by putting the chemical either in the food or in the air--not in the drinking water. Although it is possible to adjust the results to compensate for the different route of exposure, this introduces another source of possible error.

Thus, it is not too surprising that there are few studies which have shown a definite connection between a particular chemical in the groundwater and a specific toxic effect in a human population. The ones that are available have mostly been done outside of the United States. There are a number of claims of such connections in the United States, although most of these are between a contamination incident and health effects rather than a particular chemical and a specific effect. The most cited investigation is the Woburn, Massachusetts, study where a high incidence of childhood leukemia, in addition to other effects, was noted in a population exposed to contaminated well water. There is still controversy as to whether the contaminated water caused the observed effects, and there has been no determination of which of the chemicals in the water might have been responsible.

A fairly large scale study is underway in Battle Creek, Michigan, in a similar circumstance. In this case, a group of families with contaminated wells have reported a variety of health problems, and an attempt is being made to determine if these are linked to the water contamination which was uncovered a few years ago. However, there is nothing to suggest that this situation will be any more amenable to analysis than previous ones.   Thus, it is unlikely that any firm conclusions will be drawn.

Although it not possible to confidently assess health effects from water contaminants, especially chronic effects, judgments must be made in order to deal with the possibility that such effects may occur.   All current regulations dealing with toxic substances are subject to uncertainty, and the greater difficulties in assessing toxicity of water contaminants do not confer any special status to this problem.   The question is how to set enforceable levels of such toxicants.   Ideally, a total absence of contamination is desirable. Realistically, this would lead to abandoning a significant proportion, if not all, of available drinking water.

Since a total ban on all chemicals is not feasible, another approach might be to selectively ban certain chemicals.   Proposition 65 in California is a unique attempt to do just this.   However, in the absence of solid information that even minute levels cause adverse effects, and with the knowledge that large amounts of drinking water would become unavailable under such a policy, it is unlikely that this approach will be taken nationally.   Thus, society is forced to make quantitative toxicity assessments as one part of the establishment of acceptable levels.

This has been done with a limited number of water contaminants and mainly in response to acute toxicity.   These enforcement levels are known as the Interim Primary Drinking Water Standards (Table 1).   A large number of other chemicals are under review, and the initial steps toward establishing acceptable levels for these substances have been taken.   This process is being accelerated as a result of strict deadlines included in the amendments to the Safe Drinking Water Act passed in 1986.   It should be emphasized that the determination of acceptability is a nonscientific decision.   Levels that are chosen reflect both the best estimate of what exposures will lead to what amount of risk and a number of other factors such as the limits of analytical capabilities, the techniques available for decontaminating water and social, political and economic factors.

How are these levels applied in the case of common groundwater contaminants?   In the midwest United States, a chemical that is often found in groundwater is nitrate.   It arises mainly from the use of fertilizers, but there are other important sources such as sewage and feedlots.   Nitrates have the potential to pose both acute and chronic health threats.   The acute toxicity comes about through conversion of nitrate to nitrite by stomach microorganisms and the subsequent binding of nitrite to the oxygen-carrying molecules in the blood.   Toxicity has been observed only in small infants, generally less than ninety days old, and the current drinking water standard for nitrate, 10 mg/l (ppm) as nitrogen, reflects the levels at which such effects are thought to occur.

Table 1. National Interim Primary Drinking Water Standards.

Chemical Contaminants	Maximum Allowable Concentration (micrograms/liter or parts per billion)
Inorganic chemicals	50
Arsenic	50
Barium	1,000
Cadmium	10
Chromium	50
Lead	50
Mercury	2
Nitrate (as nitrogen)	10,000
Selenium	10
Silver	50
Fluoride	4,000
Endrin	0.2
Lindane	4
Methoxychlor	100
Toxaphene	5
2,4-D	100
2,4,5-TP	10
Trihalomethanes	100
Other Contaminants	**Maximum Allowable Concentration**
Coliform bacteria	1 per 100 milliliters (mean)
Radionuclides	
Radium 226 and 228 (combined)	5 picoCuries/liter
Gross alpha particle activity	15 picoCuries/liter
Gross beta particle activity	4 millirem per year
Organic chemicals turbidity	1 up to 5 turbidity units

The possibility of a chronic threat from nitrate results from a further conversion of nitrite to nitroso compounds, many of which are carcinogenic in animals. However, there is no firm evidence that this conversion occurs in humans and no good epidemiological studies relating nitrate to cancer in humans. As a result, carcinogenicity was not taken into account in the establishment of the drinking water standard for nitrate.

The situation with respect to another contaminant, arsenic, is similar in that both acute and chronic effects are possible, and the standard is set in terms of the acute effects. However, in this case, there is epidemiological evidence which suggests that arsenic is a human carcinogen. Although this evidence is not conclusive, it does lead to a cautious approach. For example, arsenic was recently found in a few wells in a community close of Michigan State University. The maximum level detected was 36 ug/l (ppb) and the standard is 50 ug/l. If the standard was accepted, this water would be considered safe. However, in view of the available evidence, the local health department recommended that the water not be used. As a result, the people involved drank bottled water.

Another common class of groundwater contaminants includes solvents such as trichloroethylene and methylene chloride. There are no established drinking water standards for these substances. Thus, there is no set approach to take if contamination is detected. On the one hand, it could be argued that a zero level be accepted until some standard is established. The other possibility is to perform a risk assessment similar to that used to establish levels (or use one that is already available) and to make judgments on this basis. In view of the difficulties in making these kinds of judgments, the first approach has often been employed and alternative sources of drinking water have been found. However, as pointed out previously, this type of solution will become increasingly difficult as the number and scope of water contamination incidents increase.

Thus, neither the presence nor the absence of standards relieves society of the necessity of making judgments. The real world presents problems of great difficulty. Very few situations fall into categories, like acute exposure, that are relatively easy to resolve. Most involve chronic toxicity and some fall in between acute and chronic. Those in between fall on a continuum of difficulty where the longer the latent period, the more poorly the available measurements reflect the actual exposure situation. In addition, the longer the latent time between the suspected cause and the probable effect, the greater the uncertainty in the applicability of animal studies. Thus, unless the incident involves a unique contaminant and a rare effect, there is bound to be residual uncertainty as to the effect a particular water contaminant episode will have on an exposed population. However, if the levels measured in the water are high enough or the effects dramatic enough, it may be prudent to act as if the connection between the contamination and the toxicity were known. There is no nice demarcation line to follow in deciding when this situation is reached. Uncertainty must be accepted and judgments made within the framework of what is known scientifically and what is acceptable socially.

CHAPTER 16

TREATMENT OF WATER FROM CONTAMINATED WELLS

Walter A. Feige
Robert M. Clark
Benjamin W. Lykins, Jr.
Carol Ann Fronk
U.S. Environmental Protection Agency
Water Engineering Research Laboratory
Drinking Water Research Division
Cincinnati, Ohio 45268

INTRODUCTION

Groundwater is a natural resource of enormous value. Three-quarters of United States cities get their water supplies totally or in part from groundwater, and more than half of all Americans rely on it for drinking water. Ninety percent of rural households have no source other than groundwater for drinking water supplies.

A number of recent studies have documented the actual or potential contamination of the nation's groundwater on a wide scale. A survey conducted by the U.S. Environmental Protection Agency (EPA) has documented that 22 percent of approximately 466 randomly sampled utilities produced drinking water containing volatile organic chemicals (VOCs) at detectable levels (Westrick et al., 1984). Other individual studies have shown the presence of pesticides in groundwater, and in recent years concern has been growing over contamination of groundwater by agricultural chemicals (fertilizers, insecticides, nematocides, and fungicides). The concern is greater, of course, when specific chemicals are identified in homeowners' drinking water wells.

Though concern in the past five years has centered on VOC contamination of groundwater, several examples illustrate the growing concern over pesticides. During the last eight years, New York's Suffolk County Health Department has examined the groundwater underlying Long Island for agricultural organic constituents and their decay products (Lykins et al., 1985). During the testing, 101 agricultural/organic compounds were evaluated, with 41 found in the groundwater. Many of these identified contaminants were present in trace quantities, but aldicarb, carbofuran, 1,2-dichloropropane, and 1,2,3-trichloropropane were at elevated levels. In addition, nitrates from fertilizer applications were present in quantities exceeding the primary drinking water standards. A 1983 study, conducted by the California Department of Food and Agriculture, stated that dibromo-chloropropane (DBCP), ethylene dibromide (EDB), and simazine were found in San Joaquin Valley wells (Litwin et al., 1983). A report submitted to the California State Water Resources Board states that more than 2000 wells statewide have been found to be contaminated with DBCP (Litwin et al., 1983).

Work conducted near Tiffin, Ohio, by EPA's Drinking Water Research Division (DWRD) has shown periodic high quantities of pesticides in surface waters associated with seasonal runoffs (Baker, 1983). Although groundwaters have not been sampled in this area, it is likely that some contamination exists. Unlike surface waters, groundwaters are likely to remain contaminated indefinitely. EPA is now proposing to conduct an extensive groundwater survey for pesticides similar to the earlier VOC survey (Cohen, 1985). Also underway and almost completed is the EPA National Inorganic and Radionuclide Survey to develop an occurrence data base for certain of those parameters.

Recently, EPA has proposed maximum contaminant levels (MCLs) for VOCs (Phase I) that are frequently found in groundwater supplies (Fed. Reg. Part III, 1985). These MCLs are expected to be in final form in the not too distant future. In addition, EPA is planning to propose MCLs for synthetic organic chemicals (Phase II), inorganic chemicals (Phase II), and radionuclides (Phase III) for contaminants that are also frequently found in groundwater (Fed. Reg. Part IV, 1985). Previous experience indicates that these Phase I, II and III regulations will have a significant impact on many water utilities in the United States.

Although the most desirable way of controlling groundwater contamination is to prevent pollution at the source, many aquifers are currently contaminated and provide the cheapest water supply to the communities they serve. Water from these aquifers can be treated to achieve quality levels that equal or exceed those being proposed by the Office of Drinking Water. EPA's Drinking Water Research Division has been conducting extensive research for a number of years into the development of treatment technologies for removing both inorganic and organic contaminants from groundwater in a cost-effective manner. Recently, this effort has emphasized the removal of volatile organic chemicals and synthetic organic chemicals such as pesticides, nitrate removal and radionuclide removal.

The research activities of the Drinking Water Research Division are categorized as pilot-, bench-, and field-scale studies. Most of the bench- and pilot-scale studies are conducted on an in-house basis, and are designed to examine the feasibility of various technologies before their cost and performance are evaluated in the field. The remainder of the research is performed via cooperative agreements with municipalities and other nonprofit institutions.

The purpose of this paper is to discuss the status of DWRD's research activities and to provide a state-of-the-art summary on the removal of organic and inorganic contaminants from groundwater.

ACTIVITIES OF THE DRINKING WATER RESEARCH DIVISION

The Drinking Water Research Division is part of the Water Engineering Research Laboratory of the Office of Environmental Engineering and Technology demonstration in EPA's Office of Research and Development. The Division consists of 44 permanent full-time employees and has an annual budget of approximately $5 million. DWRD is responsible for evaluating the various types of technologies that may be feasible for meeting the MCLs promulgated under the Safe Drinking Water Act (PL 93-523) and its Amendments. The standard research protocol followed by DWRD is to evaluate unit processes at the bench level, move the process to pilot scale for testing, and if its performance is promising, build a prototype for field evaluation. At all stages, cost and performance are key factors in moving a technology to the next phase. Thus, when a technology is tested on a field scale, cost and performance data are routinely collected.

Table 1 summarizes the treatment technologies that DWRD is evaluating for removal of VOCs, synthetic organic chemicals (SOCs), nitrates, and radionuclides from water supplies. The table indicates carbon adsorption is effective for removing both VOCs and SOCs. Packed tower and diffused aeration are best suited for removing VOCs. Ion exchange has been field-tested to show effective removal of nitrates and pilot-tested for uranium removal. Reverse osmosis (RO) has proven to be effective in the field for radium removal and pilot-tested for nitrate removals. Of the technologies that show promise and are being tested at the bench and pilot scales, conventional treatment with powdered activated carbon (PAC) is effective for removing a few of the SOCs, ozone oxidation is effective for removing certain classes of VOCs and SOCs, and certain reverse osmosis membranes and ultraviolet treatment are also potentially effective against VOCs and SOCs. Aeration and carbon adsorption are being examined for their radon removal capabilities. Each of these technologies is discussed in this paper.

Field- and Pilot-Tested Technologies

Carbon Adsorption. The Drinking Water Research Division has been conducting extensive studies on the use of granular activated

Table 1. Treatment technologies evaluated by EPA's Drinking Water Research Division for removing volatile organic chemicals (VOCs), synthetic organic chemicals (SOCs), nitrates and radionuclides from drinking water.

Technology Status	Technology	Contaminant Class or Specific Contaminant Removed
Field-tested	1. Carbon adsorption	1. VOCs, SOCs
	2. Packed tower and diffused-air aeration	2. VOCs
	3. Ion exchange	3. Nitrates
	4. Reverse osmosis	4. Radium
Pilot-tested	1. Reverse osmosis	1. Nitrates, uranium
	2. Ion exchange	2. Uranium
Promising technologies	1. Conventional treatment with powdered activated carbon	1. SOCs
	2. Ozone oxidation	2. VOCs, SOCs
	3. Reverse osmosis	3. VOCs, SOCs
	4. Ultraviolet treatment	4. VOCs, SOCs
	5. Ion exchange	5. Radium
	6. Selective complexer	6. Radium
	7. Aeration	7. Radon
	8. Carbon adsorption	8. Radon

carbon (GAC) treatment with on-site regeneration for 10 years. Most of this research has been devoted to demonstrating the effectiveness of GAC for surface water treatment (Lykins et al., 1984). More recently, DWRD has been conducting studies that incorporated the use of carbon treatment for removal of VOCs and SOCs from groundwater. GAC is being studied for removing contaminants from groundwater at Suffolk County, New York; California's San Joaquin Valley; and Wausau, Wisconsin. Each of these projects is designed to examine a different aspect of GAC application and, except for Wausau, they are intended to deal with the little understood pesticide contamination problem.

In Suffolk County, the removal of organics and pesticides, including aldicarb, carbofuran, 1,2-dichloropropane, 1,2,3-trichloropropane, and also nitrates is being studied under various flow situations. Two parallel treatment systems (one consisting of GAC and ion exchange and the other consisting of reverse osmosis) are being operated at low flows similar to home usage. The costs for these two systems will be established along with unit operating efficiency so that a larger public water supply system can be designed and tested. The results from this study will be applicable to other areas, especially in farming communities where multiple contamination of groundwater is identified.

The principal objectives of the San Joaquin Valley study are to (1) develop cost-effective design criteria for the removal of dibromo-chloropropane and other pesticides from water supplies by GAC, and (2) to improve and strengthen existing administrative guidelines and jurisdictional responsibilities pertaining to both community water systems and private wells containing the compounds of interest. Pilot studies are being conducted with GAC, and results will be compared with those from existing operating systems. Minicolumn studies as described below are taking place to predict breakthroughs. The focus in this study will be on point-of-entry GAC units that are now in use by small systems, individual homeowners, and farmers in the area.

The third site, in Wausau, Wisconsin, has multiple contaminants in its groundwater source from a nearby superfund site. GAC is the primary technology being studied, but air-stripping is being examined as a companion technology. The Wausau project is unique in that modeling techniques are being used to predict full-scale design criteria for a GAC plant. These predictions are being evaluated against the actual cost and performance associated with the building of GAC contactors on-site. The results from this study are expected to provide a useful methodology for extension to other GAC applications. The modeling techniques will be particularly useful in studying the long list of SOCs and VOCs that may be proposed for regulation under the current and future provisions of the Safe Drinking Water Act. In addition to modeling, minicolumn technology is also being studied. This methodology potentially allows investigators to acquire water from a given site and to study the performance of a small, high-pressure column in the presence of a natural water background. Results from these minicolumn experiments are expected to reproduce the breakthrough curve normally seen in pilot- and full-scale facilities.

Packed Tower. Aeration technology has proved to be especially effective for the removal of VOCs. However, the research at Wausau, Wisconsin (which also incorporates air stripping), indicates that aeration may in some cases be effective for removing compounds that have somewhat lower Henry's Law Constants than would normally be expected to be removed by this process (Hand, 1986). In this project, as with the carbon facilities, modeling is being used to predict the performance of packed-tower aeration. Air-stripping off-gas studies at Wausau are being conducted in cooperation with the American Water Works Association Research Foundation.

The Drinking Water Research Division also has an air-stripping research project currently underway at Baldwin Park, California, which is examining the removal of VOCs from a groundwater supply.

ʌɟ in Wausau, the Baldwin Park project is also examining the problems of off-gas control technology.

Two other field-scale activities dealing with air-stripping have been concluded at Brewster and Glen Cove, New York (Wallman and Cummins, 1986; Ruggiero and Feige, 1984). In the Brewster project, modeling techniques and pilot-scale facilities were used to determine the scale-up relationships to be used for full-scale, air-stripping facilities. The purpose of the project was to develop a technique that could be used by consulting engineers to adequately predict the cost and performance of full-scale facilities using pilot aeration columns. At Glen Cove, various aeration designs were evaluated for their ability to remove VOCs from drinking water. Packed column, diffused aeration, and induced-air all proved technically feasible to reduce contaminant levels to meet existing state guidelines.

Ion Exchange. Ion exchange technology has been field-tested for nitrate removal under a DWRD-supported cooperative agreement (Lauch and Guter, 1986) and has been pilot-tested for uranium (Reid et al., 1985) removal at sites in the western part of the United States.

The nitrate removal plant has been operating automatically for about three years. The 1 mgd (3.8 Ml/d) demonstration plant is located in McFarland, California, and consists of three anion exchange vessels that are designed to reduce nitrate levels to below 10 mg NO_3-N/l, the EPA MCL and also the California requirement. Currently, about 500 gpm (32 l/sec) of water is being treated and about 200 gpm (13 l/sec) is bypassed and later blended with the treated water, resulting in a total product flow of 700 gpm (44 l/sec). The blended water adequately meets the nitrate MCL and EPA's Secondary Regulation (Fed. Reg., 1979) for chloride and sulfate levels.

Bench-scale studies were initially conducted at EPA-Cincinnati for uranium removal and later pilot-tested (Reid et al., 1985). The laboratory work showed that when drinking water containing 300 ug/l uranium was passed through anion exchange resin, more than 9,000 bed volumes were treated before breakthrough was observed. The DWRD then evaluated the performance of twelve 1/4-cubic-foot (0.007 m^3) anion-exchange systems installed in New Mexico, Colorado, and Arizona at sites where uranium levels in the raw water exceeded 20 ug/l. Results confirmed the findings shown in the laboratory. Because of the high loading capacity of the anion resins for uranium, these units are well suited for point-of-use applications where on-site regeneration is not feasible. For centralized treatment, the resin may be regenerated and recycled by backwashing it with sodium chloride solution.

Reverse Osmosis. Groundwater sources in Illinois, Iowa, Florida, Texas, Wisconsin, and some Rocky Mountain states contain radium in excess of the 5 pCi/l MCL. RO treatment of radium-laden groundwater was demonstrated in 1977 when DWRD and Sarasota

County, Florida, water supply staff undertook a cooperative effort to study the operation of eight RO systems (Sorg et al., 1980). The systems were located in small communities serving a population from 39 to 15,000, and the design capacities varied from 800 gpd (3 kl/d) to 1 mgd (3.8 Ml/d). Six different manufacturers of RO systems were represented, and both the hollow fiber and spiral wound cellulose acetate membranes were used. The study showed between 82 and 96 percent Ra-226 removal for all systems, resulting in treated water that contained below the EPA MCL of 5 pCi/l.

RO technology has also been pilot tested for nitrate removal as part of DWRD cooperative agreements at three locations (Lykins and Baier, 1985; Huxstep, 1981; Guter, 1982). As mentioned previously, high nitrate levels exist in the well waters of Suffolk County, Long Island, New York, along with several SOCs (Lykins and Baier, 1985). This combination of organic and inorganic contaminants is the reason that RO was selected as one of the treatment technologies to be studied there. Seven commercially available membranes were evaluated for their rejection capabilities. Nitrate removals ranged from 75 to 95 percent, and research is continuing with one of the polyamide membranes that proved relatively efficient for both nitrate and SOC rejection. At Charlotte Harbor, Florida (Huxstep, 1981), both high pressure (265-359 psig) and low pressure (163-187 psig) RO systems were studied for the removal of several spiked inorganic contaminants, including nitrate, from a natural groundwater. The investigation showed that the high pressure system was significantly more effective for removing all substances measured. The comparison for nitrate removals, for example, was 80 percent versus 6 to 24 percent. Before ion exchange was selected for full-scale evaluation at McFarland, California, a 20,000 gpd (77 m^3/d) RO system was examined for nitrate removal (Guter, 1982). Even though the system experienced frequent electrical and mechanical failures, nitrate rejection of about 65 percent was achieved.

Promising Technologies

Conventional Treatment. Conventional treatment is unlikely to be used for treating groundwater. However, field studies are being conducted at Tiffin, Ohio, where the river source contains periodically high concentrations of pesticides because of local agricultural applications. Powdered activated carbon, added to the water normally treated only by conventional treatment, appears to be quite effective for removing synthetic organic chemicals.

Ozone Oxidation. Ozone oxidation was being studied extensively in DWRD's in-house pilot plant facility (Fronk, date pending). Controlled pilot plant ozone treatment tests were conducted on 29 VOCs in distilled water and groundwater. Results showed that aromatic compounds, alkenes and certain pesticides are well removed by ozone treatment, but that alkanes are poorly removed. Also, removal efficiency improved for the alkenes and aromatic compounds with increasing ozone dosage and for some alkanes with increasing pH. For most compounds, the efficacy of ozone was not affected by the background water matrix. Information from the literature concerning the ozone treatment of pure materials in the gaseous or liquid phase generally predicted the effectiveness of ozone in treating aqueous solutions.

Reverse Osmosis. Reverse osmosis (RO) has shown some promise in removing both VOCs and SOCs from groundwater (Sorg and Love, 1984). Most of the efforts by DWRD to date have been on a pilot- or bench-scale basis, with some limited application of reverse osmosis for the removal of organics at the Suffolk County study mentioned previously. Preliminary indications are that certain RO membranes are very effective in removing a wide range of organic chemicals.

Ultraviolet Treatment. Ultraviolet light also shows some promise for removing organic contaminants, particularly when combined with ozone. The DWRD has recently awarded a cooperative agreement to the Los Angeles Water and Power Company that deals with the removal of VOCs from groundwater using these two technologies. If successful, these chemicals will be oxidized to carbon dioxide and water, and the need to deal with off-gas control problems may be eliminated.

Radium Removal Technologies. The Drinking Water Research Division has sponsored research to study promising methods for radium removal (Myers et al., 1985; Snyder et al., 1986). The ion exchange process, with both weak acid resin and strong acid resin, was investigated. Both resins effectively removed radium from water to well below the 5 pCi/l MCL and, in most cases, to <0.5 pCi/l, representing greater than 96 percent removal. The weak acid resin in the hydrogen form also removed hardness, which was not the case for the strong acid resin in the calcium form. The maximum capacity of the weak-acid resin was about 2.3 times that of strong acid resin and much less spent regenerate per unit volume of water treated was produced from the weak-acid column than from the strong-acid column. Another part of this project was to determine the feasibility of Dow Chemical Company's Radium-Selective Complexer (RSC)[1] for removing radium from brines compared with typical groundwaters. The RSC is a synthetic resin that has a high affinity for radium. The capacity of RSC was observed to have been about 200 times greater in 450 mg/l total dissolved solids (TDS) water than in 40,000 mg/l TDS brine (51,000 pCi/dry g vs 300 pCi/dry g). However, the effect of other parameters, including calcium, sodium, and other ions, in addition to empty bed contact time needs to be determined. The application of the RSC for treatment of radium from brines is

---

[1]Mention of trade names or commercial products does not constitute endorsement or recommendation for use.

presently being studied at a small community in Colorado under a research cooperative agreement with DWRD.

Radon Removal Technologies. In response to the EPA's recent emphasis on radon in the environment, DWRD has funded a project to evaluate several treatment techniques for the removal of radon from community water supplies. Three treatment methods, packed tower aeration, diffused bubble aeration and granular activated carbon will be evaluated for the removal of radon from two community water supplies in New Hampshire. The study will compare the methods for effectiveness, costs, operation, maintenance and other related factors. To accomplish the objectives, the three treatment systems will be constructed and operated at each of the two sites. The two sites selected are trailer parks, one whose water supply serves 40 homes with an average daily flow of 4.6 gpm and an average radon concentration of 155,000 pCi/l. The other location has 56 homes, the average daily water usage is 6.3 gpm, and the average radon concentration is 40,000 pCi/l. After construction of the treatment systems, they will be operated between 6 and 12 months. At the end of this project, a report will be developed as a guideline to help state agencies and small communities select an appropriate technology for radon gas treatment in public water supplies.

SECONDARY PROBLEMS

The Drinking Water Research Division is very concerned that the technologies studied lead to no secondary problems. For example, while conducting GAC research on surface water supplies, it was discovered that dioxins were formed in the reactivation process. An extensive evaluation led to the installation of an afterburner, which was found to eliminate these dioxin byproducts when operating at a temperature of 2400°F (Miller et al., 1986).

At Baldwin Park, California, the use of high-stack dispersion was investigated as a means of minimizing the impact of VOC removal from groundwater. At Baldwin Park and at Wausau, Wisconsin, gas-phase adsorption is being investigated for removal of both VOCs and SOCs from air stripping waste gases.

The project with the Los Angeles Water and Power Company is intended to investigate the water-phase oxidation of VOCs and SOCs to CO_2 and H_2O using ozone in combination with ultraviolet light. Gas-phase oxidation of VOCs and SOCs by ozone and ultraviolet light also appears promising.

Future work will concentrate on residuals control from various unit processes such as reverse osmosis and ion exchange.

Table 2. Cost summary for selected volatile organic chemicals (Clark et al., 1984).

Volatile Organic Chemical	Capacity mgd[1]	ug/l	Percent Removal	Cost, $/1000 gal[2]		
				Tower Aeration	Diffused-air Aeration	Carbon Adsorption
Trichloro-ethylene	0.5	100	90	0.273	0.546	0.868
		10	99	0.287	0.793	0.918
		1	99.9	0.296	1.032	0.010
		0.1	99.99	0.303	1.270	0.124
	1	100	90	0.182	0.383	1.637
		10	99	0.191	0.611	1.679
		1	99.9	0.196	0.850	0.765
		0.1	99.99	0.202	1.088	0.867
	10	100	90	0.083	0.207	0.356
		10	99	0.088	0.403	0.390
		1	99.9	0.093	0.587	0.458
		0.1	99.99	0.099	0.755	0.543
Tetrachloro-ethylene	0.5	100	90	0.279	0.637	0.610
		10	99	0.293	0.935	0.660
		1	99.9	0.302	1.228	0.705
		0.1	99.99	0.308	1.486	0.805
	1	100	90	0.186	0.460	0.453
		10	99	0.194	1.752	0.502
		1	99.9	0.201	1.046	0.548
		0.0	99.99	0.206	1.296	0.651
	10	100	90	0.085	0.277	0.197
		10	99	0.091	0.514	0.224
		1	99.9	0.098	0.726	0.251
		0.1	99.99	0.103	0.905	0.313
1,1,1-Tri-chloro-ethane	0.5	100	90	0.270	0.502	1.445
		10	99	0.289	0.825	1.651
		1	99.9	0.307	1.421	1.945
		0.1	99.99	0.332	2.572	2.605
	1	100	90	0.180	0.348	1.396
		10	99	0.192	0.644	1.500
		1	99.9	0.205	1.234	1.801
		0.1	99.99	0.230	2.313	2.402
	10	100	90	0.082	0.176	0.802
		10	99	0.089	0.430	0.973
		1	99.9	0.102	0.860	1.229
		0.1	99.99	0.122	1.821	1.818
Carbon tetra-chloride	0.5	100	90	0.264	0.428	0.942
		10	99	0.287	0.531	1.021
		1	99.9	0.272	0.600	1.132
		0.1	99.99	0.280	0.648	1.340
	1	100	90	0.176	0.292	0.703
		10	99	0.181	0.371	0.775
		1	99.9	0.184	0.427	0.940
		0.1	99.99	0.186	0.470	1.063

Table 2. (continued)

Volatile Organic Chemical	Capacity mgd[1]	ug/l	Percent Removal	Cost, $/1000 gal[2] Tower Aeration	Diffused-air Aeration	Carbon Adsorption
(Carbon	10	100	90	0.081	0.133	0.408
tetra-		10	99	0.083	0.196	0.467
chloride)		1	99.9	0.084	0.247	0.550
		0.1	99.99	0.085	0.286	0.719
Cis-1,2-	0.5	100	90	0.284	0.727	2.513
Dichloro-		10	99	0.296	1.010	2.791
ethylene		1	99.9	0.304	1.281	3.153
		0.1	99.99	0.310	1.572	3.511
	1	100	90	0.189	0.547	2.156
		10	99	0.196	0.828	2.417
		1	99.9	0.202	1.098	2.760
		0.1	99.99	0.208	1.379	3.099
	10	100	90	0.087	0.350	1.735
		10	99	0.093	0.571	1.989
		1	99.9	0.099	0.763	2.327
		0.1	99.99	0.104	0.966	2.660
1,2-Dichloro-	0.5	100	90	0.276	0.587	1.286
ethane		10	99	0.285	0.749	1.465
		1	99.9	0.292	0.901	1.748
		0.1	99.99	0.297	1.054	2.322
	1	100	90	0.184	0.415	1.015
		10	99	0.190	0.568	1.177
		1	99.9	0.194	0.720	1.437
		0.1	99.99	0.197	0.871	2.980
	10	100	90	0.084	0.237	0.675
		10	99	0.087	0.368	0.820
		1	99.9	0.090	0.489	1.057
		0.1	99.99	0.094	0.603	1.566
1,1-Dichloro-	0.5	100	90	0.262	0.406	0.880
ethylene		10	99	0.265	0.448	0.963
		1	99.9	0.270	0.500	1.066
		0.1	99.99	0.272	0.531	1.243
	1	100	90	0.174	0.274	0.647
		10	99	0.177	0.307	0.721
		1	99.9	0.180	0.348	0.814
		0.1	99.99	0.181	0.371	0.977
	10	100	90	0.080	0.121	0.364
		10	99	0.081	0.144	0.423
		1	99.9	0.082	0.176	0.499
		0.1	99.99	0.083	0.196	0.640

[1]To convert from mgd to m^3/day, multiply by 3,785.
[2]To convert from $/1,000 gal to $/m^3$, multiply by 0.26412.

Table 3. Performance summary for organic technologies examined.

		REMOVAL EFFICIENCY*				
Regulatory Phase	Organic Compounds	Granular Activated Carbon Adsorption (Filtrasorb 400ᵃ)	Packed Tower Aera- tion	Reverse Osmosis Thin Film Composite	Ozone Oxida- tion (2-6 mg/l)	Conven- tional Treat- ment

VOLATILE ORGANIC CONTAMINANTS

Alkanes

I	Carbon Tetrachloride	++	++	++	0	0
I	1,2-Dichloroethane	++	++	+	0	0
I	1,1,1-Trichloroethane	++	++	++	0	0
II	1,2-Dichloropropane	++	++	++	0	0
II	Ethylene Dibromide	++	++	++	0	0
II	Dibromochloropropane	++	+	NA	0	0

Alkenes

I	Vinyl Chloride	++	++	NA	++	0
II	Styrene	NA	NA	NA	++	0
I	1,1-Dichloroethylene	++	++	NA	++	0
II	cis-1,2- Dichloroethylene	++	++	0	++	0
II	trans-1,2- Dichloroethylene	++	++	NA	++	0
I	Trichloroethylene	++	++	++	+	0

Aromatics

I	Benzene	++	++	0	++	0
II	Toluene	++	++	NA	++	0
II	Xylene	++	++	NA	++	0
II	Ethylbenzene	++	++	0	++	0
II	Chlorobenzene	++	++	++	+	0
II	o-Dichlorobenzene	++	++	+	+	0
I	p-Dichlorobenzene	++	++	NA	+	0

Table 3. (continued)

		REMOVAL EFFICIENCY*				
Regulatory Phase	Organic Compounds	Granular Activated Carbon Adsorption (Filtrasorb 400[a])	Packed Tower Aeration	Reverse Osmosis Thin Film Composite	Ozone Oxidation (2-6 mg/l)	Conventional Treatment

PESTICIDES

II	Pentachlorophenol	++	0	NA	++	NA
II	2,4-D	++	0	NA	+	0
II	Alachlor	++	++	++	++	0
II	Aldicarb	NA	0	NA	NA	NA
II	Carbofuran	++	0	++	++	0
II	Lindane	++	0	NA	0	0
II	Toxaphene	++	++	NA	NA	0
II	Heptachlor	++	++	NA	++	NA
II	Chlordane	++	0	NA	NA	NA
II	2,4,5-TP	++	NA	NA	+	NA
II	Methoxychlor	++	NA	NA	NA	NA

OTHER

II	Acrylamide	NA	0	NA	NA	NA
II	Epichlorohydrin	NA	0	NA	0	NA
II	PCBs	++	++	NA	NA	NA

*++ = Excellent   70% - 100%
 + = Average Removal 30% - 69%
 0 = Poor   0% - 29%
NA = Data not available or compound has not been tested by EPA
     Drinking Water Research Division
 a = Excellent removal category for carbon indicates compound has
     been demonstrated to be adsorbable onto granulated activated
     carbon (GAC), in full- or pilot-scale applications, or in the
     laboratory with characteristics suggesting GAC can be a cost-
     effective technology.

COST AND PERFORMANCE

All of the field studies conducted by the DWRD include cost and performance as part of the data gathered for later evaluation and extrapolation to other conditions. In addition, as bench- and pilot-scale studies are conducted, cost estimates based on the development of cost equations and cost curves from previous studies are used to predict the most cost-effective technologies on which to concentrate scarce resources. Table 2 compares the costs and performance calculations for carbon adsorption, tower aeration, and diffused-air aeration for a selected set of VOCs (Clark et al., 1984). Estimated operation, maintenance, and capital costs for reverse osmosis and ion exchange technologies are found in two EPA reports (Gumerman et al., 1979; Gumerman et al., 1985).

Some of the point-of-use technologies discussed in this paper might best be managed by a centralized or circuit rider approach. In this case, the central authority would assume responsibility for monitoring and maintaining the individual units installed on the homeowner's premises.

SUMMARY AND CONCLUSIONS

The Drinking Water Research Division is responsible for evaluating the various types of technologies that might be used to meet the MCLs promulgated under the Safe Drinking Water Act and its Amendments. Because the source water for many utilities in the United States is groundwater, DWRD is especially concerned about conducting bench-, pilot- and field-scale studies on technologies that effectively treat groundwater. Technologies being examined are carbon adsorption, aeration, ion exchange, reverse osmosis, ozone oxidation, and ultraviolet light, some at the field scale and others at the bench and pilot scales. Table 3 summarizes the various technologies examined for organics removal; and based on data gathered to date, it attempts to characterize their relative performances for both Phase I and Phase II organics. Table 4 is a similar summary for removal of nitrate and radionuclides and shows that ion exchange and reverse osmosis each result in excellent contaminant removals.

Although Table 3 only provides a general guideline for removal of compounds, several interesting trends are noted. Carbon adsorption appears to provide removal for a wide range of organics whereas conventional treatment is revealed as a poor treatment for those compounds listed in the table. Packed tower aeration manifests itself as an excellent technology for volatile organic compounds and may have application for a limited number of pesticides. Ozone oxidation appears to be a good treatment technology for certain classes of organics such as simple alkenes and aromatics as well as certain similar, but more complex organic structures. Although only a few organics have been subjected to long-term testing via reverse

Table 4. Performance summary for inorganic technologies examined.

Regulatory Phase	Inorganic Compound	Reverse Osmosis	Removal Efficiency		
			Ion Exchange	Aeration	Carbon Adsorption
II	Nitrate	++	++		
III	Radium	++	++		
III	Uranium	++	++		
III	Radon			*	*

++ = Excellent 70% - 100%
 * = Research being conducted by EPA's Drinking Water Research
        Division

osmosis, promising removals for several low molecular weight organics can be seen.

Tables 3 and 4 were generated using a variety of sources, including EPA-DWRD pilot- and field-scale studies, Henry's Law Constants for prediction of removal of some of the pesticides as well as the use of oxidative trends for predicting the removal of complex pesticides. These tables, therefore, carry the caveat that the cited removals should not be used for design purposes, but that each technology must be tested on compounds, under field conditions, before the EPA-DWRD will advocate a technology's use. Given the widespread nature of groundwater contamination, these technological evaluations should be highly useful to EPA's regulatory process and to consultants, state officials, and individual communities as well.

ACKNOWLEDGMENT

The authors acknowledge the assistance of Thomas J. Sorg, Richard P. Lauch, and the late Steven Hathaway of DWRD for the preparation of the data contained in this paper and also to Patricia Pierson for typing of the manuscript.

LITERATURE CITED

Baker, D.B.   1983.   Herbicide contamination in municipal water supplies of Ohio.   Draft Final Report, Water Quality Laboratory, Heidelberg College, Tiffin, OH 44883.

Clark, R.M., R.G. Eilers and J.A. Goodrich.   1984.   VOCs in drinking water: Cost of removal. J. Environ. Eng. Div. ASCE, 110:1146-1162.

Cohen, S. 1985. Wells in pesticide - groundwater survey. Pesticide and Toxic Chemical News, p. 4.

Federal Register, Part III. 1985. National Primary Drinking Water Regulations: Volatile Synthetic Organic Chemicals: Final Rule and Proposed Rule, 40:141.

Federal Register, Part IV. 1985. National Primary Drinking Water Regulations: Synthetic Organic Chemicals, Inorganic Chemicals and Microorganisms, 40:141.

Federal Register. 1979. National Drinking Water Regulations, 40CFR Part 143, 44:140, Thursday.

Fronk, C.A. (journal article pending) Removal of volatile organic chemicals in drinking water by ozone treatment.

Gumerman, R.C., R.L. Culp and S.P. Hansen. 1979. Estimating water treatment costs - volume 2. EPA-600/2-79-162b, Cincinnati, OH.

Gumerman, R.C., B.E. Burris and S.P. Hansen. 1985. Estimation of small system water treatment costs. EPA-600/2-84-184, Cincinnati, OH.

Guter, G.A. 1982. Removal of nitrate from contaminated water supplies for public use: Final report. (Project summary), EPA-600/S2-82-042, Cincinnati, OH.

Hand, D.W., J.C. Crittenden, J.L. Gehin and B.W. Lykins, Jr. 1986. Design and evaluation of an air-stripping tower for removing VOCs from groundwater. J. Amer. Water Works Assoc. 78(9):87-97.

Huxstep, M.R. 1981. Inorganic contaminant removal from drinking water by reverse osmosis. (Project summary) EPA-600/S2-81-115, WERL, Cincinnati, OH.

Lauch, R.P. and G.A. Guter. 1986. Ion exchange for the removal of nitrate from well water. J. Amer. Water Works Assoc. 78(5):83-88.

Litwin, Y.J., N.H. Hantzsche and N.A. George. 1983. Groundwater contamination by pesticides - A California Assessment. Submitted to the State Water Resources Control Board, Sacramento, CA, by Ramlit Associates, Inc., Berkeley, CA.

Lykins, B.W., Jr. and J.A. Baier. 1985. Removal of agricultural contaminants from groundwater. In: Proceedings, American Water Works Association Annual Conference, June 23-27, pp. 1151-1164.

Lykins, B.W., Jr., E.E. Geldreich, J.Q. Adams, J.C. Ireland and R.M. Clark. 1984. Granular activated carbon for removing nontrihalomethane organics from drinking water. (Project summary) EPA-600/S2-84-165, WERL, Cincinnati, OH.

Miller, S.E., F.L. DeRoose, J.E. Howes, J.E. Tabor, J.A. Hatchel, C.V. Sueper, D.F. Kohler and K.B. Degner. 1986. Determining the effectiveness of an afterburner to reduce dioxins and furans. EPA-600/2-86-039, WERL, Cincinnati, OH.

Myers, A.G., V.L. Snoeyink and D.W. Snyder. 1985. Removing barium and radium through calcium cation exchange. J. Amer. Water Works Assoc. 77(5):60-66.

Reid G.W., P. Lassovszky and S. Hathaway. 1985. Treatment, waste management and cost for removal of radioactivity from drinking water. Health Physics 48(5):671-694.

Ruggiero, D.D. and W.A. Feige. 1984. Removal of organic contaminants from the drinking water supply at Glen Cove, NY. (Project summary) EPA-600/S2-84-029, Cincinnati, OH.

Snyder, D.W., V.L. Snoeyink and J.L. Pfeffer. 1986. Weak-acid ion exchange for removing barium, radium and hardness. J. Amer. Water Works Assoc. 78(9):98-104.

Sorg, T.J., R.W. Forbes and D.S. Chambers. 1980. Removal of Ra-226 from Sarasota County, Florida, drinking water by reverse osmosis. J. Amer. Water Works Assoc. 72(4):230-237.

Sorg, T.J. and O.T. Love, Jr. 1984. Reverse Osmosis treatment to control inorganic and volatile organic contaminants. In: Proceedings, AWWA Seminar on Experiences with Groundwater Contamination, held in Dallas, TX, pp. 73-92.

Wallman, H. and M.D. Cummins. 1986. Design scale-up suitability for air stripping columns. (Project summary) EPA-600/S-2-86/009, Cincinnati, OH.

Westrick, J.J., W. Mello and R.F. Thomas. 1984. The groundwater supply survey. J. Amer. Water Works Assoc. 76(5):52-59.

CHAPTER 17

NATIONAL PESTICIDES IN WELL WATER SURVEY

Phyllis A. Reed
United States Environmental Protection Agency
Environmental Services Division
Chicago, Illinois 60604

INTRODUCTION

The Environmental Protection Agency has launched the National Pesticide Survey, a nationwide survey of pesticides in drinking water wells. This presentation explains the reasons for conducting the survey, how the survey will be designed and conducted, and the status of the survey planning effort at this time. The survey will be conducted jointly by the Office of Pesticide Programs (OPP) and the Office of Drinking Water (ODW). The full survey will take two years to complete, beginning in the fall of 1987 and ending in late 1989. The pilot survey will begin in late 1986.

Why is a Survey Needed?

Pesticides present in drinking water may pose dangers to human health if ingested. Since 1975, urban water systems have been required to monitor for six pesticides: endrin, lindane, methoxychlor, toxaphene, 2,4-D, and 2,4,5-T. Recent evidence, however, indicates a larger problem of pesticides in groundwater. At least 17 pesticides have been found in groundwater in 23 states as a result of agricultural practices (Cohen et al., 1986). In the last few years, studies of pesticides in groundwater have been undertaken by the states of California, Florida, Maryland, Minnesota, New York, Washington, and Wisconsin, among others. However, most of these studies have been limited to a small number of pesticides and specific geographic areas; no comprehensive nationwide study has been conducted.

For example, in California, the San Joaquin Valley and River-side County have found evidence of contamination of groundwater with ethylene dibromide (EDB), dibromochloropropane (DBCP), simazine and carbofuran. Closer to home, the central sands area of Wisconsin shows groundwater contamination from aldicarb used on potatoes.

The National Pesticide Survey is a major component of the Agency's overall effort to understand and characterize the problem of agricultural chemicals in groundwater. The survey will provide a nationwide assessment of pesticide contamination in drinking water wells and an understanding of how pesticide use and hydrogeology relate to contamination.

With adequate survey information on the concentrations of different pesticides in wells around the country, EPA can better design its regulatory programs to target pesticides of concern and to develop further regulatory initiatives. The Federal Insecticide, Fungicide and Rodenticide Act (FIFRA) gives the Agency authority to regulate the marketing and use of pesticides. Pesticides that are shown to pose potential hazards by their ability to leach into groundwater could be subject to a range of further regulatory actions, including changes in label directions, use restrictions, or suspension or cancellation of a pesticide's registration. The Agency will also use information from the survey to implement requirements of the Safe Drinking Water Act (SDWA). New maximum contaminant levels and monitoring requirements may be proposed for pesticides shown to pose a hazard in public drinking water.

GOALS OF THE SURVEY

EPA is designing the National Pesticide Survey to meet two major objectives: (1) to obtain sufficient information to characterize pesticide contamination in the drinking water wells of the nation; and (2) to determine how pesticide concentrations in drinking water wells correlate with patterns of pesticide usage and with groundwater vulnerability.[1]

The focus of the survey is on the quality of drinking water in wells rather than in groundwater, surface water, or drinking water at the tap. The survey is not designed to estimate the risk to human health resulting from pesticides in drinking water. Estimating pesticide exposure from contaminated drinking water would require a different survey and research design. The study will, however, provide substantial data to develop inferences about populations

---

[1]Groundwater vulnerability is a composite description of geologic and hydrogeologic characteristics that indicates groundwater pollution potential. Factors that affect groundwater occurrence and availability will also influence the pollution potential of an aquifer. These characteristics can be combined to estimate the potential for pollution of groundwater resources.

potentially at risk from exposure to pesticides in drinking water, and it will yield a wealth of information on the pesticides present in private and community drinking water wells.

The overall survey outputs and objectives are as follows:

(1)   Survey will establish the frequency and concentration of pesticide contamination on a national basis

(2)   Survey will examine relationships between pesticide contamination and

- hydrogeologic vulnerability
- pesticide use

(3)   Survey results will add data to examine list of *groundwater leaching* analytes

(4)   Current information is not adequate to meet objectives:

- information is state, county or area specific
- past and current analyses only cover selected pesticides
- available data are insufficient for national program

(5)   Preliminary review of selected state surveys of pesticides in groundwater shows:

- many surveys targeted vulnerable groundwater areas
- wide range of positive detections of pesticides

SPECIFIC OFFICE OF PESTICIDE PROGRAMS (OPP) OBJECTIVES

(1)   Determine the scope of the national problem of pesticides in drinking water wells to enable planning and priority setting for pesticide regulations across types of pesticides and individual compounds.

(2)   Determine the frequency and range in concentrations, and analyze the relationships between pesticide contamination, groundwater vulnerability and pesticide usage to help focus registrant monitoring requirements (both "where" to monitor and "what" to monitor for).

(3)   Ascertain the extent of contamination at levels of health concern with sufficient frequency, in association with vulnerability and usage conditions, to move directly to labelling and registration decisions.

(4)   Share survey results with states to help focus state monitoring and regulatory efforts, and state certification and training programs.

(5)   Determine statistically reliable nationwide distribution estimates of nitrates in groundwater for use in implementing the agricultural chemicals in groundwater strategy.

SPECIFIC OFFICE OF DRINKING WATER (ODW) OBJECTIVES

(1)   Identify pesticide compounds that occur at levels of health concern, and with sufficient frequency, to be considered candidates for regulation through the MCL process.

(2)   Identify the range of pesticide concentrations found in CWSs, so that the feasibility of drinking water treatment can be assessed.

(3)   Identify and analyze the relationship between pesticide contamination of drinking water wells, groundwater vulnerability, and pesticide usage to help direct compliance monitoring requirements for CWSs.

(4)   Facilitate regulatory impact analyses by building an adequate data base on the distribution of pesticide levels in drinking water.

(5)   Develop working relationships with the states in order to facilitate implementation of the drinking water program and to build state programs relating to pesticides.

HOW THE SURVEY WILL BE CONDUCTED

EPA is nearing completion of the planning stages of the project. Some additional work remains to be done in refining the research design and in coordinating activities with the many participants involved. Key components of the survey are described here.

The survey will be implemented in two steps: a pilot survey of a limited number of drinking water wells in late 1986 followed by the full survey about eight to ten months later. EPA expects to conduct the full survey over a period of two years, from the fall of 1987 through 1989.

The survey design requires the development of four major components: (1) a statistical design to select a set of wells that is representative of drinking water wells in the nation; (2) analytical methods to measure the types and amounts of possible pesticide contamination; (3) health advisories that establish the levels at which pesticide concentrations may pose a health problem; and (4) a questionnaire to collect key information to analyze additional factors affecting pesticide contamination. These four components are all at various stages of development and will be tested in the pilot survey.

Statistical Design

To test public drinking water wells for pesticide contamination, EPA will select about 500 community water systems, using the Federal Reporting Data System to identify systems with wells. The statistical design of the domestic wells side of the survey is more complex, primarily because there is no comprehensive tabulation of private (rural domestic) drinking water wells in the United States. The process of identifying and selecting representative domestic wells for sampling is organized into three stages, as follows:

Stage 1:   EPA classifies all U.S. counties using specified measures of groundwater vulnerability (obtained from the DRASTIC model) and measures of pesticide usage. From this classification scheme, EPA selects 90 representative counties.

Stage 2:   At this stage, the counties are separated further into Census enumeration districts; these districts in turn are satisfied by crop patterns and groundwater vulnerability to ensure a representative selection. Within these districts, using Census data and other sources, EPA identifies and statistically selects household clusters that use private wells.

Stage 3:   In the final step of the selection process, EPA will identify private wells and characterize their use and structure on the basis of interviews with house-holders. Over 700 private wells will be selected for pesticide sampling.

Stage 1 of the survey design and selection of counties has been completed. Work is continuing on the development of the final selection requirements for Stages 2 and 3. This survey design is summarized in Figure 1.

Analytic Methods

The water samples to be taken from the wells will be analyzed for the presence of 60 priority and other nonpriority pesticide analytes. EPA is selecting the pesticides to be analyzed on the basis of expected leaching potential, occurrence, production volume, and other considerations. Final selection of the analytes will be made when the Agency completes the analytic methods and the Quality Assurance Plan.

EPA is developing five multi-residue methods to detect and quantify the occurrence of pesticides. These methods will be completed in early 1987. Each method will detect several analytes. The methods should enable EPA, states, and industry to efficiently analyze for pesticides expected to leach into groundwater.

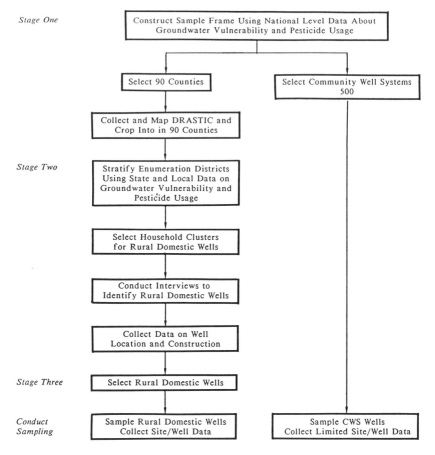

Figure 1. Sampling Design.

In addition, a method is being developed for ethylene thiourea (ETU). Methods already developed for nitrates, volatile organic chemicals (VOCs), and EDB/DBCP will also be used. Two EPA regional laboratories will act as Quality Assurance and Referee Laboratories. EPA is performing laboratory validation of these methods. This effort will validate detection limits, determine the precision and accuracy of different methods, and analyze sample preservation requirements. Work on the Quality Assurance Plan for the survey is nearing completion. EPA staff are preparing specifications for sampling procedures, sample custody, data analysis, quality control, and performance audits.

Health Advisories

Development of health advisories for the analytes is under way. Pesticide Survey Guidance Levels for 60 priority pesticides are being developed using information collected on physiochemical properties, uses, chemical fate, health effects, treatment, and existing criteria and guidelines. External review drafts of the health advisories will be available in January, 1987.

Questionnaire

The questionnaire is being designed in several stages. The Agency will first identify specific information needs, and then determine potential sources of this information and evaluate alternative sources and measurement strategies. The major categories of information to be collected are: locations of wells, well use and construction characteristics, pesticide use in relation to the well, hydrogeologic characteristics, demographic characteristics, and economic and crop characteristics.

Pilot Study

The pilot study is intended to field test the major components of the project, some of which have been developed specifically for this project. The pilot will also provide practical experience in conducting the full survey. Pilot studies are now considered essential in any large-scale or benchmark effort.

The pilot study will sample between 50 and 100 drinking water wells in six counties in three states. The states selected for the pilot will also be included in the full survey. While it is not final at this time, it is likely that either Minnesota or Wisconsin will be selected as one of the pilot states. (Since the time of this presentation, Minnesota was selected to be one of the three pilot states, along with California and Mississippi.)

SUMMARY OF ACCOMPLISHMENTS TO DATE

At this time, the first stage, statistical design, has been completed. The DRASTIC groundwater vulnerability indices have been developed, and the pesticide usage data have been compiled for all counties in the United States. County selection has occurred. The lab method development and single lab validation is on schedule, and completion is expected by the end of September. Forty-eight of the fifty-three health advisories were scheduled for completion by the end of 1986. The remaining five health advisories will be completed in 1987. Both the health advisories and the lab methods will be extremely valuable outputs of this project, separate from the survey results themselves.

LITERATURE CITED

Cohen, S.Z., C. Eiden, M.N. Lorber. 1986. Monitoring groundwater for pesticides in the U.S. Forthcoming in Evaluating Pesticides in Groundwater, Washington, D.C. See also: Pesticides in Groundwater: Background Document, USEPA, Office of Groundwater Protection, May, 1986.

CHAPTER 18

RISK CONSIDERATIONS IN PESTICIDE
PUBLIC POLICY DECISION MAKING
IN WISCONSIN

Thomas J. Dawson
Wisconsin Public Intervenor
Wisconsin Department of Justice
Madison, Wisconsin 53707

INTRODUCTION

Over the last several years, the Office of the Wisconsin Public Intervenor has been involved in pesticide use and control policy making with state and local agencies. As a result, the Public Intervenor Office has received a growing number of requests for information, advice and aid from citizens' groups and local units of government regarding pesticide-related issues.

Every year, policy makers at all levels of government are called upon to make important decisions regarding pest management and pesticide use within their respective jurisdictions. Being debated in Wisconsin local and state government forums are many of the same issues that are being debated and addressed at other local, state and national levels of government. For the most part, these issues are not new; they are not unique to local situations, and essentially they are not different from similar issues still being argued, disputed and debated at the highest scientific and public policy making levels of government. Despite this debate, these issues are no less important and no less appropriate for deliberation by local decision makers and citizens.

LOCAL DECISION MAKERS ARE ENTITLED TO REACH
THEIR OWN JUDGMENTS ON PESTICIDE USE

Because of the nature of pesticide regulation and policy making, scientists and federal and state policy makers should not have a monopoly on the pesticide policy and decision making process. It is important to adhere to the principle that the Wisconsin public has the right to participate in pesticide policy making that ultimately affects their lives and environment. Consistent with this, local decision makers such as town, village, city, and county boards must remain free to respond to the concerns of their own constituencies, conduct their own investigations, and reach their own judgments concerning the risks and benefits of pesticide use in their communities in order to protect public health, safety and welfare.

For these reasons, the Public Intervenor has acquired some related information over the last few years which pertains to the decision making needs of local governments and citizens on pesticide-related matters. This attempts to identify for decision makers some of the important issues that must be explored in local "risk/benefit analyses" before deciding whether or not to apply pesticides and, if so, under what conditions.

NEED FOR DISCUSSION OF RISK CONSIDERATIONS
IN THE BENEFIT/RISK PESTICIDE PUBLIC POLICY
DECISION MAKING PROCESS

The benefits derived from pesticide use are well known and acknowledged. Human health as well as food and forest production have been significantly enhanced, largely as a result of pesticide development and use in this country and around the world. Certain serious diseases such as malaria have been kept in check as a result of pesticides. Health, agricultural and other economic and safety threatening pests are effectively controlled today. These benefits are widely known and espoused by pesticide manufacturers, sellers and users.

In the local decision making context, however, while the many and significant benefits are brought quickly to the attention of public and local decision makers by pesticide proponents, the risks are not explained with equal fervor. In fact, all too often proponents respond to the pesticide risk issue with the claim that, because they are "registered" for use by state and federal health and environmental protection agencies, pesticides can be presumed to be "safe."

This is not to imply, of course, any sinister motives on the part of pesticide proponents. Most have the best of intentions to control pest infestations cost-effectively and prevent their consequent threats to the quality of life. In fact, many proponents honestly believe that pesticides are perfectly "safe" as a result of federal Environmental Protection Agency (EPA) and state regulatory programs. But it would be naive to assume that they can be expected to be totally informed or candid about those factors which mitigate against decisions to embark immediately on pest control programs that are predominantly dependent on pesticides.

Therefore, the "risk" side of pesticide use issues and the factors that tend to provide counterweight to the benefits argument should be examined. This paper does not take, however, an "anti-pesticide" position. In fact, it is conceded that, in many cases, a decision to use pesticides may be not only reasonable but necessary to prevent consequences that would be worse if they were not used. At least in those cases and, hopefully, whenever decisions on pesticide use are made, the final decision will be the result of a reasoned judgment that has taken all relevant factors into account, not just those that support the case for pesticide use. Only after considering all sides of an issue can local decision makers be expected to make reasonable and balanced decisions affecting their communities.

THE MYTH OF PESTICIDE SAFETY

The Regulation of Pesticides is Based on the Concept of "Reasonable Risk," Not Safety

A common public misconception continues that because pesticides are "registered with EPA" and are the subject of extensive federal and state regulation, they are "completely safe" for use. This is a myth.

Hardly any human activity, much less pesticide use, is absolutely safe. Thus, the decisions at the federal level--to allow the registration, labelling, and marketing of pesticides; the decisions at the state level--to impose conditions on the use of pesticides; and the decisions at the local level--whether to use pesticides in the suppression of insects, weeds or other pests, is not, and should not, be based on the erroneous assumption that the use of pesticides is without risk. Even if existing pesticide regulations in the United States were carried out as intended,[1] the following points about the conceptual framework of pesticide regulation should be kept in mind.

EPA is Required to "Manage" Pesticide Risks, Not Assure Public Safety

Even though the claims persist that pesticides registered by EPA are "safe," at least when used as directed, the words "safe" or "safety" are not part of the lexicon of the pesticide regulatory process.

This is because the federal law itself does not require EPA to assure public safety (FIFRA, 1978). Rather, the law requires EPA to

---

[1]There are many reasons to be doubtful that current pesticide regulatory programs are actually working as intended. This point will be discussed more fully on page 266.

allow the use of pesticides that are judged as presenting "reasonable" risks of harm in light of their intended benefits. Under the law, EPA must register a pesticide for use after determining, among other things, that "it will perform its intended function without unreasonable adverse effects on the environment" (FIFRA, 1978, Sec. 3(c)(5)(c)).

In fact, EPA regulations regard as misbranded, having illegally false and misleading statements, pesticide labels that bear "[c]laims as to the safety of the pesticide or its ingredients, including statements such as 'safe,' 'nonpoisonous,' 'noninjurious,' 'harmless' or 'nontoxic to humans and pets' with or without such a qualifying phrase as 'when used as directed'" (CFR, 1984).

Still, "[t]he public in many cases wants assurances of safety," states Edwin Johnson, former Director of EPA's Office of Pesticide Programs. "We can't give that. Because pesticides are by their very nature designed to be biologically active and kill pests and weeds, we speak in terms of relative risks, rather than 'safety'" (Johnson, 1984). Thus, Herbert Harrison, Registration Division, EPA-Office of Pesticide Programs (EPA-OPP), "contends that it is mistake to cast discussion of pesticides in terms of how safe they are." "Debate," said Harrison, "should properly refer not to 'the safest' but to the 'least hazardous' pesticides" (Anon. 1981a).

This is a crucial distinction in the policy making process. What EPA strives for is not certifications of pesticides that are "safe," but rather certifications that will create "no *unreasonable* adverse effect if used according to label direction," Harrison explains.

Risk Management is a Policy Matter, Not Solely a Scientific One

Because EPA manages and does not seek to eliminate risks, pesticide regulation is not a matter that can or should be placed solely in the hands of the scientific community for resolution. As one National Academy of Sciences (NAS) report puts it: "The purpose of pesticide regulation in this country is to protect the human population, animals, useful vegetation, natural amenities or all sorts, and property from the *'unreasonable'* adverse effects of the use of chemical pesticides" (NAS, 1980, p. 65, emphasis added). Quoting from the same report:

> [O]ur current policy of regulating the use of pesticides [is] to permit it when the beneficial effects are deemed to outweigh the hazards, but not otherwise. This policy, in turn, has required the U.S. Environmental Protection Agency to assess the beneficial effects and the risks entailed in the use of pesticides in specific circumstances (NAS, 1980, p. ix).

Risk Assessment Is Not An Exact Science

When attempting to assess pesticide risks, scientists do not always agree on the significance of certain health and safety tests (GAO, 1980). Moreover, while EPA uses benefit/risk analyses in making its decisions regarding thousands of pesticide compounds, a NAS report reveals that the procedures for developing and applying these procedures "are still evolving" (NAS, 1980, p. ix). As John Moore, Assistant Administrator for EPA-Pesticides and Toxic Substances, puts it,

> As the review of these chemicals progresses, some chemicals that were once thought to be safe will be found to be unacceptable by modern standards. . . . We should not lose sight of the fact that even new pesticides, which have been carefully scrutinized by modern science and technology and found to be safe, may at some future date be found unacceptable in light of future scientific advances (Moore, 1984a, p. 3 and 22).

Determining What Risks Are "Reasonable"
Is a Subjective Judgment

While pesticide regulation involves crucial matters of scientific inquiry and analysis, the issue of what risks are "reasonable" for the public to incur is a public policy call and, therefore, the subject of legitimate participation by citizens and their government representatives.

> The acceptability of the risk depends, of course, on the amount of the benefit to be received for taking the risk. . . . For a pesticide with tremendous benefits to agriculture or vector control, acceptance of a higher risk is more reasonable.

> It's obviously one of the most difficult parts of regulating pesticides. While we are assisted by quantitative risk analyses and extensive benefits analysis, it all comes down to a subjective judgment as to when the benefits outweigh the risks (Johnson, 1984, p. 8).

Decisions at the federal, state and local levels are based on a complex interaction of both factual and judgmental considerations. These include recognition that the introduction of toxic chemicals, including pesticides, into the environment does create both known and unknown risks to human health and the environment.

With this recognition in mind, federal, state and local decision makers necessarily must decide about pesticide use on the basis of their own judgments of what constitute "reasonable" risks to human health and the environment. The "reasonableness" of these risks is based upon a decision making process that balances imperfectly assessed risks inherent in pesticide use and the perceived benefits.

In short, pesticide policy making, like all policy making at any level of government, is not an exact science.

There Are Good Reasons to be Skeptical
That Current Pesticide Regulatory Programs
Adequately Assure Public Safety

The preceding discussion is premised, of course, on the assumption that pesticide regulatory programs actually are carried out adequately by EPA. *Assuming that current pesticide regulatory programs are carried out according to plan*, there are still going to be unavoidable risks in pesticide use that must be weighted against their benefits in order to reach a conclusion that the risks are "reasonable."

With just its name, the federal Environmental Protection Agency (EPA) inspires confidence that an army of scientists are standing guard over citizens' health and the safety of the nation's air and water (Trimble, 1980, p. 1, 9).

Yet, there are serious reasons to question whether this assumption and public confidence are well-founded.

The History of Inadequate Pesticide
Regulation Prompted Reforms in 1972

From the passage of the Federal Insecticide, Fungicide and Rodenticide Act (FIFRA) in 1947 until 1972, pesticides were registered under data requirements that "focused primarily on acute health effects and efficacy" (Johnson, 1984, p. 4). Long-term chronic health effects and "environmental fate" data were not part of the registration process.

Unfortunately, the state-of-the-art in science was still too unsophisticated to predict chemicals' long-term impact on the environment or human health. The turning point came in the late 1960s when Rachel Carson's *Silent Spring* increased scientists' and regulators' realization that use of pesticide chemicals might have broad health and environmental effects (Johnson, 1984, p. 8).

Responding to increasing evidence of environmental havoc being caused by government registered pesticides, in 1972, Congress amended the federal pesticide law, requiring EPA to register new pesticides according to the new "reasonable adverse effect" standard and to "accomplish the reregistration of all pesticides in the most expeditious manner practicable" (FIFRA, 1978, Sec. (3)(g)).

This amendment created a dichotomy between "new" pesticides and "old" pesticides that only now is beginning to be resolved. While "new" pesticides are required to have rigorous laboratory testing. . .numerous "old" pesticides, already in use, have little testing that meets current standards. To rectify this inconsistency, Congress directed EPA to review all of the old pesticides and to apply modern testing requirements (Moore, 1984, p. 2).

Despite this broad mandate, EPA remains far from achieving the directive Congress gave it to regulate pesticides to prevent unreasonable adverse effects on the environment and the public.

1976: Inadequacies in EPA Regulation of Pesticides are Found by Congress

In December, 1976, four years after the reforms of 1972, the Subcommittee on Administrative Practice and Procedure of the Committee on the Judiciary of the United States Senate issued its staff report entitled, "The Environmental Protection Agency and the Regulation of Pesticides" (USS, 1976). In its introduction, the Senate Chairman of the Subcommittee wrote:

Pesticides have played and continue to play an important role in increasing agricultural production in America, and around the world. Pesticides have proved valuable in the past, and may be needed for some years to come until safer and more effective methods or pest control are developed. But pesticides are powerful poisons also. A number of pesticides have the potential for causing such tragic effects on man and the environment as cancer, birth defects, interference with biological reproduction, and genetic mutations.

The subcommittee staff's report on pesticide regulation led to the conclusion that the Environmental Protection Agency had not assured the safe use of pesticides as mandated by Congress. As a consequence, EPA had failed the consumer and the farmer, as well as the pesticide industry.

According to the subcommittee, this regulatory agency charged with safeguarding public health and the environment was very slow to recognize and react to many warnings between 1971 and 1976. The EPA was warned and certainly should have known that testing data, submitted by industry as long as 25 years previously should not have been accepted at face value in the reregistration of thousands of pesticide products being used on farms and in homes.

The report concluded "that apparently EPA made a conscious policy decision sometime in 1973 or 1974 not to evaluate the safety testing data submitted by pesticide manufacturers. The record behind this decision is not entirely clear. What is clear, however, is that EPA had no sound basis upon which to assume that data 15, 20 or 25 years old was generally good and reliable" (USS, 1976). The opposite is more likely in light of the dramatic improvements in safety testing methodology and interpretation of testing data over the last few decades.

1980: GAO Reports that EPA Delays Seriously Hamper Pesticide Regulation

The United States General Accounting Office (GAO) tracked EPA's pesticide-related activities from 1976, when the Senate Judiciary Committee issued its report, until 1980. The GAO's conclusions included the following:

1. According to EPA officials, important health and safety studies for many of the registration standards pesticides are missing, precluding EPA from developing final registration standards and from unconditionally reregistering all pesticides (GAO, 1980, p. 8).

2. According to EPA officials, key (safety) tests required under current EPA regulations have not been performed for many of the 514 registration standards pesticides. . . EPA needs the results of these tests to make even preliminary decisions concerning a pesticide's safety and whether it should be reregistered (GAO, 1980, p. 15).

3. Also EPA is unprepared under the registration standards program to reassess the safety of the estimated 6,000 federally approved tolerances. . . Until EPA completes a thorough review of its tolerance setting procedures. . .it cannot assure the public that federally approved tolerance levels are reasonably safe (GAO, 1980, p. 23).

Use of Industry Data Impairs EPA's Regulatory Credibility

For years, critics have questioned the advisability of EPA basing its registration of pesticides almost exclusively on testing done by manufacturers or consultants responsible to manufacturers, who obviously want their pesticide products approved.

Officials in EPA's pesticide review division admit that until recently they accepted without question all the tests manufacturers submitted to support the registration of toxic substances. Only about three years ago when one of the largest laboratories in the nation was found to be conducting substandard tests, did the EPA initiate a lab audit program to monitor pesticide testing procedures (Trimble, 1980, p. 1).

The IBT Laboratories Episode

In June, 1981, top level employees of a chemical company consultant, Industrial Bio-Test Laboratories, Inc. (IBT), were indicted for falsifying reports to EPA with fraudulent mice carcinogenicity studies on the pesticides Sencor and Nemacur (Anon, 1981b, pp. 24-26).

For about 10 years Industrial Bio-Test, based in Northbrook, Illinois, conducted more than 4,000 tests for large and small chemical manufacturers to submit to the EPA in support of chemical registrations. Many of the tests were long-term animal toxicology studies, producing key information (Trimble, 1980, p. 9).

As John Moore, EPA-Pesticides and Toxic Substances Assistant Administrator, acknowledged:

Pesticide regulation is a difficult and complex task. This task, however, is nearly impossible if data on which decisions are based are not sound and accurate. Faulty data supplied by a private laboratory in recent years have raised serious concerns about the integrity of data on which government agencies, including EPA, have relied. These events have also heightened public concern over government regulation (Moore, 1984, p. 3).

Also, EPA has found serious problems with the testing methods of numerous other labs doing pesticide work submitted in support of pesticide registration.

The EPA "Cut and Paste" Episode

With respect to EPA's supposed "independent" review of industry submitted data in support of pesticide registrations, a Congressional House subcommittee in 1983 "found evidence that company submissions, which do not always stress the worst aspects of a chemical, were being cut, reassembled, and filed by EPA staffers as their own independent work" (Marshall, 1984). "Of 578 staff reviews chosen at random, one third contained some unattributed use of company charts and prose" (Marshall, 1984). Luckily, "only 29 of the questionable studies reached challengeable conclusions, possibly affecting regulations on 21 chemicals" (Marshall, 1984) and all were being fully reviewed.

Unfortunately, in the face of fraudulent or inappropriately reviewed data in support of pesticide registrations, the agency claims,

> EPA does not have statutory authority to suspend or cancel registered pesticides when inspections show that the safety tests supporting the registration are not valid. EPA can require that registrants repeat a test but, in the interim, cannot take other regulatory action, such as suspending use. Some tests take up to 3 years to complete. During this time, the public and the environment can be exposed to potentially dangerous pesticides not supported by valid safety data.

FIFRA (the federal pesticide law) does not allow EPA to withdraw a pesticide from the market solely because fraudulent or poor quality data was used to support its initial registration. According to FIFRA, Section 6, EPA can suspend a pesticide's use only if the agency determines that it poses an "imminent hazard," defined as

> a situation which exists when the continued use of a pesticide during the time required for cancellation proceeding would be likely to result in unreasonable adverse effects on the environment (GAO, 1980, p. 54).

The Risks of Several Pesticide Formulas and "Inert" Pesticide Ingredients Have Not Been Taken Into Account

Most pesticides are mixtures of substances that include (1) "active" ingredients, which are the intended agents of plant and other pest control, and (2) "inert" ingredients, which are the other intended and unintended ingredients in pesticide formulas. Generally, only the active ingredients in pesticides, rather than the pesticide formulas, including their "inert" ingredients, have been the subject of the regulatory process.[2] It is now being found that so-called "inert" ingredients present previously unassessed risks to the public and the environment.

---

[2]There are also questions about the adequacy of testing the ingredients in pesticides. Currently, most data submitted by manufacturers to EPA relate to "technical" active ingredients that come out of the manufacturing process and are mixed in pesticide formulas. Less, if any, data are submitted with respect to "pure" active ingredients or pesticide formulas that are ultimately sold and used in the human environment.

There are estimated to be 1,200 inert ingredients in pesticides marketed today, of which 50 recently have been identified by EPA as "toxicologically significant," and approximately 50 more "in the category of unknown toxicity [that] should have a high priority for testing" (Anon, 1985a, p. 42). As observed by one EPA official, "Some inerts are more toxic than active ingredients" (Anon, 1985b, p. 29).

A few of the most notable examples of unintended inert ingredients in pesticides include dioxin contamination of 2,4,5-T and dacthal, EDB (ethylene dibromide) contamination of diquat, and DDT contamination of dicofol (the active ingredient in the insecticide marketed under the brand name Kelthane). Intended inerts include preservatives, solvents and surfactants. Although the "inerts of concern [to EPA] have been cut to 40 of which the majority are active ingredients" (Anon, 1985b, p. 42), many have not passed through the registration process.

EPA Does Not Consider the Synergistic Effects of Pesticides

Exposure to many chemicals comes simultaneously through food, air, occupational exposure, and other routes. At the present time the regulations of each agency are based largely on the assumption that the exposure regulated by that agency are the only exposures. Thus, it is quite conceivable that an individual could receive several times the safe exposure even though all regulatory standards were enforced adequately. In addition, *no agency oversees the possible synergistic effects of regulated chemicals* (NAS, 1975, p. 31, emphasis added).

"Synergistic and cumulative interactions through multiple environmental insults by various routes of exposure can lead to serious illness, disease, and environmental degradation" (Anon, 1980, p. 10). For example,

Many cancers may appear as a result of exposure to several agents--that is, they may be caused by the combined effects of continued long-term exposure to carcinogens acting together in an additive or synergistic way, perhaps abetted by a genetic susceptibility, rather than from sporadic exposure to a single carcinogen acting on a genetically uniform population (Anon, 1980, p. 155).

Yet, "interactions among chemicals to which everyone is exposed--some of them cancer causing or cancer promoting--are not predictable. Additive and synergistic or even inhibiting interactions may occur with unforeseen consequences" (Anon, 1980, p. 134).

Synergistic effects of pesticides are not assessed by EPA. EPA-Pesticides and Toxic Substances Assistant Administrator John Moore admits that, although the risks of synergistic effects of pesticides is an important issue, "we don't know how to deal with it" (Moore, 1984b).

Pesticide Regulatory Enforcement
Has Been Inadequate

An October, 1981, GAO report revealed persistent problems in state and federal enforcement of pesticide regulations (GAO, 1981).

Although improvements have been made in recent years, GAO found that the public may not always be protected from pesticide misuse because EPA and the States

1. sometimes take questionable enforcement actions against violators,

2. have not implemented adequate program administration and monitoring, and

3. are approving the use of pesticides for special local needs and emergency purposes which may be circumventing EPA's normal pesticide registration procedures (GAO, 1981, p. i).

Further, the report stated, "EPA and State enforcement programs do not always protect the public and the environment because:"

1. Many enforcement actions are questionable or inconsistent.

2. Some cases are poorly investigated.

3. State lead agencies often do not share EPA's enforcement philosophy.

4. Most States lack the ability to impose civil penalties (GAO, 1981, p. ii).

The bases for GAO's conclusions included the following:

GAO's review of 2,855 randomly selected cases for the period 1975 to 1980 at 6 EPA regions and 11 States disclosed questionable enforcement actions in 491 cases, or 17 percent (10 percent for EPA and 19 percent for the States). The extent of questionable actions ranged from 5 to 80 percent for the States visited. In these cases, States either took no action or chose enforcement

actions which were minimal when compared to the sever-
ity of the violation. . . GAO's review also disclosed that
704 of the 2,855 cases, or 25 percent (8 percent for EPA
and 29 percent for States), were not investigated according to generally accepted EPA and State criteria. The
extent of the inadequate investigations ranged from 3 to
90 percent for the States visited. . . Generally, States
are more likely to resolve misuse cases by negotiating
settlements between parties involved, rather than by
taking enforcement action against violators (GAO, 1981).

GAO explained the partial reason for state inadequacies in pesticide
enforcement as follows:

In most cases, Federal pesticide environmental laws are
enforced by State departments of agriculture which have
broad responsibility to promote increased farm productivity (GAO, 1981).

Most of the circumstances surrounding pesticide enforcement in 1981
still exist today.

The Pesticide Reregistration
Process Will Not be Complete
Until Well After the Year 2000

As pointed out earlier, Congress, in 1972, required the complete
reregistration of all pesticides registered under the 1947 federal law.
"The agency has identified just under 600 basic chemicals used to
manufacture the over 45,000 to 50,000 currently registered pesticide
products" (Johnson, 1984).

Reviewing old pesticides involves two separate steps--
the development of adequate test data and the review
and assessment of that data as it is generated. As the
initial step, the agency reviewed the available data on
old pesticides and identified what important tests to
assess health effects had never been performed. EPA
then required the manufacturers of these chemicals to
perform the needed studies. . .

The second step, after receiving this data, is to
review it carefully and to take appropriate regulatory
action (Moore, 1984a, p. 3).

Fifteen years later, EPA's record for accomplishing this task
remains dismal. EPA Pesticides and Toxic Substances Assistant
Administrator Johne Moore readily admits "we haven't come to grips
with previously registered pesticides" on which many still lack an
adequate data base (Moore, 1984b, fn 38).

In early 1984, the National Academy of Sciences reported that of all pesticides there was sufficient information on only ten percent to make a full health hazard evaluation possible and that for two-thirds not even minimal toxicity information was available (Drayton, 1984, p. 91).

So, how long will reregistration take?

[B]y the end of the 1984 fiscal year, the Agency will have reviewed 90 old chemicals. . .

According to present plans, we anticipate reviewing 25 chemicals per year. At this pace, reregistration will be completed around the year 2000 (Johnson, 1984, pp. 4-5).

As of March, 1986, 124 pesticide products had interim registrations from EPA under the 1972 Congressional mandate, and "EPA has not yet completely reassessed any active ingredient and issued a final regulatory position or standard" (GAO, 1986, pp. 25, 27). In the meantime, about 50 pesticides have been entered into the EPA "special review" process because of unexpected problems or suspected hazards to humans and the environment. Some of these include re-registered products (Bowen, 1985). Examples include aldicarb (Temik), alachlor (Lasso), linuron (Lorox), and amitrole (Ehart, 1985).

In addition, the prediction that many of these chemicals will be found to be unacceptable by modern standards is coming to pass. Former EPA-OPP Director Steven Schatzow observed with respect to the few pesticides that have passed through the slow reregistration process:

[A]ctionable health and safety problems are being found for about 60% of the pesticides being reviewed for reregistration. The official said the effect of getting a great deal of up-to-date data is that the agency is bound to find some evidence of hazards that were not anticipated when the pesticides were first registered years ago. "In short, we are dealing with a legacy of previously unaddressed risks," Schatzow observed. "It appears that many products pose risks that were not anticipated in the past, and, in some cases, are only now beginning to be recognized," he concluded (Anon, 1985c).

Meanwhile, previously registered pesticides continue to be used while awaiting reregistration under the 1972 law.

EPA Lacks Resources to
Regulate Pesticides Adequately

Current federal budget cuts are crippling what pesticide reregistration and research capabilities EPA has had.

According to EPA, "At present, the pace of reregistration is a function of resources rather than the process itself" (Johnson, 1984, p. 4). A memorandum from EPA employees to the United States Senate and House Committee Chairmen has charged that pesticide research has been all but eliminated and urged that Congress stop EPA's reorganization until Congress can investigate the situation. The memorandum said:

> For example, the pesticides research program has been essentially abolished while needs continue to exist, i.e., adolescent work requirements or regulations in agricultural occupations for the Department of Labor; 2,4,5,- T/TCDD programs, toxic waste disposal, DDT exposure in Triana, Alabama, pesticide registrations of compounds by industrial submission of inadequate data with less than scientific claims for use as economic poisons, labor reentry regulations, exposure and dosimetry data, including kinetics, metabolism and toxic effects data.

"There is no way the agency can meet its FIFRA research responsibilities under the currently projected cuts, the Congressmen were told" (Anon, 1981c). Despite these 1981 appeals for more resources,

> [I]n 1984 EPA has 46 percent less purchasing power for pesticides research than it had in FY81. EPA's pesticides regulatory office, which has lost one third its purchasing power in the same three years, also suffers from, in the words of a House Agriculture Committee study, an "apparent shortage of scientific personnel." For example, it now has only one person for pathological analysis. The same is true for statistics and risk assessment models (Drayton, 1984, pp. 93-94).

The Legacy of Pesticide Problems is,
to Some Extent, a Consequence
of the Regulatory Process

The inadequacies and failures of the pesticide regulatory process are not the claimed "accidents," "anomalies," "exceptions," or "incidents" that only occasionally present theoretical risks to the public and their environment. They are inherently direct and inevitable consequences of a regulatory process that allows unreasonable risks to persist from the use of chemicals that have been "grandfathered" by registrations under past and existing laws. The following partial list of suspended, cancelled or more severely restricted pesticides (EPA, 1985), once registered by EPA and assumed by the public as "safe," is the all too familiar legacy of the inherently flawed regulatory process:

Aldrin   Kepone
Chlordane   Lindane
DBCP   Mirex
DDD (TDE) [metabolite of DDT]   Parathion
DDT   PCBs
Dieldrin   2,4,5-T
EDB   Toxaphene
Endrin   Vinyl Chloride
Heptachlor

This list contains familiar examples of pesticides that, even though they were once registered for use and consequently considered "safe," later were restricted because their hazards to public health and damage to the environment became known. The list does not yet contain other pesticides that are still registered and widely used but whose dangers have yet to be acknowledged in current or future EPA "special reviews" or registration processes. The following are examples of "suspicious" pesticides: aldicarb, alachlor, dicofol, pronamide, captan, pentachlorophenol, simazine, and diazinon.

The "pesticide-of-the-month" syndrome is no coincidence. It is the inevitable result of a regulatory process that allows the use of previously registered pesticides whose hazards have not yet been uncovered by an unfinished reregistration process. So much is acknowledged by EPA. "And through the review of existing chemicals, we are building a more complete data base and examining it closely in order to uncover remaining pesticide health and safety problems" (Johnson, 1984, p. 7).

The Myth of Pesticide Safety Persists

To sum up, all pesticides, EPA registered or not, present risks to the public and the environment. Even if it were working properly, the current pesticide regulatory scheme does not assure, and is not meant to assure, that pesticides are "safe." Rather, the regulatory scheme is intended to prevent "unreasonable" risks to the public and environment. What constitutes a "reasonable" risk for pesticide registration and use is a public policy call, based only in part on scientific analyses. The rest is subjective judgment.

In addition, the current regulatory scheme is not working nearly as well as it should. The data upon which previous pesticide registrations are based is considered to be unreliable and incomplete. Yet, pesticides continue to be used while awaiting reregistration or special review based on new data.

Despite these circumstances, the myth of pesticide safety continues to be perpetuated on the basis of the claim that pesticides have been registered by EPA.

RISK CONSIDERATIONS OF PESTICIDE USE

Decision makers should know what the pros and cons are for a proposed course of action.

Risks of the Pest Problem

In order to assess the risks and benefits of pesticide use, an initial review of the nature and risks of a particular pest problem is in order. Pertinent questions include: "What is the nature of the pest problem?" "Is the pest problem creating a mere nuisance, or more serious health and economic risks?" "If a pest is creating merely a nuisance, is it best, in light of the risks of pesticide use, not to apply? " "How serious and real are the nuisance, health, safety and economic risks presented by the pest?"

It should not be forgotten that organisms are labeled as "pests" only when they interfere with the comfort or economic and safety interests of people in a community. Pests are a fact of life and in many cases more a nuisance than economic or life threatening. Even when control measures are called for, no one should be under the false impression that pests can be completely eradicated. They often can be controlled but usually not made extinct. In any case, it is always important for decision makers and the public to understand the role of the alleged pest in the natural ecosystem of which they are all a part.

Answers to these questions are important because the risks of the pest problem have to be weighed against the risks and benefits of pesticide use, as well as the alternative means of controlling pests.

Risks of Pesticide Use

Next, it is relevant to look at the risks involved in proposed pesticide use to suppress a pest problem. In addition to considerations of the current status of pesticide regulation discussed earlier, important considerations of risks and detriments also include the following.

Does Pesticide Use Create a Nuisance in Itself?

First, even where a pest problem is a nuisance, will the odor or public concern about pesticides create another nuisance in its place? In many communities, the odor of pesticides is considered to be a nuisance. The psychological and emotional stress among members of the public as a result of pesticide usage is another important factor to consider.

Does Pesticide Use Create Economic Hardships?

Second, will pesticide use create risks of economic hardships, and to whom? In some contexts, members of a community, such as beekeepers, organic growers or property owners, can suffer significant economic hardships as a result of certain kinds of pesticide applications.

Does Pesticide Use Create Known and Unknown Risks to Citizens?

Next is the consideration of health risks as a result of pesticide use. Decision makers want to know what the short- and long-term health risks of chemicals are. Their questions are often answered with assurances that extensive studies have been made and that there are no significant known risks to humans or the environment if the chemicals are used according to label directions.

This, of course, is one of the hotly debated subjects at local as well as at state and federal levels. On this point, decision makers must recognize another crucial distinction on the issue of both *known* and *unknown* human health risks from exposure to toxic chemicals, including those classified as pesticides. No one, including the EPA, the U.S. Department of Agriculture (USDA), the pesticide chemical industry, all of the scientists in this country and in the world together, has a full and complete understanding of all the significant human health risks presented by most of the pesticide chemicals that are currently registered by EPA and are being used to suppress pest populations. Why is this so?

The "Known" Risks of Pesticides Are Inexact and Largely Unreviewed

The pesticide risk assessment process is complex, difficult and inexact. Risk assessments generally involve a two-pronged approach involving (1) extent of *exposure* to human populations, and (2) the pathological activity, or *toxicity*, of a pesticide. For example, a highly toxic pesticide to which people are anticipated to have low levels of exposure may be assessed as presenting similar risks as a pesticide with lower toxicity but to which people are anticipated to have higher levels of exposure.

Determination of the extent of exposure involves such factors as the number of people exposed, frequency of exposure, the extent of the dosage, use patterns, volume estimates, and types of exposure, such as dermal, ingestion, or inhalation exposure. Most of this information is provided within a clinical (laboratory) context. Empirical data collected on a case-by-case basis, if collected at all and properly, is generally not considered. Clinical data on pesticide exposure to humans, except as a result of accidental exposures and intentional ingestion, is generally not available. Particular susceptibility of certain individuals to pesticides, due to allergy, infirmity, illness, infancy or old age, generally is not part of risk assessment. When data are available, they are based on controlled pesticide exposure to animals.

With respect to the second prong of risk assessment, pathological activity or toxicity of the pesticide, data should be available on individual pesticides regarding their proclivities to cause *acute* (immediate) and *chronic* (long-term) ailments, the most notable chronic ailment being cancer. The proclivities of pesticides to cause certain other diseases are not contemplated or examined or are just now being considered for study. For example, study of the ability of pesticides to cause neuro-behavioral malfunctions or immune system suppression in animals or humans is just getting underway.

Although the best evidence for establishing the chronic effects of pesticides on humans is epidemiological data (studies of human ailments among populations), such data are rarely available or conclusive. Unless an illness caused by a pesticide is uniquely associated with it, the cause often goes unrecognized. Currently, chronic effects of pesticides are extrapolated from animal testing. In addition, the synergistic effects of pesticides (multiplied effects caused by the interaction of pesticides with other chemical compounds) are largely unknown and not well understood. Consideration by EPA of risks of "inert" pesticide ingredients is just starting.

According to a previously cited NAS report, current risk assessments attempt to make "worst-case" extrapolations on the adverse effects of pesticides in the environment (NAS, 1980, pp. 80-82). But as the authors of the report warn,

> With the present state of knowledge, the results of bioassays and short-term tests may be useful as comparative measures (to compare one carcinogen to another) of anticipated effects of various carcinogens at human exposure levels, but extrapolation techniques are not yet sufficiently precise or well-founded to allow us to make credible quantitative extrapolations about the anticipated frequency with which cancers will be induced in the human population by exposure to the estimated doses of pesticides. *Most important is the possibility that because of unconsidered factors, the worst-case extrapolation could actually be an underestimate* (NAS, 1980, p. 82, emphasis added).

Once risk assessments are completed on a given pesticide (questions about their reliability aside), these assessments become a part of a body of knowledge termed "known risks."

The task of pesticide risk assessment is formidable and far from complete. Risk assessments for human exposure to pesticides require extensive expertise and numerous analyses. There are over 35,000 pesticides registered with EPA which warrant risk assessment analyses. While many pesticides can be placed in categories or groups for this purpose, the number of analyses that must be done remains formidable.

Although the USDA and EPA were originally required to determine whether "reasonable risks" were being taken with respect to the original 35,000 pesticides now registered under the law, Congressional and other studies determined that the risk assessments that were being done were inadequate. Therefore, EPA was charged by Congress in 1972 with the duty of *re*registering all pesticides based on new risk assessment analyses. The process of reregistration is continuing today. Reregistration is not contemplated to be completed within the next 15 years, especially given current budget constraints at the federal level. In the meantime, the pesticides not yet reregistered continue to be used as before.

Significant Unknown Risks
of Pesticides Remain

In addition to the known, but inexact, acute, chronic and environmental risks of pesticides, it must be kept in mind that toxic chemicals, including pesticides, may be creating unknown risks to the human environment.

First, until EPA completes the reassessment and reregistration process that will extend into the Twenty-First Century, "the health and environmental risks and benefits associated with older pesticides and their uses will not be fully known" (GAO, 1986, p. 2). Therefore, "EPA. . .cannot fully assure that the public and the environment are adequately protected against possible unreasonable risks of older pesticides" (GAO, 1986, p.20).

Second, even as the reregistration process is completed, "[u]ncertain hazards represent the black box of regulatory decision making. They are typified by second- and third-order effects and interactions with other chemicals" (NAS, 1980, p. 140). There are "no immediate solutions [to the problems of fully assessing risk] given the many gaps in our understanding of the causal mechanisms of carcinogenesis and other health effects and in our ability to ascertain the nature or extent of the effects associated with specific exposures. Because our knowledge is limited, conclusive direct evidence of a threat to human health is rare" (NAS, 1983). "Uncertainties are common, are frequently controversial, and have a legitimate role in decisionmaking if that role is explained" (Anon, 1980, p. 15). In short, risks to the human environment from a toxic chemical may exist, whether they are readily observable or not.

This is the one additional idea to keep in mind in the risk assessment context. Carefully worded reports by EPA, by chemical companies, or by others often say, "there are no *known* risks to human health or the environment from pesticide X at these levels." A risk assessment that does not show a tendency of a pesticide to cause a certain ailment results in a conclusion of "no known risk" at the exposure levels tested.   This is not the same as conclusively saying that the chemical presents "no risks," or even "no unreasonable risks."   In many cases, it simply means that more detailed testing is required.

Typical examples of this phenomenon include the use, once valid under EPA registrations, of DDT, chlordane, 2,4,5-T, vinyl chloride and other pesticides, prior to the discovery of significant health and environmental risks that became known while the public was using them under the impression those pesticides were completely "safe."

Local Decision Making and
EPA Risk Assessment

Why bring these complicated matters to the attention of state and local decision makers and citizens?   They are brought because they *are* complicated, very complex, and because the capacity of the EPA to conduct reliable risk assessment and management has been questioned, not only by scientists, but by the chemical industry as well.   This is why decision makers should not be totally comfortable with merely allowing their decisions on local pesticide usage to be made by others at state and federal levels.

Does Pesticide Use Present
Risks to Pesticide Applicators?

Another consideration is pesticide exposure to applicators. Although applicators presumably use pesticides voluntarily, they are usually the most vulnerable to exposure and the health risks that accompany it.   The decision to apply pesticides should have the applicator in mind and reflect appropriate concern for his/her well-being.   For example, the regulatory status, relative toxicity and risks of a pesticide must be considered, along with the qualification, training and education of applicators, the quality of application equipment and protective gear, and the ability and inclination of applicators to apply pesticides properly and follow protective measures.   Also, applicators should be educated regarding the risks of pesticides, as well as being trained in their use.   Deference should be given to their inquiries and concerns about being assigned pesticide application duties.

Does Pesticide Use Create Risks
of Involuntary Exposure to Citizens?

Another consideration is whether proposed pesticide applications involve risks of involuntary pesticide exposure to individuals who, because of individual susceptibility or not, would find such exposure to themselves or to other nontarget property or organisms offensive. In hearings before the Wisconsin Department of Agriculture, Trade and Consumer Protection (WDATCP) in January, 1980, the Public Intervenor contended that a primary goal of pesticide policy making should be protection of

> the fundamental right of every Wisconsin citizen to be free from unwanted and involuntary contact with biocidal agents. Our citizens have a right to breathe clean air in their homes, in their backyards, on their land, and on our public roads, unmolested by pesticide sprays, even when labeled "safe" by private industry and the government.

While no right is absolute and often must be balanced against other rights and interests (both private and public), it is clear that the right to be protected against unpermitted touching and contact of persons by others, especially toxic chemicals, is a primary right that only should be overruled under the most compelling of circumstances in the public interest.

Does Pesticide Use Create
Risks to Nontarget Properties,
Organisms and the Environment?

It is a fundamental principle of Anglo-American jurisprudence that private property should be free of unauthorized and unreasonable physical intrusion by others or their instrumentalities. The law of trespass and private nuisance exists to protect this property right, regardless of the ability of the owner to prove physical or other damage to property resulting from an unauthorized intrusion (trespass), or where the enjoyment of land as a result has been unreasonably infringed (nuisance). Private property protected under trespass and nuisance includes real and personal property, including domestic animals and pets. Many property owners consider the direct or indirect spraying or drifting of pesticides on their personal or real property to be a significant infringement of their rights, and they consider such activity as tantamount to a trespass or private nuisance. Their concern is heightened with the knowledge that pesticides are toxic chemicals, once also appropriately called "economic poisons."

Decision makers are often expected to be responsive to the interests of their constituent citizens and unwanted pesticide spraying on private property is a matter of growing concern.

There is also the matter of risks, known and unknown, to nontarget wild animals, plants, beneficial insects, and local ecosystems. In some communities, the environmental quality, scenic beauty and natural wholesomeness of the surrounding area forms an especially important basis of the community's economic and social well-being, such as through the tourist trade. Potential detriments to these amenities by pesticides have to be considered, along with other more direct effects.

Benefits of Pesticide Use In the Risks Context

Another consideration is the benefit to be derived from the proposed pesticide's use and the limitations of that benefit. Clearly, pesticides are a valuable, effective, and often necessary tool for the control of insects, weeds or other pests. Without pesticides, the present levels of food production would be unknown. Certain diseases, such as malaria and yellow fever, have hardly been human health risk factors, largely to the credit of pesticides. The dollar costs of pesticides are often less than alternative means of control, and they are relatively easy to use.

As the benefits of pesticides are limited, it is important to determine the extent to which they can effectively control pests in any given situation. For example, pesticides usually may not be legitimately expected to completely eradicate a pest problem. Rather, they are useful tools of pest "control." While pesticides have the capacity to substantially reduce pest infestations, often the results are temporary; and pest recurrence could require repeated applications. Greater application frequency in response to temporary pest suppression, of course, increases the risks of nontarget and human exposure. Thus, while the benefits of pesticide usage are largely known, they should not be overestimated, nor considered outside the context of the risks they also create.

ALTERNATIVES

There are, of course, alternative options to be explored with respect to pest suppression and pesticide use.

The first may be a decision to discontinue pest suppression efforts, either by pesticide or other means. The nature of certain pest problems is such that suppression may be legitimately viewed as futile, wasteful or unnecessary. For example, a pest problem that presents a mere nuisance, without significant health, safety or economic threats, may be legitimately viewed on balance by local decision makers as not worth the costs and risks of attempting pest suppression. Recent decisions against mosquito control are examples. Public education about pests and the risks of alternative suppression schemes is, therefore, essential.

A second alternative may be to employ pest control methods by nonchemical means in order to obtain differing degrees of pest control. Mechanical, cultural, or physical treatment of pests and their environs, alone or in combination, may offer viable alternative pest control, with different levels of risks and costs to the community.

A third alternative is the judicious use of pesticides under conditions set by the local decision making body. No one would suggest that pesticides should be used indiscriminately, carelessly, or without strict compliance with EPA-approved label directions. Still, given the risks and benefits of pesticides, local decision makers may wish to consider placing additional conditions on pesticide use besides those required on the labels. For example, pesticide use may be permitted only as "a last resort" or in combination with other nonchemical means of pest control (integrated pest management).

Decision makers also may wish to establish threshold criteria. For example, applicator decisions to use pesticides can be made contingent on verification of designated pest levels or risks, rather than by ordering potentially unnecessary routine or "preventive" pesticide application schedules. Questions to be answered include: Will mere complaints about pests trigger a decision to use pesticides? Or will demonstrations of more serious pest-induced risks be required? Who will decide to give the "go ahead" on pesticide use after all relevant considerations are taken into account, and what oversight will be provided to assure that the criteria triggering pest control measures are faithfully met?

In the event that a community embarks upon the task of controlling a pest problem, whether by chemical or nonchemical methods or both, the following considerations should be kept in mind. First, any pest control program should be organized around the following components: (1) education about pesticide efficacy and risks; (2) training of qualified pesticide application personnel in use and self-protection; (3) pest preventive planning; (4) pest surveillance and monitoring; and (5) alternative and complementary pest control methods. The relative emphasis placed on each component is primarily a function of the type of pest problem. The use of each pest management method should be based upon a reasoned judgment by decision makers on the risks, costs and benefits inherent in each.

Second, it is essential that trained and qualified personnel be employed for pesticide application. They should be able to identify habitat areas and types of pests to determine and treat only the problem causing areas and species. A selective treatment program, rather than a broad based program, should be considered. Biorational pesticides, i.e., biological control agents, sex attractants and juvenile hormones are now coming on the market. They should be considered as well.

PESTICIDE PUBLIC POLICY PRINCIPLES

Local decision makers now are faced with many of the same issues and decisions as state and federal agencies with regard to pesticide use and control.

In January, 1980, the Wisconsin Public Intervenor outlined in written testimony before the Wisconsin Department of Agriculture, Trade and Consumer Protection proposed fundamental goals of the pesticide public policy making process in Wisconsin, just getting underway at that time. Not only are those goals as relevant in 1987 as they were then, but they are totally consistent in the local policy making context as well. Those goals are fivefold.

The foremost principle or goal that should be advanced in public pesticide policy making is the protection of the health and the environment of the public. It is the charge and authority of local governments to act in a manner consistent with the health, safety and welfare of their citizens. Thus, this goal must remain primary for local decision makers. The following goals are subsets of that charge.

The second goal is protection of citizens' rights to be free from unpermitted contact with biocidal agents. This was discussed previously.

Third, the citizens must have a voice in pesticide use policy making. The tradition and practice of local governments, providing public forums for local citizens to speak out and be informed on issues that affect them, must remain alive and well in the United States. It is all the more important in pesticide-related issues.

Fourth, the public has a right to know about the use of pesticides in the environment. Many citizens of Wisconsin, including decision makers at state and local levels, are unaware of the problems, inadequacies, debates, and lack of resources to regulate chemical pesticides and protect the public effectively. Myths abound regarding the "safety" of pesticides and the ability of national and state regulators to protect the public. They have a right to know the risks of pesticides and, more importantly, the right to make their own judgments as to whether those risks are "reasonable." National and state policy makers should not be the exclusive repositories of information for making this judgment. Every town board and every citizen is entitled to that knowledge.

Fifth, the burden is not on the public to prove pesticide hazards. Rather, it is on others to show the reasonableness of those hazards. No one should presume that pesticides are "safe." There exist known and yet to be discovered risks associated with their use, and these can only be weighed against the benefits and risks of not using them. Because of this reality, the burden of showing that the risks of using pesticides outweigh the risks of not using them should not be imposed on those least able to meet this heavy burden. The risks of toxic chemicals are created by those that profit from their use and are most able to pay for the studies, the assessments, the data and the facts that are necessary for the public to make its judgment. The burden should fall on those interests.

CONCLUSION: INFORMED DECISION MAKING IS CRUCIAL

Of course, all governmental decision makers seek to make their decisions based on the best information available. While they are gathering this information, particular attention should be given to the source and the perspectives of that source on the pesticide issue at hand.

This point perhaps can be made in the context of California's 1981 battle with the Mediterranean fruit fly with the pesticide malathion. Debate over the use of malathion raged for weeks. While the State's $600 million fruit industry was at stake, many local governments objected to the aerial spraying of the pesticides in their communities because of public health concerns raised by widespread exposure to the chemical.

During the debate, numerous spokespersons said that malathion was "safe." The director of Florida's Office of Entomology was reported as saying,

I'd say it's [malathion] one of the safest pesticides there is. . . We've used it in Florida for almost 30 years and we've never had the first instance of reported illness from malathion.

The USDA and the pesticide manufacturers voiced similar opinions, and a New Orleans entomologist was reported as echoing, "There has been no problem whatsoever" from ground and aerial spraying of the chemical.

Who was making these statements about pesticide safety? Were they persons charged with the obligation to protect public health or did the persons have duties or interests more attuned to killing pests and protecting economic interests? Entomologists are insect experts, not medical authorities. The USDA is charged with promoting the economic health of American agriculture. Pesticide manufacturers have the most direct economic interests.

Those who are charged with protecting health and the environment are much more reserved about touting pesticide safety and are quicker to recognize the risks. EPA officials perceive pesticides in terms of relative risks, not safety. Medical and environmental researchers, including those at the University of Wisconsin, continue to conduct research that may reveal subtle but significant long-term adverse effects on human beings exposed to pesticides.

Of course, all sources of pesticide-related information, including this one, must be examined with a critical eye. But by doing so, the quality of policy and decision making can only be enhanced.

LITERATURE CITED

Anon. 1980. Toxic Chemicals and Public Protection--A Report to the President. Toxic Substances Strategy Committee, Environmental Council on Quality, 722 Jackson Place, Washington, DC 20006, pp. 187.

Anon. 1981a. Relative risks in war on pests. New York Times, July 26.

Anon. 1981b. IBT personnel hit with criminal charges for fraud in pesticide tests. Pesticide and Toxic Chemical News, June 24, pp. 24-26.

Anon. 1981c. EPA pesticide research being scuttled, congressmen told. Pesticide and Toxic Chemical News, August 5, pp. 13-14.

Anon. 1985a. At least 100 inerts will need toxicology testing, FIFRA SAP [Science Advisory Panel] told. Pesticide and Toxic Chemical News, February 20, p. 42.

Anon. 1985b. Pesticide inerts of concern cut to 49; majority are active ingredients. Pesticide and Toxic Chemical News, March 13, p. 29.

Anon. 1985c. Health, safety problems found for 60% of pesticides reregistration. Pesticide and Toxic Chemical News, January 23, p. 4.

Bowen, D. 1985. EPA-OPP Registration Division, Room 711B, CM-2, 401 M Street S.W., Washington, DC 20460, personal communication, June 26.

CFR. 1984. United States Code of Federal Regulations, 40 C.F.R. Sec. 162.10(a)(5)(ix), 1984.

Drayton, W. 1984. America's Toxic Protection Gap. Environmental Safety (July), citing National Academy of Sciences, Toxicity Testing (1984), pp. 10-12.

EPA. 1985. Suspended, Cancelled and Restricted Pesticides, 3rd Ed. U.S. Environmental Protection Agency, Washington, DC, January, p. 29.

Ehart, D. 1985. Regulatory calendar. Proceedings of the 1985 Fertilizer, Aglime and Pest Management Conference, Madison, WI, January 15-17.

FIFRA. 1978. The Federal Insecticide, Fungicide and Rodenticide Act (FIFRA), Public Law No. 92-516 (October 21, 1972), as amended by Public Law No. 94-140 (November 28, 1975) and Public Law No. 95-396 (September 30, 1978), 7 U.S.C. Sec. 136.

GAO. 1980. Delays and unresolved issues plague new pesticide protection programs--report by the Comptroller General of the United States. CED-80-32 (February 15, 1980), U.S. Accounting Office, Washington, DC, p. 16. This report can be obtained from the GAO Document Handling and Information Services Facility, Box 6015, Gaithersburg, MD 20760, Telephone: 212/275-6241.

GAO. 1981. Stronger enforcement needed against misuse of pesticides. CED-82-5 (October 15, 1981), U.S. General Accounting Office, Washington, DC. This report can be obtained from the GAO Document Handling and Information Services Facility, Box 6015, Gaithersburg, MD 20760, Telephone: 212/275-6241.

GAO. 1986. Pesticides: EPA's formidable task to assess and regulate their risks. Report to Congressional Requesters, RCED-86-125. Resources, Community and Economic Development Division, U.S. General Accounting Office, Washington, DC 20548. This report can be obtained from the GAO Document Handling and Information Services Facility, Box 6015, Gaithersburg, MD 20760, Telephone: 212/275-6241.

Johnson, E. 1984. EPA and pesticides: An interview with Edwin Johnson. EPA Journal 10(5):4-8.

Marshall, E. 1984. EPA ends cut and past toxicology. Science 223:379.

Moore, J. 1984a. Pesticide regulation: An overview. EPA Journal 10(5):2-3.

Moore, J. 1984b. Keynote address, National Governors' Association Conference on Environmental Health Issues in Pesticide Management, San Diego, CA, September, 19.

NAS. 1975. Decision Making for Regulating Chemicals in the Environment. Committee on Principles of Decision Making for Regulating Chemicals in the Environment, National Research Council, National Academy of Sciences, Washington, DC 20418, pp. 232.

NAS. 1980. Regulating Pesticides. Committee on Prototype Explicit Analyses for Pesticides, National Research Council, National Academy of Sciences, Washington, DC 20418, pp. 288.

NAS. 1983. Risk Assessment in the Federal Government: Managing the Process. National Research Council, National Academy of Sciences, Washington, DC 20418.

Trimble, N.   1980.   Critics fault EPA testing methods.   Burlington County (NJ) Times, March 4, pp. 1, 9.

USS.   1976.   The Environmental Protection Agency and the Regulation of Pesticides.   Staff Report to the Subcommittee on Administrative Practice and Procedure of the Committee on the Judiciary of the United States Senate, Washington, DC, 50   pp.   This report and its appendix (762 pp.) is offered for sale by the Superintendent of Documents, U.S. Government Printing Office, Washington, DC 20402; Stock No. 052-07-03867-5.

CHAPTER 19

WELL CONSTRUCTION:
DRILLING, LOCATION AND SAFETY

Donald K. Keech
Michigan Department of Public Health
Lansing, Michigan 48909

INTRODUCTION

Wells installed in an agricultural community are generally drilled to meet two basic needs; either a well to supply water for drinking and other domestic users at the homestead, or to provide the larger water requirements for irrigation purposes. Actually, any well drilled in the State of Michigan for production of freshwater must meet the well construction rules promulgated by the Michigan Department of Public Health under the authority of the Groundwater Quality Control Act passed in 1965. This statute has now been incorporated into the Michigan Public Health Code as Part 127 of Act 368, P.A. 1978, as amended.

These rules cover any freshwater well that is installed, whether it is for drinking water needs, irrigation, or industrial use. Exceptions to these rules are mineral wells drilled for industrial purposes and water wells drilled to supply drinking water for public water supplies which are governed by the Michigan Safe Drinking Water Act, Act 399, P.A. 1976. The Michigan Safe Drinking Water Act regulates any well installed where the water is to be used for drinking purposes by persons other than the owner of the supply. This act covers certain supplies in the farming community such as water systems for migrant housing or grade A dairy farms (because the water is used in the production of grade A milk to be sold to the public), and perhaps other large installations where several employees are hired in a commercial venture. But the basic well construction requirements are the same, and differences only relate to isolation from known or potential sources of contamination and the need to obtain a permit for any well that may serve more than 25 persons or 15 separate connections.

The promulgated rules cover most aspects of well construction including the location, actual construction materials and procedures, details and provisions for sanitary pump installation and operation. It appears appropriate to look at the implication of each of these criteria as it relates to installation of a well in a rural location.

The basic isolation requirement for the location of a nonpublic well serving a private residence requires a 50 foot (15.24 m) horizontal separation from any possible source of contamination such as an on-site sewage disposal system including the tile field and septic tank, a barnyard, or other similar sources of contamination.

The basic horizontal isolation distance is increased to 75 feet (22.86 m) for small public water supplies such as a well serving a grade A dairy farm or migrant housing. A well serving less than 25 persons or having less than 15 connections is classified as a Type III public water supply according to the Michigan Safe Drinking Water Act and may require a permit prior to the start of drilling when permits are issued for private wells by the local health department. All other public water supply wells require a permit prior to construction, with Type II permits being generally issued by the local health department; but permits for larger Type I supplies must be obtained from the Division of Water Supply of the Michigan Department of Public Health.

In the farming community other factors, in addition to simple horizontal separation, are also important in selecting the well location. It would be wise to observe surface drainage, make a guess at the possible groundwater gradient in the area, observe areas of possible flooding, and then locate the well upgradient from possible sources of contamination and out of any drainage way or possible flood area. For example, a well could be located the required 50 or 75 foot (15.24 m or 22.86 m) isolation distance from a barnyard, but be downgradient such that runoff from the barnyard or flooding conditions could cause highly contaminated water to pond directly around the well casing.

The same principle would apply to highly fertilized areas, particularly areas where fertilizers are stored and losses could occur and leach into the groundwater. Areas used for mixing herbicides and insecticides are also vulnerable areas and are a concern from a public health standpoint. The rules require that wells be located a minimum of 150 feet (45.72 m) from sites where chemicals are stored or mixed for field application. Other concerns to consider are separation from gasoline storage areas or areas where solvents or degreasers are used and could accidently be disposed of or spilled on the ground surface. Many problems are simply the result of poor housekeeping in respect to volatile organics.

SELECT WELL SPECIFICATIONS AND DRILLING PROCEDURE

After deciding upon the proper well location, it is then necessary for the owner to determine the size of the well casing, which depends upon the capacity needed from the well and the drilling procedure to be used.

Submersible pump installations are usually the systems of choice for the smaller capacity well at a residential home or small farming installation. The use of a submersible pump requires a minimum of 4 inch (10.16 cm) casing. A determination will have to be made to use either a 4 inch (10.16 cm) or larger steel casing or approved PVC well casing. The minimum size for PVC casing is 5 inch (12.7 cm) ID pipe with a minimum wall thickness specified by the Michigan Department of Public Health standards. It should also be pointed out that there is a minimum casing depth requirement stating that each well shall contain at least 25 feet (7.62 m) of protective outside casing and that this minimum depth applies even in high bedrock areas.

It is still permissible to install a 2 inch (5.08 cm) diameter well. Generally speaking, a 2 inch (5.08 cm) well produces a smaller quantity of water, and the smaller diameter wells are not as popular for new construction as they were several years ago. Accordingly, time will not be allocated to discussing construction of smaller diameter wells; however, the Michigan Department of Public Health is available to provide assistance on proper well construction and jet pump installation that are needed for 2 inch (5.08 cm) wells.

Basically, the 4, 5, and 6 inch (10.16 cm, 12.7 cm, and 15.24 cm) diameter cased wells are installed by use of a cable tool drilling machine or rotary equipment. The cable tool rig is the typical solid tool rig with a heavy iron drill connected to a steel cable strung over a walking beam to provide an up-and-down motion. The drill bit is used to grind up the earthen materials and a steel casing is driven in the drilled hole. This procedure requires that a casing be installed as the hole is being drilled.

On the other hand, the rotary construction method uses a bentonite drilling fluid as a medium to carry the cuttings to the surface and to form a filter cake in the bore hole. This permits drilling a hole without a casing until the water-bearing formation is encountered. At that point the casing, either steel or PVC, is installed. Either drilling procedure permits the well to be finished in a sand and gravel formation by installing and developing a screen. In areas where bedrock formations are to be used, the casing is set and then the bedrock is drilled out to provide an uncased opening into the water-bearing formation.

GROUTING

The need to properly grout the annulus between the well casing and the native soil formation is becoming more evident to protect the safety of the drinking water supply and the water producing-aquifer. The basic construction rules say that an open annulus shall not be left around the well casing, and such an annulus shall be properly grouted.

However, in the past, many contractors simply used bentonite drilling fluid and cuttings from a rotary hole for grouting, or believed that a cable tool installation provides a tight casing that does not leave an annulus around the well casing and thus needs no grouting. It is the opinion of the Michigan Department of Public Health and the result of field observations that with either type of construction an open annulus can be left around the well casing which must be properly sealed to provide reasonable assurance for the integrity of the well construction.

Grouting is particularly pertinent in agricultural areas where nitrate levels may be high or pesticide/herbicide chemicals are used in the crop production that have a possibility of entering the water-bearing formation by migrating along the water well casing. Accordingly, it is the recommendation of the Michigan Department of Public Health that regardless of the type of drilling procedure used, some type of grouting material be used to seal any possible annulus around the permanent well casing. Precise procedures are currently being developed to provide uniform requirements for grouting all water wells.

Currently, there is a multitude of materials available for grouting purposes. The western bentonite used as the basic component in a drilling slurry for rotary drilling is an excellent sealant. The drilling bentonite is a Montmorillonite sodium-based clay that expands several times its dry weight volume when fully extended due to simple water absorption. The secret in proper use of bentonite as a plugging or grouting material is to place it in the well annulus before it has become completely extended, which means it must be placed while it is in a dry condition or treated temporarily to retard the swelling caused by water absorption.

There have been some very interesting developments in the bentonite field during the last couple of years. It appears that most of the major bentonite producing companies are recognizing the need to properly grout wells, properly abandon existing wells, and are working to develop grouting or sealing materials for this purpose. Today, bentonite is available in a powdered form which is used for drilling purposes, in a granular form which can be used for grouting, and also in a chunky form screened to a 1/4 to 3/4 inch (6.35 mm to 19.05 mm) size which is available for plugging and sealing purposes. In addition, many of the companies are making a pelletized bentonite which is used for sealing, but this form is considerably more expensive due to the processing, and its use is generally restricted to monitoring well construction.

Neat cement, which is simply Portland cement and water mixed at a ratio of one bag cement to no more than six gallons (22.7 l) of clean water, is a recognized grouting material that is used routinely in both the water well and oil well drilling industry.

The secret in grouting is to place the material between the well casing and the native formation to completely fill any void that may exist. In rotary construction an oversized hole is drilled and the grout material can either be pumped down a grout pipe along side the casing or placed inside the casing and forced up the annulus to assure complete filling of the annulus. This office is currently working on procedures that can be used that will minimize the time and effort and provide for universal grouting and assure that at least the upper 25 to 30 feet (7.62 m to 9.14 m) of the well casing is properly grouted. Basically, the casing should be sealed to a depth of at least 25 to 30 feet (7.62 m to 9.14 m) below the ground surface upwards to a point where the pitless adapter is attached to the well casing. The pitless adapter connection for the well casing is generally located just below frost level and provides for a watertight and sanitary connection of the discharge pipe from the pump in the well to the point of use.

It should be pointed out that if cable tool drilling procedures are used, there is still a need to properly seal the small annulus around the casing. A recent development is the use of a coarse, approximately number 8, mesh bentonite to provide this casing seal by adding the coarse bentonite around the well casing during the drilling operations. A generous amount of dry, coarse-grade bentonite is kept around the casing at all times; and as the casing is actually driven into the ground, the bentonite tends to follow the casing. It is recognized that this method is not as precise, and there is no way of determining the depth the bentonite penetrates. However, a casing installed this way was excavated and bentonite was found to a depth of at least 30 feet (9.14 m) below the ground surface. It should be pointed out that there are other special grouts being developed that may likewise be used for sealing purposes.

There are certain locations where a neat cement grouting material must be used. Neat simply means pure cement mixed with water at a rate of no more than 6 gallons (22.7 l) of water per one 94 pound (42.63 kg) bag of Portland cement. The neat cement slurry can be pumped down the outside of the well casing by use of a grout pipe to the depth required to be grouted. There are other recognized methods of pumping neat cement down the inside of the casing which forces the grout up the annulus. Neat cement is required for construction of wells where the bedrock is within 25 (7.62 m) feet of the surface due to the broken and porous nature of most bedrock at the near surface. This type of geological condition permits contamination to be transmitted directly into the drinking water well or aquifer. This is a serious problem in the cavernous limestones found in Monroe County, Alpena, and the Presque Isle County area of northern Michigan and across some of the southern part of the Upper Peninsula.

Neat cement is also the grout of choice if a well is being drilled into a flowing well area. Neat cement grouts guarantee that the casing is properly sealed in place and prevent the casing from washing out due to the high head from a flowing well. Other advantages to the use of neat cement are that it is a readily available material, is relatively easy to handle, and gives a couple hours working time for placement of the grout slurry.

Properly grouting a well casing in agricultural areas is of paramount importance to safeguard drinking water supplies from possible migration of nitrates or other chemicals into the aquifer.

IRRIGATION WELLS

The same questions should be answered by a farmer contemplating installation of a large capacity irrigation well. The casing diameter must be large enough to provide installation of large capacity pumping equipment.

One of the most important decisions is related to pump selection for irrigation wells or other large capacity wells. It is recognized that an entire paper could be written to thoroughly discuss proper pump selection; however, there are two or three things that a person should know when selecting a large capacity irrigation pump. Generally speaking, the large capacity industrial pumps are driven at a nominal speed of 1750 rpm. However, pumps are available that are designed to operate at a nominal speed of 3500 rpm, and obviously the actual pumping capacity is twice that of a pump driven at only 1750 rpm. Thus, a smaller pump can be used to provide the same capacity. However, the head is increased by a factor of four, which makes the horse power increase by a factor of eight.

The information presented in Table 1 illustrates the effect that occurs when the rotational speed of pump is changed. The figures compare a single 6 inch (15.24 cm) pump impeller or stage of identical design, but operated at two different rpms.

Generally, the higher speed pump is used because the overall purchase cost is probably less, but the cost of operation has to be much higher. The cost of operation is generally not considered of prime importance in irrigation applications since a pump is only operated a few weeks out of the year. In selecting a pump, a pump bowl is selected to give peak efficiency at the desired production and then the number of impellers or stages is determined to meet the

Table 1. Pump operating characteristics as a function of rotational speed.

Pump Characteristics	1750 rpm	Pump Speed - - 3500 rpm - -	
Volume (GPM)	150	150	300
(l/min)	568	568	1136
Head Capacity/Stage (ft)	10.7	59	42.5
(m)	3.26	17.98	12.95
HP/Stage (hp)	0.55	3.45	4.4
Efficiency (percent)	74	63	72
Peak Efficiency	76% @ 135 gpm (511 l/min)		75% @ 260 gpm (984 l/min)

desired head requirements. Each impeller has a finite head output, and the number of impellers increases as the head requirement increases. However, the pump should not be ordered until the well is test pumped and a reliable well capacity is determined.

It is important to realize that pump size must be determined in order to determine the casing size before drilling or selecting the well casing for an irrigation well. Generally, speaking, one should consider the use of an 8 inch (20.32 cm) pump for a 600 gpm (2271 l/min) unit and perhaps a 10 inch (25.4 cm) or larger diameter pump for a 1,000 gmp (3785 l/min) output. It is recommended the casing be at least 2 inches (5.08 cm) larger than the nominal pump size. This guideline indicates the need to use a 10, 12, 14 or sometimes a 16 inch (25.4 cm, 30.48 cm, 35.56 cm, or 40.64 cm) casing for irrigation wells. The owner has a choice between steel casing or approved PVC well casing. However, at the present time, there are very few wells in Michigan installed in the irrigation field with PVC well casing. It appears to be quite popular in the western states, but the toughness and general reliability of steel casing has dictated its use in Michigan.

Cable tool drilling or a special type of rotary drilling called reverse circulation are the drilling procedures generally used for irrigation well installation. Again, in the cable tool procedure, the casing is driven as the hole is drilled. The reverse rotary operation requires drilling a large diameter hole, and the casing is then slid into the hole once the drilling is completed to the required depth. The reverse rotary procedure is used when it is desirable to use a gravel pack installation since a larger diameter hole can be drilled which permits room to install the gravel. The pack material is actually a coarse graded silica sand or small gravel installed around the well screen.

Graded silica gravel pack material is selected to match the size of the native sand or gravel formation in the aquifer. Then the slot size or openings in the screen is selected based upon the size of the gravel pack material.

The grouting comments previously made likewise apply to grouting an irrigation well. Special considerations are required when a gravel pack construction is employed because the top of the gravel pack must be properly sealed to avoid migration of native fine sands or silt into the pack material. And again the annulus between the casing and the natural earthen borehole must be properly sealed with either a neat cement or bentonite grouting material.

WELL DEVELOPMENT AND PUMP TESTING

Proper development of a screen or even a bedrock formation requires special techniques by the well drilling contractor to assure that the well produces the required quantity of water without excessive drawdown and without producing fine sand or silt. Development of a well may take as long as the actual drilling, and this is an area that should not be slighted by the owner; the cost and procedures used for development should be agreed upon with the contractor prior to starting construction.

Pump testing the well can be done along with the development, but a final test should be conducted to determine the reliable capacity of the completed well prior to ordering the large capacity pump. Pumping tests are normally run between eight and twenty-four hours with precise water level measurements being made regularly during the full length of the test.

PUMP INSTALLATION

It is vitally important that a sanitary procedure be followed when installing the pumping equipment. Basically, the well casing must terminate a foot (30.48 cm) above the natural grade level, the casing must be properly vented, and all other openings into the casing must be properly sealed to prevent the entrance of surface waters or vermin into the well. Michigan now requires a vermin-proof cap for the smaller diameter installations. This cap is likewise vented to provide for equalization of pressures within the well casing. This eliminates drawing extraneous materials into the casing when the water level in the well is drawn down during pump operation. Irrigation wells likewise require special cap and venting precautions due to the nature of the large diameter pumps being used. Irrigation wells are often installed in the lowest area of the field which, from a practical standpoint, makes common sense to the farmer since this area is not available for cropping. But low lying areas are often subject to flooding, and if the casing is not extended above flood level, the flood waters will actually enter the well.

Occasionally, irrigation wells are terminated only a few inches above the ground level in an area subject to flooding. The well casing is not sealed to the pump base, and this resulted in flood waters actually entering the well and aquifer. This can be a serious problem due to possible contamination of the aquifer along with contamination of the well. It could also result in silt and sand plugging the well which could limit the well's production capacity. It is important that well casings extend a minimum of 2 feet (60.96 cm) above any known flood level and that the pump base or discharge head be tightly sealed to the water well casing and all openings properly sealed or plugged. A screened vent should be provided, terminating above the top of the well casing and pointing downward to minimize possible contamination from entering the well through the vent. It is extremely important that an adequately sized vent be used to allow equalization of the atmospheric pressure within the well casing during the pump's operation.

One final comment relating to irrigation pump installations relates to the need to install a check valve on the pump discharge. A check valve is required at this point to minimize any back siphonage from the irrigation system into the well, and it likewise minimizes the possibility of injecting chemicals or fertilizers directly into the well. The whole subject of proper installation of the chemigation equipment has been covered in printed materials from the Agricultural Engineering Department at Michigan State University, and it is recommended that these safeguards be followed when installing this type of injection equipment.

DISINFECTION

All well and pump installations are required to be properly disinfected with chlorine prior to putting the system into service. Chlorine is the disinfectant of choice because it is readily available in either liquid or solid form and produces a chlorine residual that continues to be an active disinfectant over a period of time.

It is important that the chlorine solution come in contact with all parts of the well and pump to assure complete disinfection. This requires that a method of application be used to assure that chlorine solution reaches the bottom of the well. Chlorine concentration and contact time are the two critical variables. It is recommended that the aqueous chlorine solution in the well contain a minimum of 100 mg/l (0.0134 oz/gal) chlorine, and sometimes it may be necessary to increase the chlorine concentration to a much higher level of perhaps 1500 to 2000 mg/l (0.2 oz/gal to 0.267 oz/gal) when it is difficult to obtain safe bacteriological results. Note that many times it is necessary to repeat the chlorination process several times before all evidence of contamination has been eliminated.

Contact time between the chlorine, water, and the areas being disinfected is likewise necessary. The longer the time period that chlorine is left in the well, the better the chance for complete disinfection. A minimum of two hours is required, but generally, a longer contact time of up to twenty-four hours is recommended to assure production of bacteriologically safe water.

WATER ANALYSES AVAILABILITY

The Michigan Department of Public Health provides laboratory services for water quality analyses of drinking water wells in Michigan. This work is done in cooperation with the local county health department whereby the local health department collects most of the water samples which are then sent to the Lansing, Michigan, laboratory for analyses. Water samples from the Upper Peninsula can be sent to the Houghton Lab for bacteriological and partial chemical analyses. There are three types of sample analysis available through the local health department. The basic bacteriological analysis is run to determine the safety of a drinking water supply from a bacteriological standpoint, and this generally relates to whether the well and pump installation have been properly disinfected after installation or any repair. A basic chemical examination is likewise available which includes six routine measurements for iron, hardness, fluoride, chloride, sodium and nitrate in addition to specific conductance which is useful as a general measurement of the total solid content of the water. Bacteriological and partial chemical analyses are available by making a request to the local health department.

In addition, special analyses are also available with specific approval. This includes heavy metals or other nonroutine chemical analysis such as a complete mineral analysis, gas chromatograph type analysis for the halogenated and nonhalogenated organic volatile chemicals, plus analysis for nonvolatile chemicals such as pesticides and herbicides. The special metals and organic analyses are much more sophisticated and time-consuming than the routine chemical parameters and, accordingly, there is a finite laboratory capacity for this type of analysis. This means that specific approval has to be given on a site-by-site basis to obtain this service. The Michigan Department of Public Health laboratory is currently analyzing over 500 water samples for these special types of contaminants each month, and if there is a need for such a determination, authorization can be given. The local health department can frequently assist in collecting the proper water sample.

CONCLUSION

It is recognized that this discussion has covered many topics in a general way, and an in-depth review of several of the subjects would each require an individual paper. However, additional information is available from the staff of the Ground Water Quality Control Section of the Michigan Department of Public Health. In addition, each county has a local health department that provides consultation relating to well and pump installations.

PART V

REGULATION AND REMEDIAL ACTION

CHAPTER 20

OVERVIEW OF MICHIGAN GROUNDWATER LAW

Patricia M. Ryan
Leighton L. Leighty
Department of Resource Development
Michigan State University
East Lansing, Michigan 48824

Authors' Note: This article is written to provide an overview of the case law and statutes which constitute the current body of Michigan "groundwater law." It considers the common law embodied in Michigan appellate cases on "percolating water" and "subterranean water," and the cases decided under the growing body of statutory "environmental law" which regulate groundwater. While attempt was made to include pertinent laws and cases, the cases, statutes, and regulations discussed are not exhaustive. No single case or statute dictates Michigan law on this subject, and readers are cautioned to use the information as an overview.

INTRODUCTION

The courts employ the terms, percolating water and groundwater, interchangeably; therefore, groundwater has been defined, for legal purposes, as:

> . . .Those waters which slowly percolate or infiltrate their way through the sand, gravel, rock or soil, which do not then form a part of any body of water or the flow of any watercourse, surface or subterranean, but which may eventually find their way by force of gravity to some watercourse or other body of water, with whose water they mingle, and thereby lose their identity as percolating waters. 2 Kinney, *The Law of Irrigation and Water Rights* 2150 (2nd ed., 1912).

The term groundwater lacks a single precise legal definition. Its meaning depends on the language of the traditional or pragmatic views announced by court-made law (common law).

These waters ooze, seep, or filter through the soil without a defined channel or shape. Groundwater is found below the water table in the "zone of saturation." It is water contained in the interstitial spaces of rock formations and unconsolidated sediments. Availability of groundwater is generally related to porosity and permeability of bedrock and unconsolidated materials in any given area.

The Michigan law on percolating water includes both (a) the common law embodied in Michigan Supreme Court and Court of Appeals cases and (b) the case law evolving through interpretation of the statutes enacted since the early 1970s, generally referred to as "environmental laws." This overview will address both the common law and the newer body of statutory case law.

In the common law cases decided prior to enactment of the statutory environmental laws, the courts emphasize the property right of landowners (1) to withdraw groundwaters from wells on their private property, and (2) to use such water either on their land or at some location off the land, usually quite a distance from the well. These cases typically involve the quantity of the water available, and/or taken for use rather than the quality of the water, although the "reasonable use rule" also forbids unreasonable impairment of water quality (*Schenk v. City of Ann Arbor*, Appendix 1).

The basis for this body of Michigan groundwater law is the set of real property rights that attach to the waters subjacent to the land. In Michigan, this set of interests includes the right to use percolating water subject to the "reasonable use doctrine." This doctrine permits the landowner/user to make use of the water on the land above the aquifer and requires that any use off the overlying land must neither unreasonably interfere with the rights of adjacent or neighboring landowners to the reasonable use of subsurface waters from their lands, nor result in diminution of flow of the neighboring landowners' water, nor decrease the value of the land for agriculture, pasturage or other legitimate uses. This reasonable use rule dates to the *Schenk v. City of Ann Arbor* case (Appendix 1), decided in 1917. The *Schenk* reasonable use rule for groundwaters was reiterated as the Michigan rule in 1984 in a Court of Appeals case, *U.S. Aviex Co. v. Travelers Insurance Co.* (Appendix 1).

The other general body of Michigan law affecting groundwater is found in statutes regulating groundwater quality. These statutes include the underground injection portions of the Safe Drinking Water Act, the Resource Conservation and Recovery Act, the Comprehensive Environmental Response, Compensation, and Liability Act on the federal level and the Water Resources Commission Act, the Fire Prevention Code, the Michigan Environmental Protection Act, the

Michigan Environmental Response Act, the Solid Waste Management Act, and the Michigan Mineral Well Act on the state level (Appendix 2).

In some instances, these statutes require permits for the disposal of substances which might impair the quality of the groundwater. Other statutes, such as the Water Resources Commission Act, prohibit the discharge into waters of the state of any substance that is or may become injurious to public health or to the users of the water. None of the Michigan statutes regulate removal of any quantity of groundwater available in an aquifer, unlike many of the western states which do regulate the quantity of water which can be withdrawn from an aquifer under an appropriation system for groundwater.

If there is need for injunctive or declaratory relief, the Michigan Environmental Protection Act grants plaintiffs a cause of action, allowing them a day in court, if they can establish their burden of proof. A recent statute, the Michigan Environmental Response Act, provides a means of assuring the clean up of sites by authorizing the state Department of Natural Resources to clean up sites listed on a priority index established under the Act.

There are also administrative rules pertaining to groundwater which implement a nondegradation policy (Appendix 2). These rules are currently (March, 1987) under revision by the Michigan Department of Natural Resources, the agency responsible for promulgating and enforcing the groundwater rules under the Michigan Water Resources Commission Act.

COMMON LAW PROPERTY RIGHTS IN GROUNDWATER

The Right to Withdraw and Quantity Issues

The Michigan common law property rights cases decided by the Michigan Supreme Court and the Michigan Court of Appeals do not define "groundwater." Instead, the terms used in English case law, "percolating waters" and "subterranean waters," are used to describe underground waters of the State of Michigan. Only two recent cases (*Maerz v. U.S. Steel Co.* and *Attorney General v. Biewer*, Appendix 1) use the more scientifically accepted terminology of aquifers. The early case of *Schenk v. City of Ann Arbor* makes the distinction between percolating waters and subterranean waters flowing in a defined channel or contained in an underground lake or pond, noting in *dicta* that the water from gravel deposits at issue in that case were percolating waters.

According to the *Schenk* court, percolating water is water not flowing in a defined stream and without a definite channel, while subterranean waters are those flowing in a definite channel or stream. The majority of the Michigan cases involve glacial drift percolating waters. While the depth of the wells in the cases was not discernible from the reported cases in most instances, it would be of interest to determine whether the glacial drift was the only source of percolating water considered in Michigan common law. The ease of locating water may have played a significant role in Michigan in stifling the development of a definitive body of case law establishing rights to groundwater. It is obvious from the holding in *Joldersma v. Muskegon Development Company* (Appendix 1), that the technology did indeed play an important role in the opinion of the Justices when considering the law and equities.

In *Schenk v. City of Ann Arbor*, the Michigan Supreme Court set out the qualified right of a landowner to take such percolating waters as he may collect from his own lands, even though the landowner would use little or none of the water "upon, or for the benefit of, the land from which it takes it, or for its own benefit as landowner" (*Schenk* at 81). In *Schenk*, the City of Ann Arbor had drilled test wells for the supply of municipal water; the pumping of water from those wells caused the wells of neighboring landowners to go dry or to flow diminished quantities. The city was planning to drill additional wells and construct a waterworks; plaintiffs sued for injunctive relief and damages. The evidence showed that the plaintiffs' wells had ceased or decreased flow during the test but resumed flow when pumping ceased after completion of the tests. The city presented evidence at trial which demonstrated that 4,000,000 gallons per day could be pumped without lowering the head of the underground water or the water table more than it was lowered during the tests. The trial court awarded the Schenks damages for the cost of digging their well deeper, but did not enjoin the city from pumping the water from city owned land.

The Supreme Court in *Schenk* cites English and American cases supporting the absolute ownership rule, yet rejects that rule for Michigan. The Court noted that

> [T]he letter of the law. . .affirms the right of the owner of land to sink wells thereon, and use the water therefrom, supplied by percolation, in any way he chooses to use it, to allow it to flow away, even though he thereby diminishes the water in his neighbor's wells or dries them entirely, and even though in so doing he is actuated by malice. Such a right has been held a property right, which cannot be taken away or impaired by legislation, unless by the exercise of the right of eminent domain or the police power (*Schenk* at 82).

The *Schenk* Court rejected this absolute ownership rule in favor of the reasonable user rule, which the Court adopted from a New Jersey case, *Meeker v. City of East Orange* (Appendix 1). In framing its rule the Court found that the city, as a municipal corporation, had no greater rights than a private land owner to take and use the water, yet the city could reasonably make use, for the intended municipal water supply purposes, of a large volume of water from the land the city owned. The Court did permit modification of the decree to permit plaintiff to apply for equitable relief if the city's actions damaged plaintiffs in the future.

In the next case decided, *Bernard v. City of St. Louis* (Appendix 1), the Michigan Supreme Court ordered the modification of a decree "to require defendant not to interfere with an adequate supply of water for the plaintiffs' reasonable use," while plaintiffs had to conserve water and received compensation for their damages "sustain[ed] by having to install pumping machinery and other appliances" (*Bernard* at 165). As in the *Schenk* case, the quantity of water available was at issue, not the quality. The case involved a municipal corporation (the City of St. Louis) pumping and using the water for municipal water supply and fire protection. The city pumped the water from five wells drilled on land owned by the city located adjacent to the Park Hotel, a mineral spring spa with internal distribution of mineral waters to the hotel rooms. When the city had a fire, the wells did not provide sufficient water; the wells were shut down and the lines to the Pine River, water source prior to drilling the wells and building the reservoir, were reopened. The pumping of water from wells caused plaintiffs' mineral spring flowing well to cease flow, requiring installation of pumping equipment and resulting in litigation seeking injunctive relief to preclude defendants from resuming pumping to such an extent which diminished plaintiffs' flow and to force defendants to refrain from interfering with plaintiffs' pipeline or equipment. The trial court granted the injunction.

The *Bernard* Court described the case as one of first impression in Michigan; the rule on the issuance of injunctions in percolating water cases had not been heard previously. The cases and law cited indicate the intention of the Court to grant injunctive relief on cases involving percolating waters

> only upon the clearest showing that there is immediate danger of irreparable and substantial injury and that the diversion complained of is the real cause. . . (*Bernard* at 164 citing 27 R.C.L. at 1182).

The *Bernard* Court implies that injunctions will be denied if the plaintiffs stand by while development is made for public use, or if the party makes no use of groundwater from his land.

The Court concluded that ". . .if the city makes a reasonable use of the percolating waters and the plaintiffs do not permit it to go to waste. . ." there should be ample water for all parties. The Court did uphold the injunction because

> [if] there should not be [ample water] then the plaintiffs should not be deprived of a supply of water sufficient for their reasonable use without compensation, nor should

they be required to install new machinery without compensation (*Bernard* at 163).

Injunctions are generally granted, after such "balancing the equities," only when money damages cannot make the plaintiff whole.

In the *Davison v. City of Ann Arbor* case (Appendix 1), the plaintiffs sought an injunction against the operation of municipal wells and the assessment of damages against the city for damage to plaintiff's muck lands, which they contended the city had rendered valueless for agricultural uses by pumping groundwater. This case is one of the first in Michigan in which the Supreme Court describes in detail the scientific evidence about groundwater hydrology and geology, as presented at trial, to resolve the case. *Schenk* involved economic issues and municipal supply needs in addition to the scientific evidence. In *Davison*, the proof presented to the court included testimony of a University of Michigan Forestry professor that

> . . .in the area surrounding the pumping station there was a layer of muck varying in thickness, under which there was a very thick clay cap about 15 feet at the pumping station, *which was impervious to the movement of water*, and below that, a thick bed of water-bearing gravel from whence the water from the city was pumped by the city (*Davison* at 455).

This testimony was collaborated by a civil engineer witness who also testified that tests of the plaintiffs' wells indicated no variation in flow when the city wells were flowing and when they were not, and to the minor decrease in level of the total water table due to the city pumping. Additional testimony cited included a farmer and a fruit grower, both of whom testified that there would be no decrease in agricultural value due to the city pumping.

Plaintiffs' witnesses testified to the quality and value of muck land, its need for water in the soil, and the injury to muck land by the removal of water. No witness for the plaintiffs countered the testimony regarding the impervious clay. That this persuaded the court is apparent:

> [B]ut they gave no explanation of how the surface water could be pumped from the soil by pumps which extended below a considerable thickness of clay, impervious to water (*Davison* at 457).

The Court was persuaded by defendant City of Ann Arbor's evidence and held for the city that their use was reasonable. Hence, the pumping was permitted to continue.

Ruehs v. Schantz (Appendix 1), involved lands burdened by the natural flow of water from the adjacent, dominant estate owned by defendant, who also subtiled his land, causing additional water from the tiled water collection system to flow across plaintiffs' lands. The issue was whether defendant had the right to drain his lands and "cast" groundwater or subsurface water through such tile drains to lower adjoining farms through the tile drains; or, as defendant phrased the issue, whether the drainage would unreasonably increase the amount of water that will naturally flow onto plaintiffs' lands from defendant's lands. The Supreme Court held that the additional water from defendant's tile field was a damage and a trespass to plaintiffs, and permanently enjoined defendant from constructing and maintaining artificial drainage which may discharge water onto plaintiffs' lands in excess of the natural flow of surface water.

This case establishes the rule that subsurface tile drains may not be used to drain lands artificially in a manner different from the natural surface flow. It should be noted that *Ruehs* is really a near-surface water case, involving artificial water flows, along the line of *Peacock v. Stinchcomb* and *Miller v. Zahn* (Appendix 1) cited by the *Ruehs* Court. It is not a percolating water or subterranean water case.

In *Hart v. D'Agostini* (Appendix 1), the plaintiff sued for damages incurred in securing a temporary water supply during the time that defendant was installing sewer lines, causing plaintiff's water supply interruption. Citing *Schenk*, the Court of Appeals found that the sewer benefitted the area, that it was not unreasonable to have it constructed, and that there was no usage of the water off the property, as in *Schenk* and *Bernard*. The Court of Appeals, because of these findings of the facts, held that plaintiff had no right to damages. *Dicta* in the case provides the flowery statement that "[T]he right to enjoyment of the subterranean water beneath a person's land cannot be stated in the terms of an absolute right" (*Hart* at 321). The Court relied on the Restatement of Torts Sec. 822, at p. 226, Annotation, 29 ALR2d 1357 (1953) for the rule that

> [T]he liability for interference with the subterranean water supply of a neighbor. . .depend. . .upon whether the causative activity (1) if intentional, was unreasonable, or (2) if unintentional, was negligent (*Hart* at 322).

The Court concluded the construction was obviously intentional, and was reasonable. Therefore, plaintiff was not entitled to damages.

The Court of Appeals in *Hart* distinguished *Schenk* and *Bernard* because both those cases involved partial destruction of the water table, specifically the intentional removal of the water from the supply and the transport of it elsewhere for consumption.

In *Maerz v. U.S. Steel Corp.* (Appendix 1), the Court of Appeals interpreted the Michigan rule on groundwater use to be a correlative rights rule, instead of reasonable use. This case attempts to modify the rule in *Schenk*. The Court of Appeals in *Maerz* concluded that the trial court erred in granting summary judgment to U.S. Steel when U.S. Steel's uses of the groundwater caused the loss of plaintiff's well. This well provided the only source for plaintiff's potable water supply. In other words, U.S. Steel could not continue its use of groundwater for use in its mining operation, even though such use involved no diversion of groundwater.

Maerz interpreted *Schenk* as a correlative rights rule. The specific issue of whether waste is permitted for uses of water on the land from which it was withdrawn has not been decided by the Michigan Supreme Court, but the strong emphasis on reasonable use over absolute ownership by the Supreme Court, and the trend of appellate opinions would indicate the reasonable use rule applies to such uses also.

Stidham v. Algonquin Lake Association (Appendix 1) involves both surface waters and percolating water. Defendant Association controls the lake level artificially, and one year this allegedly caused plaintiff gravel business owner's well, water from which was used to wash the gravel, to go dry. Plaintiff attempted to establish that defendant

> . . .[b]reached its duty to use the lake waters in a reasonable manner, consistent with the rights of nearby subterranean water users when, aware that lake lowerings affected the subterranean water level, defendant nonetheless obtained permits in 1978, 1979, and 1980 to lower the lake level (*Stidham* at 97).

The Court of Appeals found that the defendant did not owe a duty to plaintiff to maintain the lake at certain levels. This case is the only Michigan case recognizing the interrelationship between groundwater and surface water. However, it is primarily a surface water case.

The Quality Issue

The only Michigan case brought solely on the issue of impaired water quality is *Joldersma v. Muskegon Development Company* (Appendix 1). The Court found that plaintiff had not met his burden of proof to prove that salt water contamination of his land was caused by defendant. Other cases involving quality rely primarily on statutory authority for relief from injuries sustained and, consequently, are discussed in the Statutory Case Law section below.

Joldersma (Appendix 1) is the first Michigan case involving conflicting uses of water allegedly changed in quality (due to defendant's operation of its oil well). In this case, the plaintiff operated a celery farm, and his irrigation water, drawn from groundwater sources on the farm, became unfit for use, allegedly due to defendant negligently dumping salt water on the ground near the well. The

Court described the location of the farm and noted the surrounding landforms, including the farm, a hill, a channel filled with refuse, oil, and salt water from oil wells, and a disposal pit used for disposal of salt water and refuse or bottom sediments from the oil well, a practice in accordance with the regulations and recommendations of the department of conservation in effect at the time of disposal.

The Court acknowledged that the plaintiffs' farm was worthless for raising celery. However, it held for the defendant because

> . . .plaintiffs have failed to produce sufficient evidence for the consideration of the jury upon which they might properly find that the salt deposited was due to a subterranean flow of percolating waters containing salt from defendant's sump (*Joldersma* at 523).

Facts influencing the Court's decision included evidence that four wells were drilled on plaintiff's farm, one of which was still producing salt water being drained to one of the ditches; wells had also been drilled on nearby property; the lack of testimony to establish, with any degree of accuracy, the direction the subterranean percolating waters would take from the point at which defendant's sump is located so that the jury could properly be permitted to find that the brine from the sump was thus carried to plaintiff's land. The Court noted

> that the question of the relationship, if any, between the salt water deposited in the defendant's sump and that to be found upon plaintiffs' land becomes highly speculative. The jury cannot be permitted to conjecture (*Joldersma* at 525).

Thus, the court found that the plaintiffs had not met their burden of proof to establish a relationship between the contents of defendant's sump and the plaintiffs' damage. In light of present technology and knowledge of groundwater hydrology, the facts in this case could enable a new plaintiff to meet such a burden of proof today.

COMMON LAW NUISANCE

Common law public nuisance doctrine has been used in Michigan to abate water pollution as early as 1913. The case of the *Attorney General v. Thomas Solvent Company* (Appendix 1), affirmed the trial court's mandatory preliminary injunction ordering defendant to abate groundwater contamination based on the trial court findings of common law nuisance and statutory public nuisance under the Water Resources Commission Act. Groundwater contamination, specifically toxic chemicals found in the public water supply to the City of Battle Creek and some 80 private residential water wells, in the case arose in part from defendant's business operation involving sales of solvents and other chemicals. The Appellate Court found that the

> [S]tatus quo ante is an unpolluted environment. It is clear that the status quo in this case is the maintenance of uncontaminated groundwater and soil (*Attorney General v. Thomas Solvent Co.* at 64).

While the case has definite significance, the result did not rest solely on common law nuisance theory. It is better viewed as a statutory law case involving a strong public interest factual issue, i.e., drinking water and public health, and involving known toxic chemicals such as toluene. The scope of the case will be determined in future cases, which may indicate the strength of common law nuisance as a basis for relief in groundwater cases.

One theory of liability used in litigation involving conflicting uses of percolating waters is nuisance theory which may apply whether the case involves water quantity extraction, or whether the suit focuses on pollution problems related to water quality. The basic standard for liability under nuisance theory is the question: Was the water use or other related activity (which invaded the plaintiff's use of the enjoyment of plaintiff's land) a substantial injury, and also reasonable under all the facts and circumstances placed in evidence before the trial court? Ironically, this theory allows a water user lawfully to intentionally (Restatement of Torts definition) injure a neighbor's water supply so long as the injuries are only "reasonable" ones. *Damnum absque injuria.*

Without legislative guidelines, the focus of the law concerning "reasonableness" has been case law. The experience in Michigan has been to produce "facts" at trials to determine whether a particular use (in that specific situation or set of circumstances) is "reasonable" or "unreasonable." There are two balancing processes involved. After the facts are presented, the trial judge (or jury) must determine the reasonableness of the use in that narrow context. This first process is sometimes called "balancing-the-interests." Under the "reasonable use rule," the interests to be balanced are the rights of each landowner to use the waters of an underlying aquifer in a reasonable and ordinary manner, with the understanding that all neighbors have the same rights and expectations of land use and water use. If the "balancing-of-interests" process indicates that the use is "unreasonable," then, by definition, an unlawful "nuisance" has been established. (If the use is "found" to be "reasonable," the case ends without further deliberation.) However, once a "nuisance" has been established, the next procedural step in the case is the second balancing process. Here the finder-of-facts (judge and/or jury) determines what remedy to apply against the defendant water user-- money damages to pay for losses and/or an injunction to stop the excessive use or to modify its impact and application. This second process is sometimes called "balancing-the-equities."

A Michigan statute has attempted to cover some of the water problems left unanswered or ambiguous by the *Schenk* decision. The statute, part of the Revised Judicature Act (Appendix 2), establishes that certain actions involving waste and/or depletion of a neighbor's water supply constitute a nuisance requiring abatement and allowing damages to the injured person or persons. It is clear under the statute that the alleged nuisance use must be the cause of the injury, and the water must be used unreasonably and unnecessarily. When these facts are shown, the judgment ordering abatement must specify the daily volume the offending well may flow and must establish a reasonable time for implementation of abatement. The statute also allows reopening of the judgment on the question of reasonable use if circumstances change or if there is another equitable reason to reopen the judgment.

This statute, however, leaves the definition (and application) of "reasonable use" to the court system which defines it one-case-at-a-time on the particular circumstances of each individual suit, with only slight assistance from the doctrine of *stare decisis*. Some behavioral patterns are always "unreasonable"--e.g., total destruction of a neighbor's access to the aquifer; malicious waste of water with no economic-profit motivation; contamination of groundwater when a causal relationship is clear. These listed extremes at one end of the behavioral spectrum do not produce ambiguity or uncertainty as frequently as the "gray" situations between the extremes. The determination of "reasonable use" is a factorial process, reviewing a number of "factors" and placing subjective weights or values on each factor.

In any event, factors should be listed (either by the legislature or by the courts) to evaluate any determination of "reasonable use." Physical characteristics, conflicting uses, community economics and local politics are just a few of the many diverse factors which must be considered when attempting to determine whether a use (in the specific context) is "reasonable." Physical characteristics might include, for example, the size of aquifer, recharge history, location of wells, drawdown expectations, transmissivity and storage coefficients, recovery period, recharge boundaries, type of soil materials, porosity, permeability, etc.

In the *Jones* case, *supra*, the defendant water authority constructed 25 wells for a municipal water supply and began pumping in July, 1973.

> [T]he plaintiffs. . .began to experience loss of water, lower water pressure, excessive pumping, and other well problems. . . .The Authority then contracted to have the situation studied. The study was completed in June, 1974. It indicated that certain of the Authority's wells created an excessive drawdown (depletion) when operated together and that the problems with the private wells in the area were directly caused by the Authority's well

operation. The study took into account the indirect effects of regional pumping in determining that pumping in Lansing and at Michigan State University did not cause the interference. Subsequent to receipt of this study, the Authority continued to pump the wells identified as creating the greatest interference.

.... The trial court found the construction and operation of the Authority's wells a "reasonable use" under the provisions of 1955 P.A. 233; M.C.L. Sec. 124.281 *et seq.*; M.S.A. Sec. 5.2769(51) *et seq.* This Act permits municipalities to incorporate authorities for the purpose of acquiring, owning, extending, improving and operating sewage disposal systems and water supply systems. ...

The Michigan Court of Appeals disagreed and reversed this decision. The case was remanded to the trial court for a determination of the damages issued. The Court of Appeals found a "nuisance" to exist and employed the above noted statute:

We find that it was unreasonable for the Authority to initiate a program in complete disregard of the effects its actions had on neighboring landowners, especially where viable alternatives for achieving the same ends were available and might have been discovered and implemented had a proper study been conducted prior to initiation of the program. Plaintiffs' claim is viable under M.C.L. Sec. 600.2941(2); M.S.A. Sec. 27A.2941(2).

The *Jones* case originally gave the hope that criteria for defining "reasonable use" could be established by the courts. On remand, the case was settled out-of-court, and no criteria were forthcoming. See *Jones v. East Lansing-Meridian Sewer & Water Authority, supra.*

STATUTORY ENVIRONMENTAL LAW

The growing body of environmental statutes involves many laws primarily directed to remedy existing or potential damage to the environment in general or to a specific natural resource, such as water laws prohibiting discharges into waters of the United States without a permit, such as the Federal Clean Water Act (Appendix 2). Many of the statutes regulate discharges to groundwater by reference, others accomplish this by regulating the substances themselves. Either way, groundwater seems amply protected from degradation from any nonpermitted or nonmanifested substance or waste if the substance is "toxic" or "hazardous." These are defined terms of art under state and federal statutes and regulations, but definitions are unique to the individual statutes.

These environmental laws exist at the federal and state levels, and most are accompanied by a battery of regulations dictating quality standards, procedures, and penalties. As even the novice to environmental issues knows, these laws and regulations have spawned the creation and propagation of new agencies and expanded jurisdiction for others. Examples include the U.S. Environmental Protection Agency, and the expanded role of the Michigan Department of Natural Resources. They have also generated a body of administrative law all their own which is outside the scope of this article.

FEDERAL LAWS

The primary federal laws (Appendix 2) involving groundwater are the:

Safe Drinking Water Act (SDWA), which requires the establishment of primary standards for drinking water supplies in the United States and regulates underground injection control (UIC) for disposal of substances by injection into wells.

Resource Conservation and Recovery Act (RCRA), in particular the sections regulating underground tanks and groundwater monitoring assessment; penalties are significant.

Comprehensive Environmental Response, Compensation, and Liability Act (CERCLA or Superfund), which specifically references groundwater and establishes clean up standards.

STATE LAWS: MICHIGAN

Pertinent Michigan statutes (Appendix 2) impacting groundwater, especially preservation of the quality of groundwater include the:

Michigan Water Resources Commission Act, administered by the Department of Natural Resources, it makes violation of this Act a *prima facie* nuisance, and establishes civil penalties.

Michigan Safe Drinking Water Act, which is administered by the Michigan Department of Public Health, which also administers the drinking water portions of the Federal Safe Drinking Water Act.

Fire Prevention Code, which regulates leak detection of storage tanks containing flammable liquids, above and underground.

Michigan Environmental Protection Act, which establishes a cause of action and allows plaintiffs a day in court.

Michigan Environmental Response Act, which is remedial in that it provides for identification, risk assessment, and priority evaluation of environmental contamination sites and permits the

Attorney General to sue for recovery of clean up costs expended by the state.

Michigan Mineral Well Act, which regulates deep well injection in Michigan.

Other statutes, such as the Great Lakes Conservation Act, which establishes the Great Lakes and Water Resource Planning Commission and charges it to compile all hydrologic cycle information in the state, involve groundwater but are beyond the scope of this overview.

STATUTORY CASE LAW: MICHIGAN

Note: The cases discussed in this section include only selected cases involving groundwater and certain of the above-cited statutes. There are numerous other cases under these statutes which are outside the scope of this overview.

Attorney General v. Thomas Solvent Co. (Appendix 1), is the most significant case involving groundwater in Michigan to date. All Michigan cases were decided at the Court of Appeals level; the Supreme Court has yet to address these statutes in the context of groundwater contamination, except by exercising denial of leave to appeal, which serves to leave the lower court ruling in effect.

In addition to the *Attorney General v. Thomas Solvent Co.* case, the Court of Appeals has addressed insurance contract coverage in environmental cases in *U.S. Aviex Co. v. Travelers Insurance Co.*, which involved whether defendant Travelers was liable for clean up costs under its policy with plaintiff after a fire destroyed plaintiff's chemical manufacturing facility and water used to extinguish the fire caused toxic chemicals to seep into the ground, contaminating the groundwater beneath plaintiff's property. The Department of Natural Resources had ordered plaintiff to investigate and correct the contamination. The case opines in *dicta* "[t]he damage to natural resources is simply measured in the cost to restore the water to its original state." Whether this standard is applied in a case awarding damages remains for future cases to tell us.

The other significant aspect of the case is the reaffirmation of the *Schenk* rule:

The "[r]easonable use" rule adopted in lieu of the English rule by the *Schenk* Court permitted use by a landowner of the percolating waters as long as such use did not interfere with a neighbor's reasonable use of the water beneath his land. In adopting this rule, the *Schenk* Court clearly rejected the right of absolute ownership over percolating groundwater:. . . .*Schenk v. City of Ann*

Arbor, supra, remains the law today (*U.S. Aviex Co.* at 591-2).

The other case involving groundwater and environmental statutes was brought under Michigan Environmental Protection Act and the Water Resources Commission Act by the Attorney General, *Attorney General v. John A. Biewer Co., Inc.* (Appendix 1). This case involved judicial construction of the Water Resources Commission Act, the time from which civil penalties should be calculated, appropriate remedies for violations (which the Court held to include both injunctive relief and the scope of remedies applicable to the law regarding public nuisances), and the appropriateness of duplicate remedy (requiring hook-up to city system at defendant's cost even after defendant paid for new wells to be drilled for damaged property owners with contaminated water supply). Scientific evidence cited by the Court in reaching this decision included the depth of the plume (60 feet approximately), and the lack of a clay barrier to prevent the contaminants in the upper part of the aquifer from migrating downward in the aquifer, and the presence of chromium in a new well. The facts in this case indicate blatant violations of the Water Resources Commission Act. Toxic chemicals (chromium and arsenic) had been found in both the old wells and the new wells. The impact of this case awaits outcome on appeal.

SUMMARY

Michigan case law establishes a reasonable use rule for withdrawals of groundwaters by property owners. Quantities of water withdrawn are not without limit. It is probable that this limitation will be applied to uses on the land from which the water is withdrawn, as well as for uses distant from the well site, if and when the Michigan Supreme Court accepts a case on these facts. To date, only one Appeals Court case (*Maerz v. U.S. Steel Corp.*) has decided this issue, and all Supreme Court cases have involved off-site uses with a significant public interest, namely a public drinking water supply.

On the quality side, it seems that the groundwaters of Michigan are aptly protected by statutory and common law. The significant penalties which can be imposed under these statutes, including criminal penalties and imprisonment, serve as the most significant protective measure. Additionally, such penalties encourage prevention of contamination as a sound business practice.

Where contamination has occurred, there are remedial measures at the state and federal level to enforce clean up; on-site clean up is preferred under the Superfund regulations. While a few more cases may reach appellate courts, most activity is likely to involve administrative law and reach appeal only in rare cases. The costs and potential penalties, compared with the strength of the statutes and balancing of the equities, will preclude most challenges by alleged violators.

In conclusion, Michigan groundwater is amply protected, if existing law is enforced. No new statutes are necessary.

APPENDIX 1

LIST OF CASE AUTHORITY

Attorney General v. John A. Biewer Co., Inc., 140 Mich App 1, 363 NW2d 712 (1985), leave to appeal applied for.

Attorney General v. Thomas Solvent Company, 146 Mich App 55, 380 NW2d 53 (1985).

Bernard v. City of St. Louis, 220 Mich 159, 189 NW2d 891 (1922).

Davison v. City of Ann Arbor, 237 Mich 453 (1927).

Hart v. D'Agostini, 7 Mich App 319, 151 NW2d 826 (1967).

Joldersma v. Muskegon Development Company, 286 Mich 520 (1938).

Maerz v. U.S. Steel Corp., 116 Mich App 710 (1982).

Meeker v. City of East Orange, 77 NJLaw 623, 74 Atl 379.

Miller v. Zahn, 264 Mich 306, 249 NW 862.

Peacock v. Stinchcomb, 189 Mich 301, 155 NW 349.

Ruehs v. Schantz, 309 Mich 245 (1944).

Schenk v. City of Ann Arbor, 196 Mich 75, 163 NW 109 (1917).

Stidham v. Algonquin Lake Association, 133 Mich App 94 (1984).

United States Aviex Co. v. Travelers Insurance Company, 125 Mich App 579, 339 NW2d 838 (1983).

APPENDIX 2

LIST OF STATUTORY AUTHORITY

FEDERAL STATUTES

Safe Drinking Water Act, 42 USC 300f *et seq.*

Resource Conservation and Recovery Act, 42 USC 6901 *et seq.*

Comprehensive Environmental Response, Compensation and Liability Act, 42 USC 9601 *et seq.*

MICHIGAN STATUTES

Michigan Water Resources Commission Act, MCL 323.1 *et seq.*, MSA 3.521 *et seq.*

Michigan Safe Drinking Water Act, MCL 325.1001 *et seq.*, MSA 14.427(1) *et seq.*

Fire Prevention Code, MCL 29.1 *et seq.*, MSA 4.559 *et seq.*

Michigan Environmental Protection Act, MCL 691.1201 *et seq.*, MSA 14.528(201) *et seq.*

Michigan Environmental Response Act, MCL 299.601 *et seq.*, MSA 13.32(1) *et seq.*

Michigan Mineral Well Act, MCL 319.211 *et seq.*, MSA 13.141(1) *et seq.*

Great Lakes Conservation Act, MCL 323.51 *et seq.*, MSA 13.31 *et seq.*

Revised Judicature Act, MCL 600.2941, MSA 27A.2941.

MICHIGAN REGULATIONS

Groundwater Quality Regulations ("Part 22" Rules), 1980 AACS R323.2201 through R323.2211.

CHAPTER 21

WISCONSIN PROGRAMS ON GROUNDWATER POLLUTION LIABILITY

Thomas J. Dawson
Wisconsin Public Intervenor
Wisconsin Department of Justice
Madison, Wisconsin 53707-7857

INTRODUCTION

The terms "liability" and "pollution" have to be defined initially to evaluate state programs for groundwater pollution in general and for that of Wisconsin in particular. To environmentalists, industry, pollution victims, and perhaps to a lesser extent, to regulators, liability is the risk or actuality of paying the costs of what has been termed "groundwater pollution." For the most part, these costs are considered here from the economic standpoint, although, of course, certain groundwater pollution consequences include noneconomic costs, such as physical and emotional injury for civil wrongs, and even imprisonment for criminal violations. Even these could be said to have economic value, too. However, this consideration of "liability" for groundwater pollution in Wisconsin involves civil, criminal, and regulatory liability, such as the costs of complying with regulatory orders for clean up and monitoring.

The term "pollution" also can mean different things. "Pollution" is the noun form of the verb "pollute," which is variously defined in the dictionary to mean: to make foul or unclean, dirty; to make impure. Synonyms include "soil," "befoul," "taint," "contaminate," "corrupt," or "debase." Unfortunately, these common definitions and synonyms offer little help in defining groundwater pollution for liability purposes, but Wisconsin has attempted to do so.

HOW IS WISCONSIN DEALING WITH
THE LIABILITY ISSUE?

Wisconsin's Definitions of Pollution

For various regulatory and liability purposes, the term "pollution" in Wisconsin takes several forms. For example, in chapter 144, Wisconsin's general water protection statute, the following definitions appear:

> "Environmental pollution" means the contaminating or rendering unclean or impure the air, land or waters of the state, or making the same injurious to public health, harmful for commercial or recreational use, or deleterious to fish, bird, animal or plant life (Sec. 144.01(3), Wis. Stats.).

> "Pollution" includes contaminating or rendering unclean or impure the waters of the state, or making the same injurious to public health, harmful for commercial or recreational use, or deleterious to fish, bird, animal or plant life (Sec. 144.01(10), Wis. Stats.).

The term "waters of the state" includes "groundwater, natural or artificial, public or private, within the state or its jurisdiction" (Sec. 144.01(19)). In addition, the term "water supply" is defined as "the sources and their surroundings from which water is supplied for drinking or domestic purposes" (Sec. 144.01(20)). Various specific forms of pollution are defined. They include the terms, "garbage," "hazardous substance," "industrial wastes," "other wastes," "refuse," "sewage," "solid waste," and "wastewater."

Wisconsin administers the federal Clean Water Act and the pollution discharge elimination system (WPDES) law, which regulates the discharge of pollutants from point sources into ground as well as surface water. It includes the following pollution definitions:

> "Pollutant" means any dredged spoil, solid waste, incinerator residue, sewage, garbage, refuse, oil, sewage sludge, munitions, chemical waste, biological materials, radioactive substance, heat, wrecked or discarded equipment, rock, sand, cellar dirt and industrial, municipal and agricultural waste discharged into water (Sec. 147.015(13), Wis. Stats.).

> "Pollution" means man-made or man-induced alteration of the chemical, physical, biological or radiological integrity of water (Sec. 147.015(14), Wis. Stats.).

"Toxic pollutants" are also defined (Sec. 147.015(17)).

Two nonstatutory definitions used in Wisconsin are worth mentioning. The first is the term "detrimental effect on ground or surface water." For the purposes of Wisconsin's solid waste laws, it means "having a significant damaging impact on ground or surface water quality for any present or future consumptive or nonconsumptive uses" (Sec. NR 180.04(16), Wis. Adm. Code.).

The other nonstatutory definition that has been used in Wisconsin groundwater pollution policy parlance for years is "anti-degradation," also known as "non-degradation." Wisconsin's anti-degradation policy, stated for the purposes of the point source discharge law, is "no waters of the state shall be lowered in quality" unless it has been demonstrated that "such a change" is justified as a result of economic and social development and no assigned uses are injured by the change. Aside from the policy being ill-defined, it is not being fully enforced in Wisconsin.

Last, but not least, Wisconsin's newly enacted groundwater protection law mandates agency regulatory actions that can result in considerable economic costs to polluters based upon the exceedence of groundwater standards for various substances. These groundwater standards may be viewed as one definition of "pollution."

Within Wisconsin's definitions of "pollution," the regulations are applied in ways that create various forms of liability. This analysis is limited to the provisions in Wisconsin law that are not typically found in either federal pollution laws or their state counterparts. Certainly, Wisconsin has many other regulatory permit and licensing laws which impose regulatory costs, forfeitures, fines, equitable remedies and imprisonment, which may be deemed by the regulated community as "liabilities," but they are not being considered here.

Wisconsin's Solid Waste Laws

Wisconsin administers the federal Resource, Conservation and Recovery Act (RCRA). Beyond that, Wisconsin has several programs and laws affecting liability arising out of the disposal of solid and hazardous wastes. They include the following.

1. Solid waste facility long-term care is required. In short, Wisconsin law (Sec. 144.441, Wis. Stats.) requires new solid waste operators, as a condition of licensing, to demonstrate and assume financial responsibility for facilities 20 or 30 years after the closing of the facility, based on an election of garbage "tipping fees," paid into a "waste management fund." Newly approved and licensed facilities may not escape the fees or the responsibility until the termination of the optional 20 or 30 year period after closing. The 20 year option does not apply to mining or hazardous waste facilities. After termination of the post-closing responsibility period, responsibility is assumed by the state waste management fund into which the facility has contributed. Currently, the fund must be maintained at levels

between $12 to $15 million for the possible clean up of these facilities beyond the responsibility periods.

2.   Waste site environmental repair cost recovery is allowed under Wisconsin's "environmental repair fund" (ERF) law. This mini-version of the federal superfund imposes various fees on operating solid waste facilities to pay for environmental repair of "orphaned" dumps and facilities not covered by the waste management fund. Under this law, the Wisconsin Department of Natural Resources has broad authority, although not yet the money, to take direct actions to avert environmental pollution at sites, to repair sites, establish long-term care programs, to restore the environment, replace private water supplies and take other actions to protect public health, safety welfare or the environment.

Liability comes into play when the department seeks to recover the costs of these remedial actions.   Persons who may be held responsible for repair costs include those who are defined as an "owner" or "operator," including subsidiaries or parent corporations, who (1) knew or should have known that disposal would likely cause release of the substance in the environment; (2) if the person violated applicable statutes, rules, plan approvals or special orders; or (3) the "person's action related to the disposal caused or contributed to the condition at the site or facility and would result in liability under common law in effect at the time the disposal occurred, based on standards of conduct for that person at the time the disposal occurred" (Sec. 144.442(9)(c), Wis. Stats.).

Another subsection in this statute, however, states that no common law liability and no statutory liability, which is provided in other statutes, for damages resulting from a site or facility is affected in any manner by this section.   The law states that voluntarily assumed remedial actions may not be considered as evidence or admission of liability.

3.   Citizen-initiated administrative lawsuits regarding solid and hazardous wastes are authorized.   Wisconsin law (Secs. 144.465 and 144.725, Wis. Stats.) authorizes any six or more citizens or any municipality to petition for review of solid or hazardous waste violations.   Upon the filing of these petitions, the Wisconsin Department of Natural Resources may either conduct an administrative hearing to determine the petition's merits or immediately initiate enforcement proceedings on its own.

Wisconsin's Mining Laws

During the last ten years, Wisconsin has developed a new set of laws to regulate mining, following the discovery of extensive copper, zinc and nickel deposits, including one of the larger copper/zinc deposits in North America by Exxon Minerals Corporation. Aside from the extensive environmental impact review and regulatory approval processes imposed on mining activities, the following statutory provisions warrant particular note.

1.   Administrative awards for mining-induced private water supply damage are provided. As part of Wisconsin's mining regulatory statutes, a person claiming damage to his or her private water supply from mining can be compensated through an administrative process. Under this law (Sec. 144.855(4), Wis. Stats.), a person files a claim with the Wisconsin Department of Natural Resources and with the local municipality if an immediate alternative source is needed. The municipality must supply necessary amounts of water to replace the damaged private water supply. If, after the hearing, the Wisconsin Department of Natural Resources determines that prospecting or mining is the principal cause of the damage to the private water supply, Wisconsin Department of Natural Resources is required to order the operator to provide water to the injured person, to reimburse the municipality for the costs of supplying water, and to pay compensation for any damages unreasonably inflicted on the victim as a result of the damage to his or her water supply up to $75,000 per claimant. If, however, the department concludes that mining is not the cause of the damage, the claimant must reimburse the town for the costs of supplying water. Failure of the operator to comply with a Wisconsin Department of Natural Resources order is grounds for suspension or revocation of the prospecting or mining permit.

2.   Citizens' suits to enforce the mining laws are allowed. Wisconsin's mining law has a citizen suit provision (Sec. 144.935, Wis. Stats.) for commencement of civil actions against either a mining operator for violations of mining law or against the Wisconsin Department of Natural Resources for failure to perform nondiscretionary duties. In addition, a court must award litigation costs, including reasonable attorney and expert witness fees, to the plaintiff if he or she prevails; and the court is also required to award treble damages to any plaintiff proving damages caused by a person mining without a permit or willfully violating mining laws, permits or orders.

Wisconsin's Point Source Laws

The Wisconsin Department of Natural Resources administers the National Pollution Discharge Elimination System (NPDES) permit program in Wisconsin. In addition to the usual "liabilities" that are entailed in the administration of that program, two additional provisions in Wisconsin's point source law are worth mentioning.

1.   Wisconsin point source law applies to groundwater, as well as to discharges into surface water (Secs. 147.02; 147.015(20), Wis. Stats.).

2.   Wisconsin's point source law imposes strict liability on violators for prosecutorial and remedial costs. Under Sec. 147.23, Wis. Stats., Wisconsin Department of Natural Resources "may recover

the cost of removing, terminating or remedying the adverse effects upon the water environment resulting from the unlawful discharge or deposit of pollutants into the waters of the state, including the cost of replacing fish or other wildlife destroyed by the discharge or deposit." The term "adverse effects" includes Wisconsin's definition of "environmental pollution."

Wisconsin's Groundwater Law

In May, 1984, Wisconsin's governor signed into law 1983 Wisconsin Act 410, Wisconsin's groundwater law. The law amended existing groundwater programs, while creating new ones. For example, this law created the environmental repair fund (ERF) mentioned above, a contaminated well compensation program and a program requiring mandatory agency actions to be prompted by specifically created groundwater pollution standards. Two other programs in the groundwater law also affect liability.

1.   Mandatory prohibitions of polluting facilities or substances are required. Under Wisconsin's groundwater law, two sets of numerical standards must be established for substances that have been, or are likely to be, found in Wisconsin's groundwaters (Ch. 160, Wis. Stats.). The first set of numerical standards, called "enforcement standards," (ES) are based on either federal numbers such as EPA approved maximum contaminant levels (MCLs) or, in their absence, on a statutorily stated methodology used by the Wisconsin Department of Health and Social Services. The second set of standards, called "preventive action limits" (PALs), are percentages of the enforcement standards. For example, for carcinogens, the PAL is 10 percent of the enforceable standard; for other substances of "public health concern," the PAL is 20 percent of the enforceable standard.

When an enforcement standard is attained or exceeded, the regulatory agency with jurisdiction over that substance or facility causing the exceedence, has no discretion but to: (1) prohibit the activity or practice which uses or produces the substance; and (2) implement remedial actions with respect to the specific site in order to achieve compliance with the enforcement standard. The prohibition may not be prevented or lifted unless it can be shown "to a reasonable certainty, by the greater weight of the credible evidence" that an alternative response will achieve compliance with the enforcement standard (Sec. 160.25(1), Wis. Stats.). A wide range of remedial actions are authorized as long as they are "reasonably related in time and scope to the substance, activity or practice which caused the enforcement standard to be attained or exceeded."

In the event that the PAL is attained or exceeded, the regulatory agency with jurisdiction over the facility or substance is required to assess the cause and implement responses designed to: (1) minimize the concentration of the substance in the groundwater where technically and economically feasible; (2) regain and maintain compliance with the PAL where technically and economically feasible; and (3) ensure that the enforcement standard is not attained or exceeded. At the PAL level, the agency may not prohibit the activity or substance unless it is able to meet the burden of showing that no other alternative is available to prevent exceedence of the enforcement standard.

The liability, at least in terms of costs imposed on regulated entities, should be apparent. If a facility or substance causing exceedence of these standards is prohibited or restricted, the costs of these actions should become clear.

In Wisconsin, the pesticide, aldicarb, marketed under the trade name Temik, has been prohibited over extensive areas. The costs to growers, and less so to the manufacturer, are tangible. It also remains to be seen what effect Wisconsin's groundwater standards have in civil liability cases to help define actionable damage to water supplies. For example, people are warned not to drink or use water contaminated at or above an enforceable standard. Relevant to this question is another liability-related aspect of the groundwater law, the contaminated well compensation program.

2. Well water compensation with state subrogation to claims is provided. The groundwater law created a modest program of compensation (Sec. 144.027, Wis. Stats.). Initially created for private water supplies, it now covers some municipal water supplies (Sec. 144.28, Wis. Stats.). Unless claims are prorated according to the amount in the compensation fund, compensation is potentially available for 80 percent of the cost of repair or replacement of a well, with a limit of $9,600 for a private well, while municipalities can receive grants of up to 60 percent of eligible costs if they make claims to supply water to private polluted well owners.

Contributory negligence is not a bar to recovery, although payment of claims based on fraud or intentional contamination is barred. Contamination is defined for the purposes of this program as exceedence of the enforcement standards. Because the state is subrogated to the rights of the claimant, the liability for the award may ultimately end up in the lap of a polluter who has caused the contamination. At least that is the intent.

In addition to these laws, Wisconsin has three other statutory provisions of note that can impose significant liability on groundwater polluters. These independent statutory provisions deal with damaged water supplies, hazardous substance spills and a unique citizens suit provision.

Wisconsin's Water Supply Damage Law

Under Wisconsin's water supply damage law (Sec. 144.265, Wis. States.), if Wisconsin Department of Natural Resources finds that a "regulated activity" has caused a private water supply to become "contaminated, polluted or unfit for consumption by humans, livestock or poultry," it may, after a hearing, order the owner or operator of the regulated activity "to treat the water to render it fit for consumption by humans, livestock and poultry, repair the private water supply or replace the water supply and reimburse the town, village or city for the cost of providing water. . ." For the purposes of this law, a "regulated activity" means one for which the department may issue an order under any other provisions of Wisconsin's water pollution law or in violation of licenses, permits, orders or rules. For example, a "regulated activity" recently has been interpreted to include one for which Wisconsin Department of Natural Resources is authorized to issue orders to clean up hazardous substance spills.

Wisconsin's Hazardous Substance Spills Law

Wisconsin's hazardous substance spills law provides in part:

> A person who possesses or controls a hazardous substance which is discharged or who causes a discharge of a hazardous substance shall take the actions necessary to restore the environment to the extent practicable and minimize the harmful effects from the discharge to the air, lands or waters of the state (Sec. 144.76(3), Wis. Stats.).

The term "discharge" is broadly defined as including, but not limited to, "spilling, leaking, pumping, pouring, emitting, emptying or dumping."

The Wisconsin Supreme Court recently held responsible under this statute a subsequent purchaser of land that was found to be leaking and emitting hazardous substances from the property into groundwater, even though that person was not responsible for placing the substances on the property in the first place. (*State v. Mauthe*, 123 Wis2d 288, 366 NW2d 871 (1985)). The statute also imposes notification of discharge requirements on persons, authorizes the Wisconsin Department of Natural Resources to require preventive measures to be taken, and requires responsible persons to reimburse the Wisconsin Department of Natural Resources for actual and necessary expenses incurred in carrying out its duties under the law.

Wisconsin's 6-Citizen Complaint Statute

Last, but not least, Wisconsin's statutory framework for imposing liability for groundwater pollution includes the 6-citizen complaint law. One of the oldest of the statutes mentioned, this law requires the Wisconsin Department of Natural Resources to hold a hearing relating to alleged or potential pollution upon the verified complaint of six or more citizens filed with the department. This underutilized statute empowers Wisconsin Department of Natural Resources to remedy actual or threatened environmental pollution which, again, is defined broadly as "the contaminating or rendering unclean or impure the air, land or waters of the state, or making the same injurious to public health, harmful for commercial or recreational use, or deleterious to fish, bird, animal or plant life." Although this law has not been used recently in groundwater pollution cases, it has been used successfully in air pollution cases, particularly in difficult cases involving odor.

Besides these Wisconsin statutes on the groundwater pollution liability issue, one Wisconsin Supreme Court case should be noted concerning the issue of proximate cause in tort cases.

Collins v. Eli Lilly Company

Although not a groundwater pollution case, the Wisconsin Supreme Court case of Collins v. Eli Lilly Company, 116 Wis2d 166, 342 NW2d 37 (1984) appears to be consistent with the trend in various courts to liberalize the concept of proximate cause in tort cases.   In this one, plaintiff Collins alleged personal injury as a result of diethylstilbestrol (DES) taken by her mother.   Although defendant Eli Lilly marketed the drug along with several companies, plaintiff was not able to prove that the particular DES taken by her mother was manufactured by the defendant.   The lower courts had dismissed her case because she was not able to prove that Eli Lilly manufactured the particular DES alleged to have caused the injury.

The Wisconsin Supreme Court reversed after reviewing several existing causation theories and ultimately created one of its own to allow plaintiff a theory for recovery.   Adopting a modified "risk contribution" theory, the court held that where a defendant contributed to the *risk* of injury to the public and, consequently, the risk of injury to individual plaintiffs, each defendant shares, in some measure, a degree of culpability in producing or marketing the product that caused the injury.   The court held the company potentially liable on policy grounds.   Quoting the decision:

> We conclude, however, that as between the plaintiff, who probably is not at fault, and the defendants, who may have provided the product which caused the injury, the interests of justice and fundamental fairness demand that the latter should bear the cost of injury.   Accordingly, we have formulated a method of recovery for plaintiffs in DES cases in Wisconsin. . . .   Moreover, as between the injured plaintiffs and the possibly responsible drug company, the drug company is in a better position to absorb the cost of the injury.   The drug company can either insure itself against liability, absorb the damage award, or pass the cost along to the consuming public as a cost of doing business.   We conclude that it is better

to have drug companies or consumers share the costs of
the injury than to place this burden solely on the innocent plaintiff. Finally, the cost of damages awards
will act as an incentive for drug companies to test
adequately the drugs they place on the market for general medical use (*Collins*, 116 Wis2d at 191-192).

We are hopeful that this case, coupled with the Supreme Court's
interpretation of the hazardous substance spills law in *Wisconsin v.
Mauthe, supra*, indicates a trend toward establishing liability against
those that have not only caused injury to people and the environment, but against those contributing to the risk of such injury.
However, other issues also have to be taken into account.

POLICY OPTIONS FOR ADDRESSING LIABILITY:
ADVANTAGES AND DISADVANTAGES

Numerous major policy issues and options for addressing liability have been encountered in recent years in Wisconsin. The first
and all encompassing issue for addressing liability is whether, and to
what extent, people recognize and are willing to address the true
costs of groundwater pollution. The term "liability" is merely an
expression of actual and potential costs that are imposed by polluters
on society, on individuals or on the environment. As the Wisconsin
Supreme Court saw it in the *Collins* case, the overriding public policy
issue is on whom the costs of groundwater pollution should be
imposed. As between the innocent groundwater pollution victim,
society, the environment, or the polluter, who should bear these
costs? This public policy issue transcends many of the other issues
such as the burdens of proving proximate cause, assessing damages,
and taking remedial actions. The following issues are subsets of this
larger one.

A second issue is, what policies must be established regarding
groundwater? The policy choices include: "anti-degradation," "nondegradation," "no detrimental effect," and attainment of numerical
standards. Policy choices and definitions must be agreed upon and
adopted. Then a determination must be made as to whether the
policies of prevention, maintenance and restoration should be placed
on equal footing or be given priorities.

A third issue is to whom will the authority and assignment of
duties be given to carry out regulatory and nonregulatory actions to
achieve the policies and goals. Will one agency be assigned to
conduct these programs as a "central unit of government," or will
different agencies have concurrent authority?

Fourth, what is the extent of authority that will be given to
agencies in order to achieve the goals, and will that authority be
accompanied by nondiscretionary duties to adopt programs, take
actions and achieve goals? Unfortunately, granting an agency the

discretion to do everything is the same as granting it the discretion to do nothing.

Fifth, what about the victims of contamination who must often absorb the costs of groundwater pollution?   First, they must be defined. Are they only domestic well owners? Are they municipal well owners?   Are they businesses?   Are they farmers?   Are they animals and plants? Is the role of state and federal government to help victims, especially those that have no real recourse for remedy or compensation?

Sixth, who's going to pay for the administration of regulatory, compensation, research or other programs?   How will they be paid for?

Seventh, what regulatory and nonregulatory incentives for achievement of goals will be provided to polluters?   What proper mixture of carrot and stick is right, both with respect to preventing and to remedying damage that has already been done?

Eighth, on whose shoulders will the burden of proof fall when pollution occurs? Will it fall on victims to show harm when drinking water guidelines or standards are violated? Will it fall on regulatory agencies to show actual or imminent harm from pollution before they can take necessary actions?   At what levels of contamination will presumptions work in favor of victims and regulatory agencies and place burdens of proof on polluters?

Ninth, is an adequate monitoring program established to prove that what is not known is not necessarily bliss?   If groundwater pollution is not searched for, it only follows that too often it will not be found.   What kinds of groundwater monitoring are needed? Are adequate standards established to assure its reliability?   And again, who's going to pay for it, and how?

Tenth, how will one deal with the contamination that has already been caused and cannot be prevented, versus ongoing polluting activities and future sources of contamination?   Questions arise concerning who will be held accountable, what standards of conduct will apply to past and present actors, and the allocation of society's burden versus that of past and present polluters.

In light of these issues, the options start to become clear. The costs of groundwater pollution can be borne by victims, society, the environment or the polluters, or some combination thereof.

Pollution can be defined ambiguously or specifically. Although ambiguous definitions of pollution, such as "rendering unclean or impure" can be enforceable, they tend not to be enforced. Although numerical standards can provide necessary triggers for establishing liability, the political fight over what those numbers will be is considerable.

Although an argument exists to make one regulatory agency accountable for the protection of groundwater quality and the assessment of liability against polluters, it may be equally healthy to create institutional rivalry between regulatory agencies with concurrent jurisdiction over groundwater.

While groundwater pollution may call for the granting of extensive authority to agencies to deal with it, political and economic pressures can thwart the exercise of that authority. So, while enforceable nondiscretionary duties requiring agencies to take specific actions should accompany extensive grants of authority, the difficult balance to strike is defining the actions to be taken while maintaining flexibility to respond to different groundwater pollution circumstances.

Are groundwater pollution victims expected to be left to an antiquated and ineffective tort system to compensate them, or will Twentieth Century concepts for compensating them finally be implemented? Options include the development of liberalized toxic tort and proximate cause standards; authorizing regulatory agencies as well as courts to replace damaged water supplies; compensating for incidental damages; and ordering restitution in appropriate cases.

Regulatory, compensation, research and other programs can be paid for either through tax dollars or through fees imposed on activities that are known to cause groundwater pollution problems. Both regulatory and nonregulatory incentives for the achievement of goals should be provided to polluters, but with no illusion about the fact that there are always going to be bad apples in the best barrel of responsible corporate and other actors.

The burdens of proof imposed on plaintiffs and regulators can be lightened on the basis of sound public policy grounds, left alone for the courts to evolve, or statutorily constricted back to Seventeenth Century standards.

One can go on assuming there is no groundwater pollution problem where it hasn't been looked for, or necessary monitoring programs to better discover and respond to groundwater pollution can be developed.

Lastly, predominantly reactive programs can be enacted that are doomed to condemn groundwater supplies to perpetual and unacceptable contamination, or serious preventive measures can be taken now that will save billions of dollars in the long run.

THE RATIONALE BEHIND THE OPTIONS
CHOSEN IN WISCONSIN

Except with respect to the two Wisconsin Supreme Court decisions whose formal rationales are explained, the rationale behind legislatively adopted policies and laws may not have been expressed. Although most state and federal laws may begin with good ideas, their finally adopted details are molded by power and influence, not by purely reasoned judgment. For example, the Wisconsin ground-water law was adopted because it ultimately had the support of environmentalists, industry and agriculture. That was not because any one of those groups placed a proposal on the table that was instantly agreed on by the other two. A two year process of bargaining, negotiating, trading, arguing, heart-wrenching, and power brokering molded the Wisconsin groundwater law. It is what it is because of the political pressures that were brought to bear on it. Industry got something; environmentalists got something; and agriculture got something out of it. The same process occurred with regard to all of the other laws that were described.

In short, these laws are not either stronger or weaker because of the counterbalancing influence wielded by the political and economic interests in the policy-making process.

IMPACT OF THE LIABILITY PROVISIONS
OF THE GROUNDWATER PROGRAM

The answer to this question is that some are working while others are not. Despite the limited progress exhibited in the *Collins* case, the tort system may well be the poorest for compensating victims and society, and imposing liabilities on those responsible despite cries about the "liability crisis" to the contrary. While the insurance industry is quick to heap notoriety on multimillion dollar judgments which are needed to respond to the costs of multimillion dollar damages to people and the environment, the injury to many more groundwater pollution victims goes largely unreported, much less compensated. Homeowners report that their properties have been rendered unmarketable by groundwater pollution, mothers decry children who have been drinking toxins from household taps, small businessmen's livelihoods are threatened, farmers are on the brink of bankruptcy because their cows have been killed or injured by contaminated well water, retired couples truck water several miles a day in plastic bottles to bathe in and to drink. These are the people who can't afford lawyers' fees, or can't even interest lawyers in contingent fee contracts or *pro bono* work. These are the ones who can't afford chemical and hydrogeologic tests or expertise to prove causation even if they can get a lawyer. They are the ones who are told by bureaucrats that the role of government is to regulate, not serve or compensate. They are the ones that have no meaningful access to the tort system, much less a chance of being compensated even if they did. As the insurance and polluting industries talk about the "liability crisis," one should not forget the perhaps larger "compensation crisis." The tort system is not delivering nearly the compensation it should to deserving victims.

Recent interpretations of the Wisconsin Hazardous Substance Spill Law and the Water Supply Damage Law offer promise for quicker administratively ordered replacements of contaminated wells. And although the Wisconsin contaminated well compensation program has replaced dozens of wells in Wisconsin, it will not survive unless a way is found to fund it in the long term, preferably based on fees imposed on polluting facilities and substances.

The bottom line question is, "Are most groundwater polluters being held liable for the costs imposed on individuals and society under Wisconsin programs?" The answer is probably "no." First, they must be found. Then the obstructions to holding them accountable, without depriving them of fairness and due process of law, must be broken down.

ADVICE FOR OTHER STATES

First, although groundwater pollution costs government, society, consumers and industry, unfortunately, those costs are largely being borne by the wrong individuals and the wrong segments of society. The costs of pollution should be borne by those who cause and directly benefit from it. That, after all, is the basis for imposing liability. The way to do that is to identify those costs, such as by establishing monitoring and compensation programs and reimposing those costs back on polluters who can then either insure themselves or pass those costs on to their consumers and customers who benefit from the polluting activity. Therefore, costs of pollution should be deflected back into enterprises as readily as they were externalized into society.

Define pollution carefully. That definition will be an ally or enemy. Start to determine the costs of groundwater pollution by first monitoring drinking and other groundwater supplies. Take away from government the discretion to be ineffective by defining triggers when actions should be taken and by specifying the actions to be taken when the triggers are pulled.

Don't wait for the federal government to deal with the issue of groundwater pollution or its concomitant liability questions. By the time the federal government develops numerical contaminant standards, it may be too late or they may not be strict enough. Start now, drawing on the progress that has been made at the federal and other state levels.

Don't seriously limit liability for groundwater pollution under the guise of "Tort Reform" or under any other. Robert Reno wrote in a recent *Newsday* article, "The question a heavily insured society must confront is: At what point does the prevalence of insurance against risk promote the incidence of loss?" A related question is, "At what point does the limitation against risk of liability promote the incidence of loss?"

The answers to groundwater pollution and its costs to victims and society are prevention, remediation and compensation. It is not appropriate to shift the costs from risk takers onto pollution victims. Tort liability claims should be unnecessary, not futile. The way to reduce the risks of groundwater pollution liability is to reduce the risks of groundwater pollution in the first place. The way to reduce the risks of exorbitant tort claims is to provide a more responsive system to compensate victims and society so as to pre-empt the need to resort to the tort system.

In the meantime, it is unconscionable to even suggest that the "answer" is to shift groundwater pollution liability costs to pollution victims and society, which is the essence of "tort reform." If the risks of the activities and products causing groundwater pollution (including those unknown at the time they are created) are truly worth taking, then those who created those risks must assume their costs and their real price. The liability "problem," if there is one, is how, not whether, liability will be imposed, people compensated and the environment restored.

Pollution victims usually don't like lawyers and the tort system any more than the insurance industry and polluters do. Most victims simply want to be made whole again. They seek out tort lawyers only as a last resort after polluters, insurers, and government have failed to protect and compensate them for new wells, water supplies, loss of property values and other damages. Their chances of receiving recovery are already limited by the existing tort system without heaping on them the additional costs of "tort reform." Even successful plaintiffs often are not fully compensated because of attorneys' fees and expenses.

Lastly, business and industry must become a part of or partner in the solution, instead of remaining the problem. The answer to groundwater pollution and the compensation crisis is not the creation of another problem called "tort reform." The problem of groundwater pollution liability will be corrected by solving the problem of groundwater pollution. Industry must be willing to take responsible stands and offer constructive alternative solutions. Limiting liability doesn't help. Fingering irresponsible polluters does. Supporting business and property purchaser right-to-know laws does. Actively supporting regulatory agency staffing and resources does. Working with environmentalists and victims on real solutions does. Offering effective and efficient compensation systems for actual damages does.

CHAPTER 22

COUNTY LEADERSHIP FOR GROUNDWATER CONCERNS

Donald J. Brown
Western Michigan University
Science for Citizens Center
Kalamazoo, Michigan 49008

THE ACTORS

Who's responsible for groundwater quality and its protection? Everyone knows the alphabet soup that usually comes back in response to that question! Environmental Protection Agency, Department of Natural Resource, U.S. Geological Survey, Michigan Department of Public Health, U.S. Department of Agriculture, the local health department, the city, township, or state. The DNR's *Groundwater Strategy for Michigan* (DNR, 1981) identified four state commissions, four state departments, 37 divisions and units, eight federal agencies and six types of local agencies which have diverse and sometimes overlapping roles and responsibilities in groundwater management. Where then is an individual citizen or members of a community concerned about private or local water quality to break into this maze?

A COUNTY FOCUS

The originators of the Southwest Michigan Program saw the county as an appropriate and manageable sized governmental unit to address the problem. County health departments are the first point of contact for many people with drinking water concerns. It is the county that issues well permits, sites septic systems, tests drinking water, issues health advisories and in general is the public's first contact in instances of public health concerns. In most cases, county personnel work closely enough and frequently enough with constituent clientele to enjoy a degree of confidence not necessarily held by representatives of larger state or federal agencies.

Typically, counties have at least a small staff trained in water quality assessment and public health protection. Hence, the public health departments of member counties have been the focal points of much of this program.

GOALS AND OBJECTIVES OF THE SURVEY PROGRAM

The purpose of the Southwest Michigan Groundwater Survey and Monitoring Program is to safeguard the public health by using systematic methods to identify and protect the quality of groundwater in the region. Funded by a W.K. Kellogg Foundation grant, the Program operates a full-scale project in five counties and a more limited program in five additional counties. Program goals and objectives are the following.

Goals

(1)   establish a computerized geologic data base sufficient to identify and map the significant aquifers in the region

(2)   establish a computerized water quality data base characterizing baseline or background water quality

(3)   provide training in each member county to enable establishment and continuance of the data bases

(4)   conduct an extensive program of chemical analysis of groundwater

(5)   provide training in each member county to enable sanitarians to interpret and apply geological and water quality data to public health and water quality protection

(6)   cooperate with all appropriate entities to produce a model system and program for state-wide and broader application

(7)   develop and execute a public education project focusing on groundwater and public health

Objectives

(1)   design, develop and deliver necessary training

(2)   create, test and install necessary software

(3)   evaluate, select and establish a data base management system, graphics, mapping and other supportive software and provide necessary interfaces

(4)   integrate existing data into both data bases

(5)   establish uploading and downloading capabilities to a selected mainframe computer and demonstrate data transfer

(6)   provide necessary technical expertise to member counties

(7)   establish or facilitate mechanisms for continuance of the groundwater program

(8)   extend appropriate aspects of the program beyond the original service district

(9)   produce and distribute a groundwater guide and participate in public education activities relating public health and groundwater quality protection

(10)   provide extensive public notification of the outcomes of the project to facilitate broad utilization of its methods

ACCOMPLISHMENTS

It is not possible in a short paper to cover all aspects of the Survey. An outline of central activities, however, would be as follows. Sanitarians select well records believed to contain reliable information, and the locations of the wells are confirmed. A unique identifying number is assigned to the well, spatial coordinates are determined and its location computer plotted. Sufficient wells are selected, typically one hundred per township, such that a basic understanding of the township's aquifers can be inferred when geologic cross sections are plotted from the well records. Cross checking with previous aquifer studies is done when they exist. Approximately 24 wells are selected so as to provide chemical sampling that represents the important segments of the township's aquifers. Site visits and permission to sample are arranged with the well owners and scheduled. Site inspections provide essential information on the water system and well environment for rating the vulnerability of the well to various sources of contamination and its appropriateness for follow-up testing for organic contaminants, pesticides and herbicides.

Following a prescribed protocol, testing of pH, temperature and conductivity are done and a sample drawn for inorganic analysis. Results of the inorganic analysis and site evaluation contribute to a decision to return for further sampling. Over two thousand samplings covering 35 parameters (Table 1) will be done in the five full-study counties. Results are compared to maximum contaminant levels prescribed by statute and to recommended maximum contaminant levels. Computer generated, but personalized, letters inform well owners of the sampling results, health implications and follow-up

Table 1.   Inorganic water quality parameters and levels of detectability (Anon., 1985).

PARAMETERS, DETECTION LIMITS (mg/l), AND MCLs (mg/l)

pH, temperature, conductance

membrane filter coliform bacteria	(1 per 100 ml)*	
chloride	(250)	
fluoride	(1.5 - 2.4)	
hardness (calcium carbonate)		
nitrate	(10)	

aluminum	0.03	manganese	0.01	(0.05)
antimony	0.10	mercury	0.05	(0.002)
barium	0.005 (1.0)	molybdenum	0.005	
boron	0.10	nickel	0.10	
calcium	0.10	phosphorus	0.10	
cadmium	0.01 (0.01)	potassium	0.02	
chromium	0.01 (0.05)	selenium	0.05	(0.01)
cobalt	0.01	sodium	0.10	
copper	0.005 (1.0)	thallium	0.05	
iron	0.05 (0.30)	zinc	0.005 (5.0)	
magnesium	0.01			

alkalinity	1.00	lead	0.003 (0.05)
ammonia	0.05	silica	1.00
arsenic	0.002 (0.05)	sulfate	0.66 (250)

---

*Maximum contaminant levels (MCLs) and suggested MCLs, where established, are indicated in parentheses.

indicated, if any.   The inorganic chemical data base constructed serves as a primary source for interpreting the hydrochemistry of the aquifer.

Sampling for organic contaminants, pesticides and herbicides (Table 2) is done only at sites selected according to a vulnerability rating scheme which includes local land use and history as well as physical characteristics of the site.   All instances of apparent contamination are confirmed and followed up on according to health department protocol.

Templates and computer programs have been created by the Survey and are used to record the contents of well records (Table 3), inorganic/physical data (Table 4), organic analyses, and other well site information needed by the health departments for management

Table 2. Organic water quality parameters and levels of detectability.

Purgeable Halocarbons (0.001 mg/l)

Methylene Chloride   1,1,1-Trichloroethane
1,1-Dichloroethylene   Bromodichloromethane
1,1-Dichloroethane   Trichloroethylene
t-1,2-Dichloroethylene   Dibromochloromethane
Chloroform   Bromoform
1,2-Dichloroethane   Tetrachloroethylene

Purgeable Aromatic Hydrocarbons (0.001 mg/l)

Benzene
Ethylbenzene
Toluene
Styrene
Xylene

Chlorinated Pesticides (0.001 mg/l)

Aldrin   Heptachlor
Chlordane   Heptachlorapoxide
DDD   Heptachlorcyclohexane
DDE   Lindane
DDT   Isophorone
Dieldrin   TCDD
Endosulfan   Toxaphene
Endrin

Acid Herbicides (0.001 mg/l)

2,4-D
Silvex

Miscellaneous Pesticides (0.01 mg/l)

Alachlor
Triazine Herbicides
Atrazine
Simazine

---

purposes (Table 5). A data base management system (Condor 3) allows any of the data to be regrouped, retrieved, compared or displayed in any form desired. For instance, records for wells at certain depths, located in specific aquifers, containing certain concentrations of chemicals, drilled before or after a specified date,

Table 3. Well Log File.

```
      Local Health Department WELOG File:          Pilot Project          v1.0
*****************************************************************************

LOCATION:   W.NO 230400301   (ctttssxx)          LOC.E 1913900   LOC.N 461300
            ADD 7312 Old River Trail             JUR
            Quarter Section  QS SW NE SW         Well Head  ELE         840

WELL:       W.DPTH 155     CMPL.DATE 08/17/84    RIG 3           W.USE         1
CASINGS:    C.SIZE 4.0 in. C.DPTH 87 ft.         C.HGTH 1        (+, - surface)
SCREEN:     S.TYPE none    S.DIAM .00 in.        S.LNTH .0       S.BETW       .0
PUMPING:    SWL 50         PG.LVL 50             PG.HRS .0       PG.RATE       0
WELL HEAD:  CMPL.TYP 1     GROUTED 1             GROUT.TYP 4     GROUT.DPTH   87
CONTAM:     CNTM.DIS 75    CNTM.DIR E            CNTM.TYP SEP
PUMP:       PUMP.IN 1      PUMP.DPL 1            PUMP.CAP 10     g.p.m.
OTHER:      DRY.HOLE 0     DRILLER# 0            PERMIT 574      ETC ow plugged

FORMATIONS: S  12 BOT  16   S  10 BOT  21   S  31 BOT  51   S  65 BOT  155
            S   0 BOT   0   S   0 BOT   0   S   0 BOT   0   S   0 BOT    0
            S   0 BOT   0   S   0 BOT   0   S   0 BOT   0   S   0 BOT    0
            S   0 BOT   0   S   0 BOT   0   S   0 BOT   0   S   0 BOT    0
            S   0 BOT   0   S   0 BOT   0   S   0 BOT   0   S   0 BOT    0
            S   0 BOT   0   S   0 BOT   0   S   0 BOT   0   S   0 BOT    0
            S   0 BOT   0   S   0 BOT   0   S   0 BOT   0   S   0 BOT    0
            S   0 BOT   0   S   0 BOT   0   S   0 BOT   0   S   0 BOT    0
            S   0 BOT   0   S   0 BOT   0   S   0 BOT   0   S   0 BOT    0
            S   0 BOT   0   S   0 BOT   0   S   0 BOT   0   S   0 BOT    0
```

Table 4. Inorganic File.

```
     Local Health Department Partial    CHEM    File: Pilot Project    v1.1
*********************************************************************************
LOCATION:  W.NO  230400301    (cctttssxx)    LAB1 MSU    LAB2 MDPH    LAB3 SNELL
           L.NAME                                        COL.DT 01/20/86

*Values ppm except Con. (MMHOS)   *Not Detected as -det.level   *Not Tested as 0.00

Alkalinity      300.000      Aluminum     -0.030      Ammonia        0.100
Antimony         -0.050      Arsenic       0.007      Barium         0.100
Boron            -0.010      Cadmium      -0.010      Calcium       82.000
Chloride        -10.000      Cobalt       -0.010      Chromium      -0.010
Conductivity    553.000      Copper       -0.005      Fluoride       0.400
Hardness        278.000      Iron          0.900      Iron.2         0.300
Lead             -0.050      Magnesium    29.000      Manganese      0.020
Mercury          -0.050      Molybdenum   -0.010      Nickel        -0.100
Nitrate          -0.030      pH            7.360      Phosphorus    -0.100
Potassium         1.200      Selenium     -0.050      Silica        23.000
Sodium            5.000      Sodium.2    -10.000      Sulfate       48.000
Detergents       -0.010      Thallium     -0.050      Zinc           0.100
TDS               0.000      C         V   0.000      C         V    0.000

          Note Apple Orchard              Note2
```

C and V are "spaceholders" for additional parameters and their numerical values.

Table 5. Administration File.

```
   Local Health Department WQADM File    Groundwater Survey Project    V.3.0
 *******************************************************************************

 LOCATION:  [W.NO]_____  (cctttsssxx)              [JUR]_____  (cctttssdddxxx)
            [LOC.E]_____ [LOC.N]_____               [TR]_____  [Q]_____
            [UTM.E]_____ [UTM.N]_____               [TWSP]_____
            [ADD]_____                        [OF.NAME]_____
 OWNER:     [OL.NAME]_____                         [O.CITY]_____
            [O.ADD]_____                          [OTHER]_____
            [O.ST]_____ [O.ZIP]_____

 Health Department Activity:
 *******************************************************************************
            Service Category [SC]____      Activity [AA]___ [AB]___ [AC]___
            Permit App. Date [PD]____      [FEE]____
            Fee Paid Date [FPD]____
            Provider Code [PC]____         Init. Inspect. Date [IID]____
            Service Time [ST]____          Permit Issue Date [PID]____
            Permit Exp. Date [PXD]____     Permit Status [PS]____
            Last Activity [LA]____

 WELL
 ACTIVITY: Completed [CMPL.DATE]____       Log Recorded Date [WLD]____
 COMMENTS: [COM]____

 *******************************************************************************
```

constructed in specified ways or other information can be retrieved and the contents of the records compared or statistically examined. Graphic display (Figure 1) and mapping programs (Figure 2) facilitate data interpretation. All of these program-tested procedures and computer applications are described in detail in the Survey's *User's Guide* (SCS, 1986).

As a result of the Survey, water quality will be known at the region of entry into an aquifer, throughout the aquifer and at discharge points from the aquifer. With continued monitoring, deviations from this baseline water quality will become apparent and early detection of possible pollution made more likely. Reduced cleanup costs and less extensive pollution should become possible. The identification of vulnerable recharge areas should facilitate initiation of needed aquifer protection efforts.

HIGH TECH/LOW TECH BALANCES

After thorough consideration, the program opted for using the considerable power and versatility of modern microcomputers instead of minis or mainframes which are much more costly and carry much more severe personnel commitments. Dramatic recent expansion in the range of software designed for micros contributed to the decision, and similar conclusions were embodied in the recently released Michigan *Statewide Groundwater Data Base Strategy* (DMB, 1986). Elegant but costly and nonessential technologies have been avoided in favor of methods which can be utilized and economically supported by health departments or other local governmental units. Sophisticated data treatment and research follow-up utilizing the data bases can be accomplished in research institutions.   However, graphing, plotting and statistical functions important in applying the data from the Survey to the needs of the county and state are integral parts of the program.

Topo maps, sample bottles, boots, pencils and hours in the field remain the low tech essentials of a sanitarian's day. But computer-generated letters to well owners, maps, records and reports make the days easier and more productive.

AN ACTION TEAM

Delivering such a complex program has absorbed much of the staff time of the Science for Citizens Center at Western Michigan University and would have been impossible without the eleven technical consultants and several students employed by the Project.

Also invaluable has been the input of a Science Advisory Group representing state and federal agencies and university disciplines. County health departments have committed sanitarians and supervisory

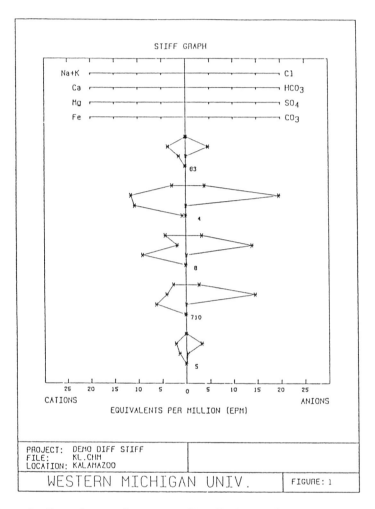

Figure 1. One of several water quality display options.

time as well as physical facilities. In some cases, sanitarians began with no computer experience and no computer available to their section. Now they all can do the essential computer work, run many of the applications programs and are beginning to feel comfortable with geologic and water quality interpretations. Individualized as well as group instruction has made this possible.

The Michigan Department of Public Health routinely contributes to portions of the inorganic analyses, provides occasional checks on results and follows up on any reported instances of contamination.

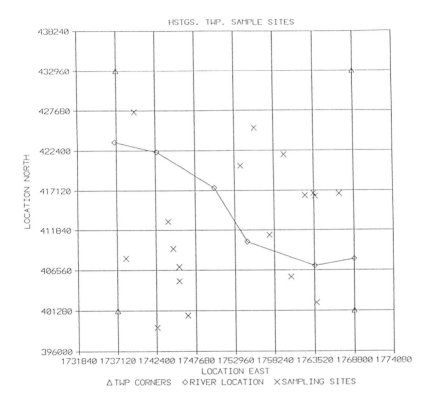

Figure 2. Location map.

All other analyses are conducted by private laboratories and the Michigan State University Animal Diagnostic Laboratory. Western Michigan University's Geology Department,its Program in Water Resources and Contaminant Hydrology, its Center for Statistical Services and Center for Social Research have provided essential services.

A FOUNDATION ON WHICH TO BUILD

The Project has demonstrated the possibility of developing groundwater expertise at the county health department level and has produced the necessary training program and materials to extend the program to other counties.   A computerized data base of perhaps 20,000 well records and of over 80,000 chemical test results soon will be in place along with the mechanisms for their continued expansion.

Originally covering five counties, the program now encompasses nine with work also being done in a tenth county and several more considering affiliating with the Project. The *Statewide Groundwater Data Base Strategy*, published in April, 1986, by the Department of Management and Budget Office of Management and Information Systems (DMB, 1986) recognized the Southwest Michigan Program as an appropriate model to be adapted and supported statewide. In September, an Advisory Board was established to develop an implementation strategy. The Strategy derives from and builds upon the earlier *Groundwater Protection Initiatives* (CCEP, 1984) and *Local Government Roles* (DNR, 1985) statements. Hence, while a great deal remains to be determined concerning the details of establishing and supporting the system, Michigan clearly is moving forward in groundwater management, and county governments have a major role to play.

LITERATURE CITED

Anon. 1985. National Primary Drinking Water Regulations: Part IV. Synthetic Organic Chemicals, Inorganic Chemicals and Microorganisms. Fed. Reg. 40:141.

CCEP. 1984. Groundwater Protection Initiatives. Cabinet Council on Environmental Protection, Lansing, MI 48909.

DMB. 1986. A Statewide Groundwater Data Base Strategy. Office of Management and Information Systems, Department of Management and Budget, Lansing, MI 48909.

DNR. 1981. Groundwater Management Strategy for Michigan: Agency Roles and Authorization in Groundwater Management. Water Quality Division, Michigan Department of Natural Resources, Lansing, MI 48909.

DNR. 1985. Local Roles in Groundwater Protection and Management. Local Groundwater Advisory Group, Michigan Department of Natural Resources, Lansing, MI 48909.

SCS. 1986. Southwest Michigan Groundwater Survey and Monitoring Program Users Guide. First Revision. Science for Citizens Center, Western Michigan University, Kalamazoo, MI 49008.

CHAPTER 23

RURAL GROUNDWATER CONTAMINATION: IMPACTS OF AND POTENTIAL BENEFITS FROM LAND USE PLANNING AND ZONING

Lawrence W. Libby[1]
Department of Agricultural Economics
Michigan State University
East Lansing, Michigan 48824

Jessica T. Kovan[2]
Department of Agricultural Economics
Michigan State University
East Lansing, Michigan 48824

INTRODUCTION

It is the contention of this paper that the behavioral roots of groundwater contamination are reasonably well understood. Contamination is an unfortunate side effect of reasonable and predictable actions by rational people in pursuit of valid objectives. It is essentially a residual of human response to incentives that define the broad "terms of trade" in society. Thus, any reduction in contamination will require adjustments in those incentives in ways that will appropriately alter human behavior.

There is certainly much to be known about the sources, fate and effects of substances that pollute groundwater. Contamination is a highly diverse and complex phenomenon that requires careful analysis. Understanding the physical and biological character of the problem and its related threats to human health are essential

---

[1]Present address: Department of Food and Resource Economics, University of Florida, Gainesville, Florida 32611.

[2]Present address: Institute of Water Research, Michigan State University, East Lansing, Michigan 48824.

precursors to effective policy action. Treatment of contaminated water supplies requires the coordinated effort of engineers, biologists and other scientists. But avoiding future contamination is first and foremost an institutional problem.

This paper focuses on the institutional nature of groundwater contamination and its reduction. Special attention is given to institutional devices available to local governments to adjust the incentives which produce groundwater problems.

SOURCES OF CONTAMINATION

The facts of groundwater problems vary significantly from place to place. This diversity itself is a source of great difficulty in designing consistent area-wide solutions. The same land uses produce very different groundwater impacts from one area to another. This physical and biological complexity is acknowledged but will not be pursued further here. Others have done so (Bouwer, 1978; Pye, 1983; Kittleson, 1986). For simplicity and to accommodate the general institutional approach of this paper, it is assumed that the primary sources of contamination in Michigan are waste disposal sites (including sewage treatment systems), underground gasoline storage, and farms.

Waste

Land continues to be the primary medium for waste treatment in Michigan and the United States. Water passing through decomposing waste can transmit organic and inorganic pollutants directly to the aquifer. Outflow from a conventional or land disposal system for municipal waste or even a home septic system may carry nitrates to the water supply. Kittleson and others at this conference have documented many instances of waste-induced contamination of groundwater.

The obvious options are to find a medium other than land for disposing of potential contaminants, to impede water flow through the waste, or to locate waste facilities away from recharge areas where leaking to the aquifer is at least unlikely. There are various policy instruments to encourage human action consistent with those options.

Farms

Potential contaminants from farms include nutrients, pesticides and other toxic organics, and salinity. Phosphates and nitrates are residuals of fertilizers. Nitrates can also be leached from concentration of animal waste or decomposing plant residue. Intensive commercial agriculture is increasingly dependent on organic pesticides.

Application and particularly storage of these materials is a potential source of contamination (see Pratt, 1985).

Better information on the links between farm practices and groundwater problems can help the farmer make appropriate production decisions, particularly if those practices affect the farmer's own water supply. A rational person would likely try to avoid putting him or herself at risk. Perhaps the farmer is using too much fertilizer or pesticide, creating problems for groundwater at personal cost. But much of the farm-related contamination is likely the unexpected byproduct of the farmer seeking to do the best job possible within the rights that define land ownership. Other water users experience the unfortunate impacts of those rational actions. Institutional change is essential.

Gasoline Storage

An increasingly bothersome source of groundwater contamination is the uncharted network of gasoline storage tanks laced throughout the countryside. Some of these tanks are abandoned, no longer functional, but continue to leak water soluble contaminants. Benzene, a suspected carcinogen, is the most damaging constituent of gasoline. By sheer number of contamination sites, the gasoline storage system is the most troublesome contamination source in Michigan and in other states. Lack of information on number and location of these tanks makes control difficult.

THE INSTITUTIONAL CHARACTER OF GROUNDWATER QUALITY

The physical and chemical properties of groundwater are matters for careful analysis to understand the attributes of the resource. Physical factors define movement of water vertically or laterally within the aquifer and the relationship to soil and other land features. Chemical properties are the link to human health. Understanding human risks associated with chemical change in water is a formidable challenge. The essential purpose of groundwater quality policy is to alter individual behavior affecting those chemical properties. Quality changes provide a gauge to management. But the institutional properties of groundwater create the real challenge for effective policy design.

Exclusion Cost

Groundwater is basically an open access resource available to the surface land owner who can capture its services. In most states, including Michigan, quantity limits are established through common law definition of "reasonable use." Only when someone contests the reasonableness of another's rate of withdrawal are the lines between

users drawn more sharply. Those limits could be established--
allocation of supply is technically feasible and has been accomplished
in some states (see Carriker, 1985). There has not yet been political
demand for groundwater allocation in Michigan.

The quality attributes of groundwater are also open access
resources but much less easily assigned. A land user whose actions
on the surface affect water quality is reducing supply as much or
more than is a person withdrawing from the aquifer, but the link
between action and water availability is less easily defined. Exclusion
costs for attributes of groundwater quality are much higher than for
quantity. The cost is particularly high when contamination sources
are diffuse or nonpoint. Government has the authority to exclude
those who unreasonably use the quality attributes and has done so
through regulation. But cost of administering such exclusions can
become prohibitive and politically unacceptable when sources and
magnitudes are uncertain. It is known, for example, that farm
applications of nitrogen fertilizers and pesticides can under some
circumstances contaminate groundwater. The cost of excluding all of
these potential contaminators would be very high. Even if high risk
farm practices could be defined and regulated, and the political cost
of doing so were within reasonable bounds, enforcement would be a
problem. A key policy challenge is to reduce ambiguity as to source,
but groundwater quality will remain a high exclusion cost good.

Information Cost

Secondly, groundwater quality is characterized by high information cost. Reassigning rights to water in ways that will retain or
increase quality depends on availability of information on those
actions that damage quality and their subsequent links to human
health. Further, individuals whose actions affect quantity or quality
lack information on other users or even the consequence of their own
use. Human perception of the risks associated with groundwater use
affects their own actions, and their relationships to other users.

There are three elements to individual risk: (1) how land use
activities affect the groundwater; (2) how an individual's health is
impacted by the contamination; and (3) the risk preference of an
individual. The information available on these aspects will determine
how contamination episodes will impact community members. Two
neighbors who have both been warned that particular irrigation
activities will likely contaminate their wells may react differently.
One neighbor continues the polluting practices while drinking the well
water; the other immediately stops the polluting activity and also
buys bottled water. Knowledge of this risk and information variable
is essential in establishing policy emphasis, for the emotional and
physical effects of contamination will impact some individuals more
than others.

Existence Value

Many people place a certain intrinsic value on the very existence of a clean water supply. It is part of the general feeling of security associated with a well ordered society. This value goes beyond water use or concern for potential health impacts to the very existence of the resource itself. Fear about the possibility of losing that resource in the future, of sacrificing even the option for future use, creates real stress that may be translated into support for policy to avoid contamination. Value placed on the resource will be larger for those systems which are rare or threatened in some manner. As general awareness of the threat of widespread contamination of groundwater increases, the existence value which individuals place on that resource will heighten as well.

Incompatible Use

The effect one person's use of a good has on another's use can vary in many ways. When that person's actions deny consumption by another, the two uses are incompatible. Hence, an incompatible use good is one which has the aspect of two or more uses or users which are not congruous (Schmid, 1978). Groundwater has this quality. Varying quality requirements from the same aquifer can create conflicts between users, and it is not feasible for two users to both consume the same water. Two users can *share* an aquifer while requiring different quality if the contamination plume does not interfere. Two such different needs are not always incompatible. Generally, however, an aquifer should not be used as a waste dump by one and a drinking water source by another. Capture and use of those quality attributes by one user makes them unavailable to others. If an aquifer is of the quality which can be used for irrigation, but not for drinking, then the water available for drinking has been limited by its quality. Recognizing incompatible uses of the groundwater will focus attention to the immediate problem of conflicting interests in the resource.

Groundwater quality has a variety of attributes, including incompatible use properties, existence value characteristics, high information costs, and high exclusion costs. Interdependence among these attributes establishes the setting within which contamination occurs. In most contamination episodes, ignorance and uncertainty are rampant among all parties involved. Further, one must fully understand these characteristics to be able to predict how policy alternatives will affect the behavior of individuals, and thus alter the resulting performance.

THE FEDERAL AND STATE CONTEXT FOR LOCAL POLICY

Groundwater quality policy seems to be one of those areas where tremendous uncertainty combined with substantial health risk have produced the incentive for institutional innovation. In the absence of clear information on presence or absence of hazard, there is the incentive to reduce the possibility of serious irreversible problems (Milon, 1986). Objective analysis of management alternatives would place weights on various sources of risk (based on probability distributions) and true uncertainty. Spofford et al. (1986) identify four sources of system-wide uncertainty in groundwater quality: a) sources of contamination, b) transport of contaminants in soil, c) transport of contaminants in the saturated zone, and d) economic losses associated with contamination.

Federal Level

There is no single statute or program at the federal level to protect groundwater. In fact, the federal system has been labeled a fragmented mess (Getches, 1978). There are special purpose laws, however, targeted at parts of the problem. Agriculture's contribution to groundwater problems are addressed as part of the Clean Water Act amendments to the Federal Water Pollution Control Act of 1972. The only implementing strategy is voluntary installation of management practices to reduce chances of contamination. Groundwaters are not specifically included.

Waste disposal in landfills is addressed by the Resource Conservation and Recovery Act with related incentives for recycling of some waste products. There are standards and procedures for waste handling and disposal. The Federal Insecticide, Fungicide and Rodenticide Act requires pesticide labeling and testing and registration of new chemicals. The Safe Drinking Water Act of 1974 established "maximum contaminant levels" for certain contaminants of community water supplies. The Comprehensive Environmental Response, Compensation and Liability Act creates a funding mechanism for cleaning up chemical spills of other hazardous waste threats. These programs are summarized in Appendix I.

Recent federal policy initiatives seek to target groundwater protection efforts to avoid the inefficiency of treating all supplies and contaminants the same. The 1980 Groundwater Protection Strategy differentiates between Class I "special" groundwaters, Class II "current and potential drinking water supplies" and Class III "not potential sources of drinking water." These priorities, then, will drive implementation of "superfund," hazardous waste management rules and any other opportunities for federal level regulation. The assumption is that the social losses associated with "irreplaceable" or "ecologically vital" water sources in Class I are greater that those in II or III. The practical difficulty is the collection and analysis of sufficient information, given the uncertainties noted above, to arrive at a defensible classification scheme. Standards have been defined (US EPA, 1984). Again, implementation is limited to those situations where a federal permit is required. EPA reports that classification may be lower priority than more immediate pollution issues such as pesticide use.

The 1986 amendments to the Safe Drinking Water Act require states to develop programs to protect existing wellheads serving community water systems. States are also encouraged to define and protect recharge areas. The Sole Source Aquifer Demonstration Program encourages state development of comprehensive groundwater quality management plans. Potential for state policy development affecting water supplies is significant, though availability of federal funds will likely influence success of the program. Management may also include point of use treatment of groundwater as an apparently cost-effective approach to dealing with groundwater contamination (Raucher, 1986).

State Level Groundwater Policy in Michigan

Authority over groundwater quality programs in Michigan is derived from a conglomeration of state laws. These statutes take groundwater into consideration in many different forms, though none are specifically targeted at groundwater. Listed and briefly described below are some of the major laws affecting groundwater.

Act 64, P.A. 1979--Hazardous Waste Management Act: This law specifies the engineering of hazardous waste landfills and provides for the licensing and regulation of individuals engaged in the use of hazardous waste.

Act 243, P.A. 1951--Servicing of Septic Tanks, Seepage Pits, or Cesspools: This law regulates the licensing and bonding of the servicing and cleaning of septic tanks, seepage pits, or cesspools.

Act 245, P.A. 1929--Water Resources Commission Act: This general water pollution law regulates discharges to groundwater and grants permitting for certain discharges.

Act 307, P.A. 1982--Environmental Response Act: This Act is Michigan's version of the federal "Superfund" program.

Act 368, P.A. 1978--Part 127, Groundwater Quality Control: Authority for monitoring water supply and quality.

Act 399, P.A. 1976--Michigan Safe Drinking Water Act and Administrative Rules: Protects public health by providing for supervision and control over public water supplies.

Act 423, P.A. 1984--Underground Storage Tank Registration: The registration of particular underground storage tanks is required under this law.

Act 641, P.A. 1978--Solid Waste Planning and Management Act: This law regulates the management of solid waste.

There has not been a law established to specifically deal with groundwater quality and protection in Michigan, such as requiring uniform standards.

LOCAL POLICY FOR GROUNDWATER PROTECTION: EMPHASIS MICHIGAN

The real hope for effective groundwater protection in Michigan and most other states is at the local level. Federal and state guidelines, coordination and, most importantly, financial support are essential. But the instruments for effective, supportable policy to reduce contamination are at the city, township, county and multi-county regional level. The essential task is to create incentives that guide private land use decisions in ways that protect vital recharge areas. Land use planning and controls are by law and convention primarily matters of local discretion. They have been directed at valid public purposes before and must be again. When sources of incompatibility of use and high information cost go beyond local boundaries as they often will, regional, state or federal guidelines will be needed.

While the potential for local action is great, the record is skimpy at best. A recent inquiry by the American Planning Association (DiNovo and Jaffe, 1984) concluded that the midwest is behind other regions in local protection of groundwater. The major population centers of the region are close to the Great Lakes or other surface water sources for municipal needs. The more rural areas that have depended on groundwater lacked the finances or sense of urgency to act. Frequency of contamination in recent years and generalized awareness of the issue has moved it up the agenda.

Land Use Planning

The first place for considering the importance of groundwater recharge areas and wellheads is in the comprehensive planning process. In fact, defensive regulation or control must be built on the base of logic and foresight contained in the land use plan.

The characteristics of groundwater are often poorly understood by the public. There is little general awareness about the connections between human activities and groundwater quality. Further, the specific objectives of a community will depend upon the physical, economic, and social characteristics of the region. A community may choose a nondegradation policy of groundwater management. However, where a community has other objectives which encourage development, nondegradation may be impractical. The State of

Wisconsin, for example, originally established a policy of nondegradation for their groundwater resource. They have since then noted, "in recent years. . .the state has moved toward the recognition that some contamination of groundwater is almost inevitable in modern, industrial society" (Yanggen and Webendorfer, 1984).

Good data are prerequisite to sound planning and policy. In some cases, information on the physical character of groundwater is enough to demonstrate the wisdom of protecting recharge areas even without regulation. In Meridian Township near East Lansing, for example, the planning director reported that despite absence of a specific protecting ordinance, a major developer accommodated the need to protect the recharge zone. The sensitive areas were well defined and need for protection was sufficiently understood in the community. The Center for Remote Sensing at Michigan State University is developing the capability to combine and display data in ways that highlight areas to be protected. Also, a detailed data system is being developed by several southwest Michigan counties with a substantial grant from the Kellogg Foundation (Science for Citizens Center, 1985). There is nothing approaching a uniform groundwater data system for Michigan, though the importance of such a system has been recently acknowledged by the Great Lakes and Water Resources Planning Commission.

The essential need in local land use planning that includes groundwater protection is the definition of sensitive areas--recharge zones and areas of influence for operating wells. More complete classification of aquifers in a larger area would be useful as well. Land use patterns in the community should be guided in ways that accommodate groundwater protection. Obviously, some uses are consistent with recharge, others are not.

Planning is necessary but seldom sufficient for protecting sensitive areas. There are several devices available to virtually all local governments to assure that land development is responsive to the importance of the groundwater resource. These have been summarized succinctly by American Planning Association (DiNovo and Jaffe, 1984) in a table included as Appendix II. The essential options are discussed briefly here.

Regulation

The basic existing power of local governments to implement the good intentions of a land use plan is zoning and related regulations. Zoning is an exercise of police power by which government may divide the communities into districts and restrict land uses within those districts on behalf of the health, safety and general welfare of residents.

1. Conventional zoning applies to new land uses in a community. Current uses cannot be regulated out of existence but

become nonconforming inclusions in the ordinance districts. That may be insufficient to avoid some future problems. Agricultural sources will likely be poorly handled by conventional zoning since agricultural zones tend to be least restrictive and do not deal with specific crops grown or cultural practices.

The real advantage of zoning is that it exists and has been used by some communities for nearly 50 years. It is understood, generally accepted and enforcement procedures are in place. As local action, it is (or can be) responsive to local needs. However, there is little experience with groundwater provisions to conventional ordinances. Survey of all regional planning commissions in Michigan turned up just a few examples.

Springfield Township in Oakland County has very recently established a zoning ordinance which addresses the effects of new developments on groundwater quality. The zoning ordinance is based on extensive mapping of the physical characteristics of the township which helped define the sensitive aquifer areas. New developments must adopt "site plan review criteria" developed by the township. Thus, higher quality of the groundwater is maintained.

Tri-County Regional Planning Commission in Lansing has established a suggested model groundwater ordinance for local governments in the counties of Ingham, Eaton and Clinton. The City of Lansing began preparation of a draft groundwater recharge ordinance. Recharge areas within city limits were identified. There was some disagreement among the experts, in this case, a consulting firm and Michigan State University Department of Geology, as to where the recharge sites really are. No real implementation steps have been taken as yet. Officials report that the project is currently of low priority.

Meridian Township in Ingham County has what may be among the most complete inclusion of groundwater considerations in a Michigan local zoning ordinance. A Conservancy District in the Meridian ordinance includes as a declared public purpose of land regulation the protection of groundwater recharge areas. It is an "overlay" zoning ordinance in that special conservancy restrictions are added to those already established for conventional districts. Permitted uses include recreation facilities, golf courses, general agriculture (no obvious restrictions), and unpaved parking areas. Others are permitted by special use permit, supporting the generally accepted uses noted above. Meridian Township is experiencing major development pressure and making a genuine effort to avoid damaging their important groundwater supply. Data necessary to enforce groundwater provisions are now available, and the ordinance is fully operable.

Outside of Michigan, the city of Crystal Lake, Illinois, recognized that the Crystal Lake watershed was being contaminated by agricultural practices and urbanization. The city amended its zoning ordinance to protect the watershed through such means as preserving the natural groundwater flow and drainage, limiting development densities, restructuring water supplies and prohibiting septic systems in sensitive areas. These efforts have been met with strong public support (DiNovo and Jaffe, 1984).

2. Specialized zoning devices. Local regulations for groundwater protection are likely to rely on such specialized flexible zoning techniques as conditional use, overlay and cluster zoning (Yanggen and Webendorfer, 1984). Conditional uses are allowed in a given district only after assurance that groundwater resources would not be damaged. Overlay zoning (as in the Meridian case above) adds restrictions specific to protection of groundwater, beyond those contained in the conventional ordinance. The overlay protection could apply to ecologically sensitive areas that do not coincide with conventional districts. Cluster zoning permits increasing density in some areas to reduce it in others. All these specialized zoning techniques are designed to augment or embellish a conventional ordinance to avoid land use patterns that would threaten groundwater supply.

Dade County, Florida, has perhaps the tightest groundwater recharge protection ordinance in the nation. Restrictions are based on potential for dilution within the wellhead area of influence. Nonresidential property owners must submit binding covenants incorporating restrictions against hazardous substances (US EPA, 1984).

3. Regulate the polluter. An alternative, or supplement, to protecting recharge areas is to regulate actions that may cause contamination. Residential septic systems, for example, are a major source of nitrate contamination. The Southeast Michigan Council of Governments identified several discrepancies in septic procedures for Genoa Township and recommended corrective actions and information needs for avoiding future problems. In some areas, on-site disposal systems may be prohibited altogether. As always, defensible regulations require dependable data. Household hazardous substance disposal is specifically prohibited in Briley Township in Montmorency County in northern Michigan. Some local governments have regulations against use of hazardous substances. Underground storage tanks are prohibited by some communities and tightly controlled in others. Briley Township established a permit system for fuel storage tanks.

Agriculture is a potential nonpoint source of groundwater contamination. Production practices have generally not been regulated, however. The only programs to influence farm practices have been voluntary, with cost sharing as further incentives. "Right to Farm" laws in many states seek to further establish farmers' rights to engage in reasonable and generally acceptable farm practices. Defining accepted practices could protect the farmer's implicit right to contribute nitrates to the water supply and force other water users to find other sources. Alternatively, right to farm could sharpen the farmer's obligation to avoid damaging the water supply. Presumably, one criterion for approving reasonable, acceptable farm practices could be demonstrated impact on groundwater. Farm practice regulations in Michigan are not anticipated in the near future, but the point has been made.

Acquisition

Local governments have the authority to spend public funds to acquire land for important public purposes. Sensitive recharge areas are in that category. Public land comes off the tax rolls; there will be maintenance cost in addition to original purchase expense; but once land is purchased, control is certain and complete. It may be the cheapest option available. When farms are the primary nonpoint source, acquisition of recharge areas may be the most effective approach. Governments also have power of eminent domain. They can acquire land even when there is not a willing seller. A valid case could be made for using oil and gas revenues available under Michigan Land Trust for such a clear public purpose.

Purchase may focus on just the development rights or other selected rights to recharge zones, rather than full rights. Initial cost would be less and land would stay on the tax rolls, with options restricted. The owner would continue to use the land in ways consistent with protecting recharge zones. Perhaps what is needed is a variant of the Conservation Reserve Program (CRP) of the 1985 Food Security Act wherein erodible land is drawn out of crop production through a bidding process. In CRP, the federal government essentially acquires the "erosion right" from the farmer, thus securing conservation goals by leasing discretion from the land owner. Similarly, the right to permit nitrate leaching, or the potential for farm-based contamination, might be bid away from the farmer. The owner would retain ownership but could not use farm practices that could pollute groundwater. Bidding would be limited to farm land overlying sensitive recharge areas.

Transfer of development rights and easement purchase are more complex variations on the acquisition theme. The former involves controlling development in sensitive recharge areas and permitting development in other areas. Through an administered market, those who have the right to develop must reimburse those landowners who do not have the right. The latter grants public rights to "use" certain land attributes while leaving other rights in private hands.

Finally, a cross compliance strategy might be employed. Again using soil conservation policy under the 1985 Food Security Act as a model, perhaps the farmer whose land overlies recharge areas should be required to eliminate potentially damaging land practices as precondition for eligibility for Farmers Home loans, commodity price supports and other income supports from USDA.

In all groundwater protection policy, the goal is to shift discretion in some way, to avoid land actions that contaminate groundwater. That discretion may be bought, taken through regulation or influenced through education or improved data on returns to pesticides and fertilizer.

CONCLUSIONS AND RECOMMENDATIONS

The institutional character of groundwater quality absolutely mandates public action. There are few situations where the polluting action itself creates incentives that are translated back to the polluter and are sufficient to change private behavior. Exceptions are cases where the polluter destroys his or her own water supply, or as in farming where application of "too much" nitrogen or pesticide results in contamination. Unlike some areas of resource policy where private rights may be defined and private trades encouraged, exclusion and information costs are prohibitive for groundwater quality. The open access nature of the quality attributes requires direct limits on private action in the interest of the broader public.

Despite high information cost, defensible adjustments in the rights and incentives for users of groundwater quality require considerably more groundwater and related land data. There are Constitutional and other legal limits on arbitrary restrictions to individual rights, and all water users are entitled to equal protection. Data generation should be sharply focused on those aspects contributing to effective policy. While more basic understanding of the physical, chemical and biological linkages in the groundwater ecosystem would be interesting, the more immediate need is to define the dimensions of aquifers and recharge zones.

Despite the meager record of achievement thus far, local policy for groundwater protection shows the greatest promise for immediate protection. The authority and procedures for guiding private use of land are well established. The central purpose should be to divert potential contaminants away from sensitive areas. Specialized and flexible zoning, particularly overlay zoning, may be the most effective technique.

The fourteen regional planning commissions have the essential capability of encouraging consistent local action across jurisdictional boundaries. Aquifers obviously do not follow political lines. The tri-county model ordinance for local government is particularly commendable. The data project in southwest Michigan, facilitated by the Kellogg Foundation, shows great promise as well. This is the type of effort necessary for effective local action. Data and modeling capability need not and cannot be reproduced in every region, however. There are great economies of scale in centralizing the more sophisticated modeling and data display in a place such as the Center for Remote Sensing. It is felt that a set of planning grants for groundwater quality management, directed to those planning regions that can document both urgency and technical capability and to an appropriate centralized modeling facility would be money well spent and a symbol of commitment by the Governor and legislature to confront Michigan's water quality problems. Perhaps the good intentions of the Federal Groundwater Strategy and Safe Drinking Water Act amendments will be backed with federal funds. Miracles are possible, but one should not count on them. Aquifer classification would definitely require expensive data collection.

Local governments are better able to protect recharge areas than control polluters. Regulation against such specific polluting activities as waste disposal must be state and federal. To get serious about the most damaging agricultural practices will take state action as well.

Finally, further education of Michigan people in the science and policy of groundwater quality is essential. Education should increase the individual's awareness of the effectiveness of altering land management practices to decrease contamination, and alleviate much of the uncertainty which engulfs the issue. By changing the individual's perception of the problem, information can be used as a tool to encourage public participation in groundwater protection policy. The Meridian Township experience discussed above is a case in point. Local groundwater programs should provide a central point of contact for information. County Extension can help in that. A concentrated effort to provide consistent well documented technical and policy information about groundwater will decrease the uncertainty and risk felt by the public. It is known that zoning and other public controls are still rejected on ideological grounds by some people, even when controls are local. But it is also predicted that better clarification of one's self interest can influence ideology.

A successful program of public education was developed by the East Michigan Environmental Action Council for nine townships in western Oakland County. The program began after discovery of two illegal toxic waste disposal sites in 1979 generated considerable citizen anxiety. The program was based on the belief that "effective public education for groundwater protection must be carried out by local citizens." A groundwater calendar was used to relay information on basic hydrological principles, past contamination incidents, and the relationship of groundwater protection to property values. Emphasis was on the fact that citizens and local governments could take action to prevent groundwater pollution (DiNovo and Jaffe, 1984). On the other side of the state, the West Michigan Shoreline Regional Development Commission developed a list of known, suspected, and potential groundwater contamination sites as part of a water quality planning program. A three-part classification system was then established enabling planners to pinpoint groundwater pollution problem areas. By overlaying maps of aquifer vulnerability with groundwater use, a composite map was developed which systematically determined relative priorities for a proposed groundwater management plan. This information was packaged and communicated to local citizens and officials (DiNovo and Jaffe, 1984).

There is real urgency to the groundwater quality issue in Michigan. Extent of contamination of private wells is still largely unknown or unacknowledged. The economic future of the state very much requires a widespread sense of security among citizens, businesses and visitors. Capability for response exists, but must be focused. Conferences such as this are positive steps in that direction.

LITERATURE CITED

Bouwer, H.   1978.   Groundwater Hydrology.   New York, NY: McGraw-Hill.

Carriker, R.   1985.   Water Law and Water Rights in the South.   A Synthesis and Annotated Bibliography, Southern Rural Development Center, Mississippi State University.

DiNovo, F. and M. Jaffe.   1984.   Local Groundwater Protection: Midwest Region. American Planning Association, Chicago, IL.

Getches, D.H.   1978.   Controlling groundwater use and quality: A fragmented system. Natural Resources Lawyer 17(4):623-645.

Kittleson, K.   1986.   The groundwater problem in Michigan: An overview. Paper presented at the Conference on Rural Groundwater Contamination: Assessment of Needs, Strategies for Action. October, Hickory Corners, MI.

Milon, J.W.   1986.   Interdependent risk and institutional coordination for nonpoint externalities in groundwater.   Paper presented at American Agricultural Economics Association annual meeting, July, Reno, NV.

Pratt, P.F, R.A. Olson, H.G. Alford, R. Back, T.L. Carson, H.H. Cheng, D. Curwen, J.M. Davidson, K.H. Deubert, J.E. Dewey, G.R. Hallberg, J.C. Hayes, C.S. Helling, W.A. Jury, D.R. Keeney, T.L. Lavy, R.A. Leonard, L.A. Licht, H.S. Lillard, R.B. Russell, P. Nowak, V.G. Perry, D.H. Pelkington, K.G. Renard, G.O. Schwab, A.L. Sutton, V.H. Watson, N. Webb and J.B. Weber. 1985. Agriculture and Groundwater Quality. Council for Agricultural Science and Technology, Report No. 103, May, Ames, IA.

Pye, V., R. Patrick, and J. Quarles.   1983.   Groundwater contamination in the United States. Philadelphia, PA: University of Pennsylvania Press.

Raucher, R.S. 1986. Regulating pollution sources under a differential groundwater protection strategy. Paper presented at American Agricultural Economics Association annual meeting, July, Reno, NV.

Schmid, A.A. 1978. Property, Power and Public Choice. New York, NY: Praeger.

Science for Citizens Center. 1985. Groundwater Survey. Vol. 1, No. 1, Summer.

Spofford, W., A. Krupnick and E. Wood. 1986. Sources of uncertainty in economic analyses of management strategies for controlling groundwater contamination. Paper presented at American Agricultural Economics Association annual meeting, July, Reno, NV.

US EPA. 1984. The Groundwater Protection Strategy. U.S. Environmental Protection Agency, Washington, DC.

Yanggen, D.A. and B. Webendorfer. 1984. Groundwater protection through local land use controls. Department of Agricultural Economics, University of Wisconsin, Madison, WI (preliminary draft, 1984).

APPENDIX I

Existing Federal Groundwater Protection Programs

Program	Scope: What Resource Is Protected?	Differential Protection of Groundwater?
SDWA Underground Injection Control 40 CFR Parts 144-146	Aquifers which could provide public water supply; groundwater with less than 10,000 mg/l total dissolved solids	Yes--Class V wells can degrade up to MCLs vs. no degradation for class I, II, III
RCRA 40 CFR Part 264	Uppermost aquifers	No increase in hazardous waste constituents or no violation of MCLs
RCRA 40 CFR Part 257	Current underground drinking water sources and groundwater with less than 10,000 mg/l total dissolved solids	Maintain drinking water standards
UMTRCA	Uppermost aquifers	Yes--No increase in Mo or U. Radium 226/228 up to 5pCi, Gross Alpha up to 15pCi
TSCA PCB Regulations 40 CFR Part 761	Groundwater (undefined)	No--No release of PCBs to any groundwater

Duration of Control?	Regulatory Mechanisms?	Waiver/Variance Provision?	Monitoring and Remedial Action Approach
During well operation, but presumably forever	Classes I to III: Design Standards Class V: non-endangerment of under-ground drink-ing water supply	Yes--exempted aquifers can be designated	Specified in Permits
30 yrs after closure for disposal facilities	Design and performance standards	Yes--Risk-based alternate con-centration limits on case-by-case basis	Monitoring and remedial action required
Unspecified--regulatory scheme im-plies forever	Design and performance standards	None	None
Design objec-tive of 200-1000 years	Design standards--liners required	None	Case-by-case decisions
20 yrs after closure	Design and location standards	None	Monitoring required

Existing Federal Groundwater Protection Programs (continued)

Program	Scope: What Resource Is Protected?	Differential Protection of Groundwater?
CWA Construction Grants 40 CFR Part 35	Uppermost aquifers with 3 classes based on current and potential uses	Yes--Protection to levels set by three classes
FIFRA Pesticide Policy	Groundwater	No--Maximum advisable level based on no effect on 10kg child drinking 1 liter/day
Nuclear Wastes Policy Act	Groundwater (undefined)	Yes--Some aquifers become part of disposal site
CERCLA National Contingency Plan	Aquifers which could provide public water supply or use by more than one person using scoring system	Yes--Decisions on a case-by-case basis

Source: U.S. Environmental Protection Agency.   Groundwater Protection Strategy for the Environmental Protection Agency. (Final Draft). May, 1984. (Reprinted from DiNovo and Jaffe, 1984, p. 53).

Duration of Control?	Regulatory Mechanisms?	Waiver/Variance Provision?	Monitoring and Remedial Action Approach
Unspecified	Best practicable waste treatment technology	None	Case-by-case monitoring to demonstrate compliance
As long as the pesticide is registered	Controls on or prohibition of use of specific pesticides	None	Monitoring can be required of registrants if groundwater contamination is of concern
10,000 yrs	Design and performance standards	None	None
Unspecified	Case-by-case decisions	Case-by-case decisions	Case-by-case decisions

APPENDIX II

Sensitive Area Protection Techniques

Technique	Legal Framework
Purchase of watershed lands	Local government home rule authority
Retention of all publicly-owned open space properties within the watershed	Governmental prerogative
Restriction/denial of infrastructure construction and/or major capitol improvements (roads, sewers, etc.)	Planning, enabling acts, and other laws
Purchase of development rights	Local government home rule authority
Transfer of development rights	Amendments to zoning ordinance establishing transfer districts

Major Advantages	Major Disadvantages
Provides the potential for complete control of the site purchased	Acquisition and long-term maintenance of parkland can be costly
No new costs; formal designation can lead to a savings of in-lieu payments and other carrying costs of tax-default properties	Some public properties may, because they are too small or too isolated, create a need for some complementary direct acquisition
Roads, sewers, and water mains are essential for intensive urban development; the control of types and locations of facilities can protect resources without the necessity of land purchase or regulation	Control of certain types of capitol improvements (such as county roads and drains) is not within local government powers; financing or major public improvements may also be difficult
Provides the potential for simultaneously keeping property on tax rolls and controlling land use	Acquisition costs can be high and landowners can elect nonparticipation in purchasing programs
Compensates owners of environmentally sensitive lands with public values without necessitating public purchase; allows for preservation of large tracts	In order for development rights to be marketable, development pressure and limited availability of land are needed; this situation may not be present in all communities

Sensitive Area Protection Techniques (continued)

Technique	Legal Framework
Reduced density zoning in conjunction with mandatory clustering	Amendments to zoning ordinance to allow cluster development as an option
Regulation of watershed by governmental health, conservation and planning agencies	Special districts included in zoning ordinances or special-purpose ordinances
Conservation/scenic easements	Legal agreement (easement) between the landowner and the organization receiving the easement
Property tax incentives to encourage large land holdings	Amendments to local laws; certain need complementary or enabling legislation at the State level
Performance zoning	Amendments to zoning ordinance adding performance standards

Source: Greenberg, Meyland, and Tripp. Watershed Planning for the Protection of Long Island's Groundwater. (1982). Coalition for the Protection of Long Island's Groundwater. Reprinted by permission. (Reprinted from DiNovo and Jaffe, 1984, p. 100).

Major Advantages	Major Disadvantages
For an individual site plan, reduces impervious surfaces and increases open space areas; the process encourages creative site plans that take natural resources into account	Cluster development incentives (increasing densities on part of the site) may not be attractive to developers
During development reviews, allows for special consideration of the resource in question; low-cost to the community	It may not be possible to comply with regulations and still make reasonable use of the parcel
Can provide significant property and federal income tax benefits to the landowner; provides scenic natural areas without land purchase	Requires voluntary consent of the landowner; may have piecemeal effects--difficult to implement open space or conservation plan
Tax reductions may diminish the tendency of open land to be developed	There is no assurance that the properties will not be developed
Leads to an objective review of the impacts of a proposed development; encourages innovative site plans which reduce negative impacts	It may be difficult to develop quantified, objective standards; it may be difficult to convince some local officials to use objective technical standards in the site plan review process

PART VI

STRATEGIES AND ASSISTANCE

CHAPTER 24

INDUSTRY/AGENCY PERSPECTIVES ON STRATEGIES
FOR PROTECTING GROUNDWATER

NATIONAL AGRICULTURAL CHEMICALS ASSOCIATION
--Thomas J. Gilding, Director, Environmental Affairs

It is a pleasure to have this opportunity to present the agricultural chemicals industry's views on strategies for protecting groundwater. Although my comments will primarily address the aspects of pesticides and groundwater, the overall strategies certainly apply to any human activity having the potential for affecting the quality of groundwater.

Good management principles dictate that problems be clearly defined before effective strategies can be identified and implemented. Since the data base on pesticide detections in groundwater is very limited, the extent of the problem from a spatial perspective and health significance is primarily undefined. With recent expansion of monitoring activities in both the private and public sectors and the extraordinary advances in analytical chemistry, some pesticides in certain hydrogeological conditions have been found in groundwater in concentrations most commonly in the range of 0.1 to 10 parts per billion (ppb). Even though more pesticides will undoubtedly be found with this expanded groundwater monitoring, can true problem definition ever be achieved? Extensive groundwater monitoring has extreme resource constraints, and a true scientific definition of the health problem, i.e., risk, is frustrated by a strong public concern about chemicals in groundwater.

The "pesticides in groundwater problem" needs to be defined for the purpose of identifying the basis from which groundwater protection strategies for agricultural pesticides can be formulated. As understanding and characterization of the overall problem improves, it is essential that strategy formulation and implementation focus on those areas which are scientifically based and determined to be most significant in risk to health and the environment.

The National Agricultural Chemicals Association (NACA) Groundwater Protection Program is a two-step, phased approach having the overall emphasis and objective of preventing or minimizing the potential for pesticide residues to reach groundwater. The term "two steps" refers to both nonregulatory and regulatory components. The "phased" descriptor simply recognizes the commitment to start now with what is presently known, changing as future technology and information dictate. The program objective of protecting groundwater will not change, just the strategies for obtaining it.

The nonregulatory portion of the program consists of three distinct, but inter-related components. These are:

1. Define and implement an effective risk communication program. Recognizing the importance of communication in the overall groundwater risk management process, NACA has greatly expanded its attention and resources to groundwater education and communication. Awareness and appreciation of the need for protecting groundwater must be increased among the pesticide users and industry, especially those responsible for research and product development, registration, and marketing. More specific information will be communicated to the pesticide user as technology and understanding allow.

2. Promote development and use of scientifically sound prediction technology to identify the why, when and where factors that influence pesticide movement through soils and into groundwater. The objective of this technology application is to use computer modeling to identify those optimum combinations of pesticide properties, application site characteristics, and use practices to ensure that a pesticide stays in the soil zone where it is designed to perform its intended function and acceptably degrade. This technology has great potential to: (1) pesticide manufacturers in designing new products or modifying existing formulations; (2) Environmental Protection Agency (EPA) in the regulation of pesticide risks to groundwater; and, hopefully, (3) pesticide users.

3. Promote definition and dissemination of sound crop, soil and water management practices associated with pesticide application to minimize those situations which promote chemical movement. Although the current level of knowledge about the transport process of pesticides needs to be better understood before specific practices for specific pesticides for specific application locations can be defined, there are certainly some common sense, good housekeeping considerations which pesticide users can use in the storing, handling and application of pesticides. Storage and handling are mentioned because there are indications from monitoring programs that spills during mixing and loading operations of pesticides prior to application, rinsewater discharge from application equipment,

and back-siphoning into water sources may be major factors behind pesticide detections reported in groundwater.

EPA and states can make modifications to application practices for pesticides as part of their regulatory programs. Examples of this type of action would be changes in rates of application, times of application and restrictions on where a pesticide can be applied.

These latter government actions are part of the regulatory portion of the NACA Groundwater Protection Program. The main objective is to help define and support reasonable and effective regulatory strategies which EPA and the states can use under the authority of the Federal Insecticide, Fungicide and Rodenticide Act (FIFRA). EPA, as FIFRA is currently written, has the authority to regulate pesticides for their risks resulting from getting into groundwater. This authority, however, is general in scope and not defined in substance. Upon recognizing the importance for having a groundwater protection program defined in FIFRA, NACA and a coalition of 41 environmental, labor and consumer groups, known as the Coalition for Pesticide Reform, negotiated proposed amendments to FIFRA on groundwater. The groundwater amendments contained the following requirements:

Notification and Monitoring -- Registrants are required to report the detection of their products in groundwater to the Administrator and Governor of the state involved within 15 days. In order to adequately assess any detection of pesticides in groundwater, the EPA Administrator is given authority to require affected registrants to conduct any necessary groundwater monitoring.

Groundwater Residue Guidance Levels -- As a guide to EPA and states in developing appropriate regulatory action, the FIFRA amendment would require EPA to establish health-based numbers called Groundwater Residue Guidance Levels (GRGLs). The GRGLs would actually be the maximum containment level if one exists under the Safe Drinking Water Act, or if one does not exist, the health advisory under the same Act. EPA is also presently in the process of issuing health numbers for the National Survey on Pesticides in Drinking Water Wells which could be adopted as GRGLs.

In establishing GRGLs under FIFRA, the Administrator would be required to take into account the level of the pesticide in drinking water at which no known or anticipated adverse effect on the health of persons may occur, the need for an adequate margin of safety, the nature of the toxic effects of the pesticide, and the validity, completeness and adequacy of the available data. If required chronic, oncogenicity, reproduction or teratogenicity data are absent or incomplete for a pesticide, the GRGL shall be at the analytical limit of detection.

Site Specific Response -- If a pesticide residue is detected above its GRGL, the amendment requires the Administrator to take specific site response. In cooperation with the Governor of the state, the Administrator must (1) act to ensure that continued use brings and retains the pesticide concentration below the GRGL, or (2) ensure that people do not consume drinking water that exceeds the GRGL. There is also opportunity for public petition to modify or terminate action taken by the Administrator. The amendments would require EPA to distinguish between groundwater that is drinking water or potential drinking water. It also gives the Administrator the authority to initiate a special review when a pesticide is detected in groundwater that is neither drinking water or potential drinking water to determine whether the pesticide could cause unreasonable adverse effects on the environment.

Scope of Registration Actions -- The registration actions that the Administrator could take under the proposed FIFRA amendments are:

1. Limitations on the pesticide use and locations of use, considering factors such as environmentally sensitive areas and conditions.

2. Limitations on the rate at which the pesticide is applied.

3. Limitations on the time or frequency of pesticide use.

4. Limitations on the method of pesticide application, storage, handling or disposal.

5. Administration requirements, including additional registrant reporting, label changes or changes in pesticide classification to reflect specific groundwater concerns.

The amendments also recognize the importance of the public being informed on the detections of pesticides in groundwater by setting up public information files.

In conclusion, the above Groundwater Protection Program, both regulatory and nonregulatory, can be viewed as containing near term strategies, the basis for these strategies being dictated by present understanding of the problem, the state of the art of technology, and existing social/economical and political constraints. These near term strategies, however, reflect a sincere commitment to groundwater protection using what is available in today's arsenal. From a long-term perspective, the strategies described will undoubtedly lead to pesticides with enhanced nonleaching and degradation properties, while still maintaining their desired effectiveness. The methods for applying pesticides, as well as the volumes required, will all collectively contribute towards the desired groundwater protection objective.

MICHIGAN ASSOCIATION OF COUNTIES
--Ann Beaujean, Legislative Assistant

Local units of government have a critical role in the protection of their groundwater supplies. Despite the fact that there has been some confusion in this area over who's ultimate responsibility it is to protect groundwater, local units can and have accepted this very important challenge.

The Southwest Michigan Groundwater Survey and Monitoring Program is just one of many programs operating today in Michigan. At present, Barry, Berrien, Calhoun, Eaton and Ingham Counties are implementing the project to its fullest extent while Allegan, Ionia and Jackson Counties are implementing component parts of the project. Livingston County has just recently been added as an associate member of the project.

Other local groundwater protection programs have also had outstanding success in Michigan. The Tri-County Regional Planning Commission's Recharge Area Management Project has had excellent results in parts of Ingham County. This program's goal is to identify and manage groundwater recharge areas so that once these areas are found, they can be managed properly in order to protect the local groundwater supply. The local government role in these projects has been quite varied. In some cases, counties have contributed money and staff time to the project. Counties and local governmental units can also make significant contributions to groundwater protection in many other areas.

The Tri-County Recharge Area Program has identified five ways local governments can protect groundwater recharge areas. These methods include: (1) public acquisition, (2) purchase of development rights, (3) conservation easements, (4) transfer of development rights, and (5) zoning.

Other regulatory options locals have within their power include identifying underground storage tanks, promoting livestock waste ordinances and regulating the land spreading of septage. Locals also have a variety of nonregulatory options at their discretion. These include promoting waste reduction strategies such as recycling programs and the reclaiming of waste oil. These projects not only prevent groundwater pollution, but they also save valuable landfill space. Counties and local units of government can also encourage the use of agricultural best management practices in order to protect both surface water and groundwater.

Local governments are not the only organizations that can protect local groundwater supplies. Organizations like the Michigan Association of Counties can help protect groundwater by providing educational forums to local officials in order to increase their understanding of the importance of groundwater and the options they have within their power to protect this valuable resource. Information is also provided in monthly newsletters concerning dates, times and locations of household hazardous waste collection and disposal programs.

Local governments and their association have and will continue their efforts to protect Michigan's groundwater supplies. Despite their hard work, however, locals still need state and federal assistance if they are to continue their efforts with success. In fiscal year 1987, the State has appropriated $500,000 for the Groundwater Compliance Grant Program. Actions like this are extremely important if local units are to implement "groundwater protection strategies."

This basic dollars and cents issue is critical at this time. The recent elimination of federal revenue sharing has left local units of government in a desperate situation. Many units may be considering the elimination of local groundwater protection programs in favor of saving basic services like police and fire protection, although these choices may be short-sighted.

With this in mind, the Michigan Association of Counties encourages the State to continue to support and expand operations like the Groundwater Compliance Grant Program. Collectively, a good job can be accomplished in protecting the groundwater supply. Local governments want, and in many cases are the best units of government, to achieve success is this area. They will, however, need the continued financial assistance of the state and federal governments if they are to continue these projects.

THE GREAT LAKES AND WATER
RESOURCES PLANNING COMMISSION
--James A. Koski, Chairperson

Groundwater is an unseen but integral part of the hydrologic cycle. It provides the water that sustains the base flow of streams during dry weather. Many cold water streams originate as springs fed by groundwater. Groundwater is seldom studied and taken for granted as always available "good" water, yet some 30 public water supplies and hundreds of private wells have been contaminated to the extent of being unacceptable drinking water supplies. According to a U.S. Geological Survey report (Estimated Use of Water in the U.S. in 1980), groundwater supplies 43 percent of the state's population. However, groundwater accounts for only 4 percent of the total water used in the state, because most water supplies for large urban areas are from surface water.

Groundwater withdrawals for irrigation constitute about 40 percent of the total water used for irrigation, and this use is projected to increase. Groundwater withdrawals can impact other uses of the aquifer and nearby surface water by reducing groundwater flow and the level of the water table. Groundwater withdrawals can reduce stream flows beyond acceptable levels since the base flow of streams is sustained by groundwater. This, in turn, impacts the assimilative capacity of the stream and the quality of the water.

Given the magnitude of this issue area, the committee focused on the following five areas:

1. Michigan's standards for groundwater, how they protect drinking water supplies and how the Departments of Public Health and Natural Resources function to protect the resource.

2. Efforts to protect groundwater from inadequate handling and storage of hazardous materials.

3. Improved enforcement of groundwater protection regulations.

4. The role of local government in groundwater management.

5. Groundwater data management.

Groundwater quantity issues were raised for consideration. However, because of the time constraints, complexity of the issues and lack of data, quantity issues are not addressed in this paper. Groundwater quantity issues are significant and merit timely consideration through the continuing planning process.

The Philosophy Statement and 17 Recommendations describe strategies for addressing needs in each area.

PHILOSOPHY STATEMENT

The goal in groundwater protection and management is protection of groundwater quality for current or potential drinking water purposes or other uses which require a higher degree of protection. This goal can be met through: (1) moving away from commercial and industrial waste discharges into the groundwater; (2) preventing contamination at a potential source through reformulation of products, education and implementation of precautionary programs; and (3) timely permitting and vigorous monitoring activities.

An effective monitoring program is necessary. The programs implemented at the state and local government levels can no longer be reactive to contamination problems. The State of Michigan must move toward addressing issues and preventing groundwater pollution. By so doing, clean-up activities which have been enormously expensive will be minimized. The financial resources saved can be used for other priority programs such as expanding the compliance role of local government monitoring efforts in key program areas.

An effective monitoring program is necessary to achieve nondegradation of groundwater. Programs should be designed to prevent groundwater degradation from local background groundwater quality.   At the point of detected degradation, corrective action should occur.   Programs should not be designed to allow contamination up to maximum contamination level (drinking water) standards.

RECOMMENDATIONS

The Groundwater Protection and Management Committee makes the following recommendations.

Standards for Groundwater Protection and
Program Functions to Protect Groundwater

1.   *The rules in Part 22 of Act 245 should be amended to define nondegradation and be promulgated by September, 1988.*

Rationale: Nondegradation needs to be defined more specifically.   A nondegradation rule should be developed for groundwater discharges.   This rule must provide appropriate safety factors and flexibility in order to incorporate new data on a substance into the formula.   This process would provide a methodology for establishing a discharge level consistent with nondegradation policy.   A violation of the discharge requirements would cause the regulatory program to take action against the discharger to correct the action and/or cease operation.

2.   *The Water Resources Commission should develop a policy to implement the goal of moving away from commercial and industrial waste discharges into groundwater.*

Rationale: Recognizing that groundwater contamination can result from activities at commercial and industrial facilities, regulatory agencies should give priority to moving toward eliminating industrial and commercial groundwater discharges.   This policy must take into consideration alternatives available to the facilities, including reformulation of products for waste reduction, improved pollution incident prevention practices and other methods of disposal available.   This policy would cause facilities to reevaluate waste management practices in terms of environmental impacts, production efficiency and cost/benefit implications.   This policy will carefully balance environmental protection and economic development considerations.

3.   *Steps should be taken to ultimately consolidate oversight responsibility for all groundwater protection and drinking water programs within one state agency.   In the interim, a management committee with representatives from MDNR, MDA and MDPH groundwater programs and other agencies having impact on groundwater programs should be established.   This committee should meet biweekly and discuss policies, issues and programs that impact groundwater supplies and facilities siting.*

*The management committee should be established immediately
by a memorandum of understanding between the departments.*

Rationale: Coordination and communication among programs are
critical to effectively protect groundwater. In the long term, oversight of all groundwater management programs should be located in
one agency. In lieu of this arrangement to address the immediate
groundwater issues and needs, a higher level of coordination is
required. This process has been initiated by joining efforts to implement Act 307 activities. However, more coordination and communication are necessary to develop a comprehensive, effective and
preventive groundwater management program.

Groundwater Protection from Leaking/Inadequate Storage and Handling of Petroleum and Hazardous Materials

4.   *A program should be developed to protect groundwater from
      leaking/inadequate storage (including pipes) and handling of
      petroleum and hazardous materials. This program should
      address underground storage tanks and improved PIPP imple-
      mentation.*

Rationale: Tanks are 23 percent of the known sites by "point
of release" that contribute to groundwater contamination, yet current
storage standards and handling regulations are inadequate to protect
the groundwaters. Using a state and local government partnership,
sites can be inspected and monitored regularly by certified local
agencies. Consistent enforcement and technical assistance can be
provided by state agencies.

5.   *Standards for construction, installation and leak detection of
      underground storage tanks should be developed by the Michigan
      Department of Natural Resources (MDNR) in consultation with
      the Michigan State Police (MSP), Fire Marshall's Office, and
      the Michigan Department of Public Health (MDPH), Water
      Supply Division. The MDNR should present these draft rules to
      the Administrative Rules Committee as soon as possible.*

Rationale: Since underground storage tanks (UST) account for
17 percent of the "points of release" for known contamination sites
in Michigan and over 100,000 tanks are registered, the contamination
potential from UST is great. Expedient efforts to regulate these
potential sources of contaminants are critical to preventing future
groundwater contamination problems.

The program may include a statewide permit to install certain
tanks with funding support for certified local health departments for
permit issuance and construction inspection. Tank installation site
approval would take into consideration a minimum isolation distance
from current or potential drinking water sources.

6.   *The State Fire Marshall's program for installation, inspection, maintenance and removal of underground storage tanks needs to be adequately staffed and funded.*

Rationale: Currently, about 5 percent of the State Fire Marshall's field staff time is spent on hazardous materials storage. This program should be enhanced to provide staff for inspection of new installations and certification and training for local agencies to carry out installation and removal efforts for small tanks. Though this program provides inadequate environmental protection, it would remove abandoned tanks and address installations of new tanks as preventive groundwater protection efforts until new rules under Act 245 are developed.

7.   *The State should provide funding for certified local agencies for administering state standards for storage and handling of hazardous materials.*

Rationale: Such funding would provide a mechanism for local monitoring efforts for PIPP facilities and storage tanks and focus state level efforts on permitting and investigating compliance activities.

8.   *The Departments of Natural Resources, Agriculture and Public Health together with other appropriate local interests should work to develop a strategy to address groundwater degradation from fertilizers and pesticides use. This strategy should be developed by January, 1988.*

Rationale: Although agricultural and food-related sources of contamination constitute only 3 percent of known groundwater contamination, evidence from other studies indicates that nitrate and other chemicals in groundwater are one of the unaddressed problems looming on the horizon. Efforts to carefully evaluate the scope of quantitative and qualitative problems and develop preventive practices for the future will assist Michigan in maintaining the quality of life desired for the next century. The recommendations of the Strategy for Improved Pesticide Management in Michigan, December, 1985, should be a component of this effort.

Improved Enforcement of Groundwater Protection Regulations

9.   *The State should adequately staff current groundwater protection and drinking water supply programs so that existing laws and rules can be adequately enforced. This lack of enforcement should be remedied by tying staffing and funding resources to this need beginning with fiscal year 1987-88.*

Rationale: Due to lack of sufficient staff for investigative and enforcement efforts, enforcement procedures are not fully implemented. Michigan has good environmental laws; however, stronger enforcement efforts would provide a deterrent to activities which may result in groundwater degradation.

Prevention efforts cannot occur without the proper combination of a "carrot and stick" approach to groundwater and drinking water supply protection. One of the key ingredients to groundwater protection is monitoring. The state lacks resources to do adequate site monitoring. Self-monitoring alone is not sufficient to protect the environment.

10. *Increased consideration should be given to delegating, with funding and state oversight, groundwater protection enforcement functions to certified local agencies.*

Rationale: Such a partnership effort would enhance groundwater protection enforcement functions through increased monitoring and compliance activities. Some programs cannot be handled effectively by the state. For example, storage tank installation or PIPP compliance could more easily be accomplished by certified local agencies since limited resources would be effectively placed.

11. *Administrative agencies should have the authority to impose administrative fines on chronic violators in order to improve groundwater protection enforcement, including drinking water supplies. These agencies should work with all interest groups to develop an administrative fines rules package for groundwater protection laws by spring, 1988.*

Rationale: The ability to assess administrative fines for chronic violations of permit requirements, statutes and rules in a clearly defined system would do much to deter such violations by creating an economic disincentive for the violators. These fines should be administered in a consistent manner statewide with oversight responsibility from respective department directors. Administrative fines would be one method of addressing groundwater protection law violations without removing the potential for civil actions.

Administrative fines are intended to benefit the majority of businesses that accept their responsibility to maintain a clean environment by assessing monetary penalties against the minority of businesses that chronically violate regulations and pollute the environment. Administrative fines would serve to place all businesses on an equal footing by denying irresponsible businesses the market advantage they currently achieve by avoiding some of the compliance costs of groundwater protection statutes. Administrative fines serve to remind violators of their responsibilities in a tangible manner.

12. *Administrative agencies should have in-house counsel directly responsible to the department director. The financial resources to implement this program should be part of the fiscal year 1988-89 budget.*

Rationale: In-house counsel responsible to a director is a critical step to enhanced groundwater protection, including drinking water supply enforcement efforts. By being housed in a department, counsel services would be available to assist in proper investigative case development and to establish department priorities for cases referred to the Attorney General for prosecution.

The Role of Local Government in Groundwater Protection

13.   *A consistent program for private well permitting should be instituted statewide. Part 127 of the Public Health Code should be revised by fall, 1988, to provide for a state well construction permit program for private wells administered by the state or certified local health departments.*

Rationale: The current well construction program does not contain a uniform permit requirement and, consequently, permitting is inconsistent across the state and in many places inadequate. By implementing a statewide permit program, consistency and proper well construction for public health and environmental protection will be enhanced. This system would also provide more accurate well logs and data from new wells which could be used to better define the aquifers and to make sound decisions about projects which impact groundwater.

14.   *Encourage utilization of certified local agency staff in appropriate review, monitoring, inspection and construction permit activities such as PIPP programs and petroleum and hazardous materials storage facilities and handling practices.*

Rationale: Current staffing for these programs at the state level is not adequate. By utilizing certified local agencies to close these gaps in groundwater protection and management, more control over potential future sources of contamination will be obtained. Preventing contamination would redirect remedial action dollars from cleanup to protection programs.

15.   *Local units of government with state technical assistance should be encouraged to consider groundwater management in planning and zoning as part of on-going planning and zoning decisions.*

Rationale: Examples of groundwater planning and management considerations would be zoning and public acquisition of aquifer recharge areas or municipal well field areas of influence to enhance protection of the resource.   Such techniques require accurate groundwater management data. By providing access to such groundwater management data, the state can enhance locally implemented groundwater protection techniques.

Groundwater Management Information

16. *The MDNR, as lead agent, should institute a statewide ground-
     water data strategy. The implementation and completion of a
     data management system, including general mapping of aquifers
     and water quality characteristics should be funded and com-
     pleted as soon as possible.*

Rationale: Though much data is available from well logs and
sources of groundwater quality and quantity information, little has
been done to link accurate data or put it into a usable form. If such
data is linked into a computer system, much information would be
available to decision makers. Access to this information should lead
to more knowledgeable and protective decisions for activities which
impact groundwater. Using this information, economic development
can occur which enhances and protects the environment.

This strategy would facilitate analysis of aquifers in terms of
areal extent and volumes of water stored data for addressing water
quantity issues such as water rights and user conflicts.

17. *Local government must have computer access to groundwater
     management information. The state program should provide
     some funding and training to develop local expertise in using
     groundwater data for local decisions.*

Rationale: Decisions are based on the information available. If
groundwater data is available in a usable form to local officials, it
can be used to make better decisions on projects which have the
potential to impact groundwater. Since decisions are made daily, the
information must be current and readily accessible to provide a basis
for preventing groundwater contamination.

CHAPTER 25

FEDERAL AND STATE
ASSISTANCE FOR ACTION

Assistance with problems associated with groundwater contamination or its protection is provided through various federal, state,
and local governmental units or agencies as well as the private
sector. Representatives from some of these groups were requested to
present information on their agency's role in aiding the public
concerning groundwater contamination.

FEDERAL ASSISTANCE FOR ACTION

U.S. Geological Survey
Norman Grannemann

The U.S. Geological Survey (USGS), which was established in
1879, is dedicated to collecting, analyzing, interpreting, and dissemin-
ating information about the natural resources of the nation. The
mission of the Water Resources Division of USGS is to collect data
and conduct research concerning the nation's water resources. This
is accomplished, in large part, through cooperation with other federal
and nonfederal agencies by: systematically collecting data needed to
determine and evaluate the quantity, quality, and use of the nation's
water resources; conducting analytical and interpretive water resource
appraisals that describe the occurrence, availability, and the physical,
chemical and biological characteristics of surface and groundwater;
and conducting supportive basic and problem-oriented research in
hydraulics, hydrology, and related fields of science to improve the
scientific basis for investigations and measurement techniques and to
understand hydrologic systems sufficiently well to quantitatively
predict their response to stress, either by natural or human causes.

The water resources program in the Michigan District of USGS
is planned and funded cooperatively with 40 units of local government
as well as state and federal agencies. If a proposed project is
mutually advantageous to the Geological Survey and an agency or
local governmental unit, the Geological Survey enters into a formal

cooperative agreement to provide needed hydrologic information. In most cases, costs are shared equally between the Geological Survey and the cooperator.

The USGS publishes an annual series of reports "Groundwater Data for Michigan," which includes groundwater data collected for each calendar year. Additional cooperatively funded studies designed to relate land use to ground and surface water quality and quantity are now underway in Kalamazoo and Grand Traverse Counties. Deteriorating groundwater quality at some places in these counties is likely related to the use of fertilizer. Pesticides in water are also being studied. The Michigan District also operates a network of water quality stations at which the chemical and physical quality of surface or groundwater are monitored to establish a base against which future water quality data can be compared and against which the effects of new and additional stresses can be evaluated.

USDA Soil Conservation Service
Daniel A. Smith

The Soil Conservation Service (SCS), a nonregulatory agency of the U.S. Department of Agriculture, works with local farmers and land users to conserve the soil and water resources in a wide variety of circumstances. Historically, as an agency, SCS has not had a role in groundwater resources. However, since groundwater problems generally are an expression of what is happening on the land surface, how groundwater becomes contaminated and how such contamination can be prevented is of strong interest. By controlling pollution on the surface, groundwater contamination problems may be reduced or eliminated.

The majority of the agency's work is conducted at the county level where approximately 75 percent of the staff is located. SCS works cooperatively with other groups and agencies at this level to get conservation on the land. The job is far too large for any single group to handle alone. By combining resources and working cooperatively with such groups as soil conservation districts, Cooperative Extension Service, Public Health, American Soil Conservation Society, the Michigan Department of Agriculture, and the Michigan Department of Natural Resources, the SCS can begin to address the problems at hand.

The SCS is ready to assist the people of Michigan to conduct needed soil and water conservation measures, provide help in the form of planning assistance, offer technical and engineering assistance, and give information and technology transfer and educational assistance. Cooperation (people) plus Assistance (information) equals Action (solutions). The county action team is already in place and the SCS can make the best use possible of this team to accomplish the needed action.

STATE ASSISTANCE FOR ACTION

Michigan Department of Natural Resources,
  Groundwater Quality Division
Daniel O'Neill

The Groundwater Quality Division (GQD), Michigan Department of Natural Resources (DNR) is charged with administering the federal and state laws written to protect the groundwater from contamination. The Division is divided according to its two main functions to fulfill this charge.

The first function, administration of laws designed to prevent contamination, is conducted by the permit and compliance sections. Permits are drafted with restrictions on discharges and contain standards for wastewater treatment, lagoon construction, land application of waste materials and monitoring usable aquifers in regions where discharges occur. Plans are reviewed and approved for solid waste management and pollution containment. Drafting of rules and regulations to prevent contamination from underground storage tanks and oversight of septage waste disposal are also administered under these sections.

The second function is the clean-up of existing contamination, which is carried out by the remedial action section in coordination with the hydrogeological section. Under Act 307, Michigan Environmental Response Act of 1982, as amended, contaminated sites are identified and ranked according to priorities related to risk associated with the contamination. Clean-up and control measures are implemented in accordance with Act 307 and the federal Superfund program.

In addition, the GQD coordinates activities in the Pollution Emergency Alert System (PEAS). This system provides a means for reporting incidents of pollution such as illegal dumping of waste or accidental spillage of contaminated materials.

Michigan Department of Agriculture, Environmental Division
Christine E. Lietzau

The Michigan Department of Agriculture's (MDA) Environmental Division is responsible for several programs which benefit groundwater resources. The Clean Water Incentives Program is jointly sponsored between the Michigan Department of Agriculture and the Department of Natural Resources to control nonpoint sources of water pollution. A pilot demonstration program was begun in 1985 wherein grants up to $35,000 were offered by the MDA to Soil Conservation Districts to showcase their efforts to control nonpoint source pollution on a watershed basis. Nine projects were funded. This pilot was the forerunner of the larger scale Clean Water Incentives Program which began October 1, 1986. Four rural and two urban projects were selected for funding with planning grants of up to $50,000 through the DNR. The program will also be jointly administered between the two departments. Implementation dollars should be appropriated by the Legislature to be used beginning October 1, 1987.

Optimization of Agricultural Water Management Using Water Modeling is being funded by the MDA through Michigan State University, Center for Remote Sensing. The primary purpose of the project will be to develop a statewide projection of Michigan's agricultural water demands, providing the first quantitative foundation for agricultural water policy development statewide. More importantly, the project will provide decision makers with the capability to simulate the impact of their decision on anticipated water shortages.

Michigan Department of Public Health
Donald K. Keech

The Michigan Department of Public Health (MDPH) offers three distinct areas of assistance to users of groundwater as a source of drinking water. First, laboratory services for both bacteriological and chemical determinations are available to all residents of Michigan as a free service. Second, Michigan Environmental Response Act, Act 307 of 1982, as amended, provides for interim response or remedial action to assure safe drinking water whenever a drinking water well is contaminated from a point source such that use of the water could threaten human health. The third area relates to expertise and consultation available for all types of water well construction, pump installation, or groundwater development problems.

Water Analyses Availability

The MDPH provides laboratory services for water quality analyses of drinking water wells in Michigan, in cooperation with local county health departments. Three types of sample analysis are available through the local health department. The basic bacteriological analysis determines the safety of a drinking water supply from a bacteriological standpoint. This generally relates to whether the well and pump installation have been properly disinfected after installation or repair. A basic chemical examination is also available which includes six routine chemicals: iron, hardness, fluoride, chloride, sodium and nitrate along with specific conductance, which measures the total solid content of the water. Bacteriological and partial chemical analyses are available upon request to the local health department.

In addition, special analyses can also be performed with specific approval. Tests include: heavy metals or other nonroutine chemical analysis such as a complete mineral analysis, gas chromatography for halogenated and nonhalogenated organic volatile chemicals, plus analyses for nonvolatile chemicals such as pesticides and herbicides. A limited laboratory capacity is available for special metals and organic analyses. Therefore, specific approval has to be given on a site-by-site basis to obtain this service. The Michigan Department of Public Health laboratory is currently analyzing over 500 water samples monthly for these special types of contaminants. Local health departments frequently assist in collecting the proper water sample.

Michigan Environmental Response Act

The Michigan Environmental Response Act, Public Act 307, P.A. 1982, provides for the identification, risk assessment, and priority evaluation of environmental contamination sites. The Michigan Department of Natural Resources has the responsibility to develop a numerical risk assessment of each site to determine the relative present and potential hazards posed to the public health, safety, welfare, or to the environment by each identified site. The Michigan legislature then provides funding for interim response actions, including detailed site evaluations and other remedial measures to protect public health. Final response actions are designed to obtain a permanent remedy to clean up the contaminated site.

If a drinking water well becomes contaminated with organic compounds or compounds with established maximum contamination levels, an interim response is dictated to guarantee the users a safe potable drinking water. Consequently, many users receive bottled water for drinking purposes. For a very few highly contaminated sites, hauled water systems are installed to meet domestic needs. A more permanent interim response results in installation of new wells, extension of water mains, or occasionally installation of a complete new water system to serve areas where wells have been impacted by groundwater contamination.

These state funds are only available to innocent parties whose wells have been impacted by groundwater contamination from a recognizable or what is believed to be a point source. The general and quite ubiquitous nitrate contamination in rural areas does not qualify for this assistance. The majority of the groundwater contamination problems receiving state assistance relate to contamination from organic chemicals such as solvents and cleaners like trichlorethylene, or perchlorethylene, or contamination from gasoline components such as benzene and toluene.

Consultation, Technical Assistance

The staff of the Groundwater Quality Control Section has developed a wide range of technical expertise regarding all types of groundwater problems. Consultation is available for problems ranging from the development of a well in an area where water is difficult to find to all types of well construction or pump installation problems. This section also administers the state-wide well construction code requirements, which encompasses both the construction of the well and installation of the pumping equipment. Handout sheets and diagrams showing proper well construction and pump installation along with copies of the Groundwater Quality Control Act, Part 127 of Act 368 of Public Acts of 1978, as amended, are readily available to the public from both the Michigan Department of Public Health and each local county health department office.

CHAPTER 26

LONG-RANGE PLANNING AND PRIORITY AREAS
OF GROUNDWATER MANAGEMENT

Alan G. Herceg
District Conservationist
USDA Soil Conservation Service
Centreville, MI 49032

David R. Mumford[1]
Environmental Health Services
Mid-Michigan District Health Department
Stanton, MI 48888

Pete Vergot
Cooperative Extension Service
Michigan State University
Paw Paw, MI 49079

Lois G. Wolfson
Institute of Water Research
Michigan State University
East Lansing, MI 48824

Private industry, environmental groups, federal, regional and local governmental agencies and university personnel concur that communications are a top priority in any groundwater management program. However, the communications require vast improvement to establish a united front addressing groundwater quality problems. It is rather fragmented at this time, with individual groups and agencies or units of government pursuing separate goals.

---

[1]Present address: Michigan Department of Public Health, Bureau of Environmental and Occupational Health, Division of Occupational Health, Saginaw, Michigan 48607

The communication relationships that need to be strengthened include various levels of organization. They are: 1) agencies to agencies at the local, state and federal level; 2) agencies to action groups--those in positions to promote and precipitate action that will support the activities of various agencies; 3) agencies to local units of government--the information that is available from agencies is not readily available to local units who often receive the burden of responsibility to implement any actions; and 4) agencies to the general public--evidence suggests that the public is receiving insufficient information on groundwater quality issues from governmental agencies. Communications should be improved to the general public, not just to those involved in agricultural production.

In addition, a higher level of communication from the public to the units of government and their agencies must be promoted. Units of government need to know the concerns of the public and where they are willing to put forth efforts and tax dollars. The public must also keep in touch with the regulatory agencies that are making policy.

Educational strategies on a local grassroots level must be devised to define, understand and prevent groundwater contamination by suggesting alternative methods for prevention, correction and treatment. Utilizing the resources of all agencies in a cooperative effort is essential.

If communications are going to be improved, so must the quality of information. This represents another dimension of the current groundwater needs in Michigan. One very useful type of information would be uniform statewide hydrogeologic mapping. Many user groups indicated that, currently, very few such data bases are available in Michigan. Excellent information is available through the soil surveys to a depth of 60 inches but not below that level. That type of mapping must be given priority on the list of information needs. Current and future water well records could serve as a background to concentrate efforts on utilizing current parameters as indicators for future testing.

In addition, statewide and county-specific mapping to identify where groundwater problems are located should be made available. Statewide identification of the vulnerable recharge areas is needed, along with directions to assist the public in finding these new sources of information. Currently, units of government, action groups, the public, and even some agencies are at a loss as to how to secure information that may be available. By utilizing the data obtained from hydrogeological studies, groundwater mapping and aquifer vulnerability, a more comprehensive informational package could be provided for land use planning and zoning officials. Furthermore, information on groundwater impacts from surface activities must be brought to the attention of rural and agricultural land users and planners. Training should be accelerated in agencies that are working with rural land users who request water quality information.

After communications are improved, and better information is made available, other priorities need changing, and an integrated approach to water management should be conceived. Unless the priority of groundwater protection is elevated, little change will occur. Agencies at all levels, local units of government, rural residents, and agricultural land owners must have similar priorities and work together. Because Michigan is a water-rich state, the supply is taken for granted as rural residents turn on the water faucets, farmers turn on their irrigation systems, and industry flips a switch. Unless the current complacent attitude changes, progress in protecting as well as improving groundwater quality will be very slow.

An integrated water management effort must also be developed wherein groundwater activities would be integrated into ongoing programs developed through universities and the state Departments of Natural Resources and Public Health. Programs such as irrigation scheduling, soil fertility recommendations, pesticide application recommendations, home owner recommendations for lawns and gardens, and program development in school systems through science classes should have an integrated groundwater component related to water quality. These groundwater initiatives should include training of local personnel working in counties and townships.

Water resource specialists are needed to develop educational programs and support for the agencies involved in this initiative. Possible help could include some groundwater modeling on microcomputers, news releases generated by the state for local communities and also continuous support of household hazardous waste programs and collection days.

The groundwater quality issue should be incorporated as part of the total nonpoint source pollution program in Michigan. Some of the actions that could be taken are:

1. Develop a permanent subgroup of the Great Lakes and Water Resources Planning Commission (GLWRPC) to address and coordinate water quality activities in the State of Michigan. It should be composed of a cross section of agencies, units of government and the public so all segments have input. When the GLWRPC has completed its mission, this new group would be able to continue water resources planning and implementation.

2. Develop a packet of information to direct groundwater quality concerns to the right sources of information. This packet would be available in local offices such as planning commissions, the Cooperative Extension Service, Soil Conservation Service, and Department of Natural Resources field offices.

3. Build agricultural and environmental coalition and support groups. In addition, further education is needed by those concerned with rural, agricultural and environmental issues.

4. Refocus or realign some of the traditional conservation practices of the Soil Conservation Service and accelerate the transfer of information that shows the benefits from those practices to the groundwater and not just to surface water.

5. Emphasize research and modeling to insure that credible recommendations come from the Cooperative Extension Service and Soil Conservation Service. These two agencies must increase their efforts in integrated pest management, fertilizer recommendations, and irrigation water management. With respect to statewide groundwater mapping, a joint effort is needed on the part of the state and the U.S. Geological Survey (USGS) so that the information can be obtained quickly.

7. Educate the rural landowners as to their responsibilities to groundwater and the environmental consequences of their actions. Unless that component is conveyed to the land users, the profit issue will not permit change in what is occurring. There has to be some other influence and reason stronger than the dollar to encourage landowners to change to more environmentally sound practices. That may be a long-term process, but it should be started immediately.

8. Refocus the priorities of agencies and units of government in the state. The best way to influence priorities at the local level would be to provide study grants and compliance grants. Implement studies and identify the problem areas and vulnerable sites. Provide funding for personnel at the local level and to those agencies that offer technical assistance to the rural land users. If these rural land users are convinced that they need to change their practices and that they have to go to a higher level of management, then some additional technical assistance will be needed.

Many groups including local groundwater committees that are county-based, the Cooperative Extension Service, U.S. Soil Conservation Service, Public Health Department, drain commissioners and many others have begun to look at the groundwater issues at the local level and have produced reports or have presented programs in those areas. It is hoped that they can expand those efforts and start to share information among agencies, to share the network that they have and programs that they are putting on for their particular clientele, and involve the county clientele in their needs assessments, awareness programs and county initiative.

Communication is the primary key in these efforts and should be increased along with the activity level in projects. These recommendations call for an increased working relationship of the aforementioned agencies in information sharing and program development of groundwater initiatives.

(Editors' note) These recommendations were compiled during the Conference on Rural Groundwater Contamination: Assessment of Needs, Strategies for Action, held at the Michigan State University Kellogg Biological Station at Hickory Corners, Michigan, October 20-22, 1986.

INDEX